The
Living
Landscape

How To Read
and Understand It

Patrick Whitefield

Permanent Publications

Published by

THE QUEEN'S AWARDS
FOR ENTERPRISE:
SUSTAINABLE DEVELOPMENT
2008

Permanent Publications
Hyden House Ltd
The Sustainability Centre
East Meon
Hampshire GU32 1HR
United Kingdom
Tel: 01730 823 311
Fax: 01730 823 322
Overseas: (international code +44 - 1730)
Email: enquiries@permaculture.co.uk
Web: www.permaculture.co.uk

Published in association with

Permaculture Association (Britain), BCM Permaculture Association, London WC1N 3XX
Tel: 0845 458 1805 Email: office@permaculture.org.uk Web: www.permaculture.org.uk

© 2009 Patrick Whitefield

The right of Patrick Whitefield to be identified as the author of this work has been
asserted by him in accordance with the Copyrights, Designs and Patents Act 1998

Designed and typeset by Tim Harland

Cover photo by Penny Rose

Line drawings by the author

FSC
Mixed Sources
Product group from well-managed
forests and other controlled sources

Cert no. SGS-COC-2953
www.fsc.org
© 1996 Forest Stewardship Council

Printed in the UK by CPI Antony Rowe, Chippenham, Wiltshire

All paper from FSC certified mixed sources

The Forest Stewardship Council (FSC) is a non-profit international
organisation established to promote the responsible management of
the world's forests. Products carrying the FSC label are independently
certified to assure consumers that they come from forests that are
managed to meet the social, economic and ecological needs of
present and future generations.

British Library Cataloguing-in-Publication Data

A catalogue record for this book is available from the British Library

ISBN 978 1 85623 043 8

Contents

For my mother, Peggy Vickers

About the Author

Elizabeth Jacobs

Patrick Whitefield was brought up on a smallholding in Somerset and trained in agriculture at Shuttleworth College. After working on farms both in Britain and abroad he came back to his home county, where he still lives. In the 1980s he bought a wildflower meadow to save it from destruction and lived there in a tipi for most of that decade, developing an intimate relationship with the natural world. In 1990 he became a teacher of permaculture, a profession which brings together his farming background, his immersion in nature and his lifelong passion for the landscape. For the past twenty years he has kept a notebook of landscape observations which form the basis of this book. He is also the author of *Tipi Living, Permaculture in a Nutshell, How To Make A Forest Garden* and *The Earth Care Manual.*

Foreword

by Ben Law

I first met Patrick Whitefield about 17 years ago. I was attending a permaculture design course at Dartington in Devon. Patrick had attended the previous year and I was fortunate enough to witness the beginning of his now renowned permaculture teaching career. Permaculture was in its infancy in Britain, with courses beginning to be adapted from the Australian model to the more temperate climate of this island.

I remember clearly hanging onto every word as Patrick spoke. His enthusiasm and passion for the landscape shone through and I knew I was in the presence of a man who was speaking from experience. His examples of landscapes were so clearly portrayed I could have been walking through them. It is this portrayal and journey through the landscape that Patrick has now managed to transfer so vividly into *The Living Landscape*.

Patrick's background is tied to the landscape. He farmed in Africa, and lived a low impact lifestyle in a tipi, where he developed many craft skills. He has tended the land by gardening and nature conservation, become involved in green politics and dedicated his life to the research and teaching of permaculture design. He is also a celebrated author of four other books which demonstrate a life long commitment to discovering genuinely sustainable, low impact ways of living.

One of Patrick's defining qualities is his love for the British landscape. Here he shares a lifetime's knowledge about the myriad interactions that go to make up the fascinating and varied landscapes we see all around us. He takes us on a journey through landscape formation, from rocks, through soil to vegetation and the intricate web of interactions between plants, animals, climate and people which form the landscape around us. This is not only useful for the permaculture and landscape designer, farmers and smallholders, it will appeal to a far wider audience who appreciate our countryside and want to learn more.

For me, a walk through the landscape with a friend is stimulated and enhanced by the changing patterns of land use; a pause beside an earthbank to contemplate its evolution, the surprise of finding a plant outside of its usual landscape comfort zone and the joy of feeling in touch with the history of landscape evolution, past and present. The ability to read the landscape can be compared to moving from black and white to colour, such is the impact it can make on what you might once have thought was just a walk in the countryside. Patrick's book can open your eyes. Through *The Living Landscape* he inspires people to reconnect with the land as a living entity and develop an active relationship with nature and the countryside. He invites us all to actively engage with nature and experience it first hand; he understands that only a relationship with an actual place can start to make that connection real.

For me the connection with my local landscape is like that of an extended family. Familiar plants mark changes in the year, trees that I visit to forage and migratory species help me reflect upon my impact on the land. The knowledge of reading the landscape brings us within it. We become part of it, rather than remaining excluded as outsiders, and through this process begin to care for our surroundings as mutual dependence becomes apparent. *The Living Landscape* offers an opportunity to step through this veil. In his clear and accessible style, Patrick has opened the door to engaging with the landscape. He offers the reader not just the breadth of his knowledge but the first steps on a journey to become part of our landscape again.

Ben Law
woodlander, author, teacher and builder
of the famous 'Woodland House'
www.ben-law.co.uk

Introduction

I've always been fascinated by the landscape. When I was a student at agricultural college I remember a friend saying, "It's not safe to take a lift with Patrick. He 'farms' as he drives." I don't know if people use the word 'to farm' in that sense any more. It meant to observe and assess the land and decide what should be done in each field. It goes back to the days when most farms employed a lot of workers and the job of the farmer was to think and make decisions rather than tolabour on the land. Farming in that sense is one aspect of landscape reading.

As it turned out I never became a farmer, although I did work on farms for a number of years. When I acquired some land of my own it was a nature reserve rather than a farm and so I learnt to read the landscape from a somewhat different perspective. Eventually I became a teacher of permaculture, and both of these perspectives are relevant in my present work, in fact they've merged into one, more holistic view of the landscape. Permaculture is all about creating productive landscapes which work in harmony with nature and are thus truly sustainable. It's a way of designing such landscapes, whether gardens, farms, villages or towns. Observation is the absolute bedrock of permaculture design. We can only design a truly harmonious landscape if we have a good knowledge and understanding of what's already there before we start.

I originally intended this book to be a permaculture workbook, covering the all-important observation stage of the design process but my enthusiasm for the subject couldn't be bound by such a utilitarian plan. My notebook started to fill with all sorts of observations which might not be directly relevant to permaculture design but were just too interesting to pass by. I began to visualise my reader not so much as a designer but more as someone who spends time outdoors, whether for enjoyment or in the course of their job, snuggling down with the book in an armchair on a winter evening. But in the end I realised there may not be that much difference between the useful and the simply fascinating. Who can say which piece of information may be of use and which will not? In fact the kind of thinking which divides things up into the useful but boring and ugly on one hand and the useless but interesting and beautiful on the other belongs to an age which is rapidly passing. It was an idea born in the industrial revolution and its time is over. So I hope that all of you who read this book will both enjoy it and find it useful.

Its focus is on the rural landscape. This is not to say that I think the urban landscape is less worthy of attention, but as I've lived almost all my life in the country I don't know much about urban landscapes and I'm not the right person to write about them. In fact that job has already been brilliantly done by the late Oliver Gilbert in his book *The Ecology of Urban Habitats*. (See Further Reading on page 319.)

Nor is this book mainly about the history of the landscape. Many people assume that landscape reading and landscape history are one and the same. Perhaps this is because so many excellent books have been written about the history of the landscape while the broader field of landscape reading has been largely ignored. I have included a chapter on landscape history, as it's clearly impossible to understand something without knowing about its past. But the history of the landscape is just one aspect of it, the human influence. The other influences – rock, soil, climate, plants and animals – also play their parts, and the landscape we see is the result of the ever-changing interplay between all these.

The result is a very complex picture which can only be broadly sketched in a single book. There just isn't space to go into all the complexity. In fact if I were to try to do so the clarity of the picture would be lost, so I've often simplified. For example, I've sometimes had to ignore regional variations and make broad statements which are true for the country as a whole. I hope I've never simplified things to the point of distorting the truth but everything you read in the pages that follow must be taken with the added comment that actually it's more complicated than that. In any case my intention is not to tell you exactly what's going on in the landscape around you. It's to open a door on a new way of looking at that landscape and understanding it. The descriptions I give of individual landscapes are no more than examples and you may find something quite different in your own locality.

I've tried as much as possible to write from personal experience but inevitably quite a lot of research has been necessary to fill in the gaps. Since landscape reading is such a wide field this has meant casting my net wide and getting snippets of information from many different books and papers on a variety of subjects. But I must mention a special debt of gratitude to Oliver Rackham, the great authority on the history of the landscape. His books and face-to-face teaching have been a source of both inspiration and information. Special thanks also go to my sister, Cristina Crossingham, who has carefully read each chapter and made many useful suggestions.

The Landscape
and
How To Read It

We visited Blawith Common on a sunny day in August. It lies on a rounded hill, just south of Coniston Water in the English Lake District. It's not high by local standards and a walk to the top made an easy afternoon stroll for my wife, Cathy, and me. The deep green of the bracken was set off by the occasional splash of purple heather and grey rock. On the way up we passed a tarn, deep blue under the cloudless sky, though it had a dark grey and bottomless cast despite the bright light. From the top the view of the lake and surrounding mountains was a classic landscape in the aesthetic sense, the kind of view that's reproduced so often on posters and postcards that it needs no description.

On the way down I began to see that the common itself is no less fascinating, though in a different way. You could almost call it a microcosm of the British landscape. I don't mean that it's typical – our landscape is so varied that there's no such thing as a typical scene. I mean that within the space of a few hundred metres you could clearly see how the landscape is formed by four main factors: the rock, the soil, the climate and living things.

The rock which underlies the landscape determines the landform. Hard rocks are slow to erode and they make hills while soft, erodible rocks make valleys. Here in the Lake District all the rocks are hard. The narrow valleys have been carved out with difficulty by the forces of erosion, perhaps initially along a line of weakness in the rocks. Blawith Common is a hill because it's made of hard, strong rock.

The kind of rock is also a big influence on the second factor, the soil, and the soil in turn has a strong influence on the vegetation. On Blawith Common the effect of the soil on the vegetation is dramatic. Most of the hill is covered in a deep, unbroken bed of bracken. Bracken is a very competitive fern. Given a free rein it can out-compete other plants, both by shading them and by releasing a poisonous chemical which inhibits their growth. On the common the continuous coat of bracken was only broken in the low hollows and on the high spots, near where the bare rock came to the surface.

One thing which bracken doesn't like is shallow soil. As you get closer to the exposed rock the soil gets thinner and here heather has the competitive advantage over bracken. Although it's not such a vigorous and competitive plant it tolerates shallow soil, so here it out-competes the bracken. Surrounding each patch of bare rock was a purple halo of blooming heather, a bright relief to the monotonous green of the bracken. Another thing bracken can't stand is constantly wet feet. In the poorly-drained hollows it gave way to grass and rushes or, where it was very wet, to bog myrtle. These three kinds of vegetation, bracken, heather and the grassy mix, faithfully mapped out the differences in the soil without any other plants confusing the picture. At least they did on the higher part of the hill. Lower down on the eastern slope, near where the open common abuts on stone-walled fields, some young trees were growing up through the bracken.

This is where the third factor came in, climate. The warm, sunny day we were enjoying isn't typical of the local climate. The Lake District is not just the wettest part of England, it's also one of the windiest. The shelter given by the hill was just enough to make the difference between survival and death to a young tree seedling. The critical time for a young tree which is trying to get established in bracken is the autumn. When the bracken fronds die down they can swamp the seedling and kill it just as a gardener kills weeds with a mulch. The quicker a tree can grow the sooner it can get its head above the bracken and the fewer autumns it has to run the gauntlet of being mulched to death. On the open hill the odds are stacked too far against them but on the lower slopes the occasional one succeeds. When we were there, in 1993, there was a scattering of juniper, birch, eared willow, hawthorn and rowan. These are typical pioneer trees of the harsh uplands, well able to withstand the cold, wet climate and poor soils formed on the hard, old rocks. Perhaps by now they've turned into a small wood.

Landscapes are full of small-scale changes in climate such as the difference between the top of Blawith Common and the sheltered eastern side. Microclimate is the name given to the distinctive climate of a small area and this was a classic example of a difference in microclimate. Although the regional climate has a big influence on the landscape, the influence of microclimate is often much more visible because you can see the difference in one glance. Microclimate can also be very important from a practical point of view as it affects the productive potential of different parts of the land.

Standing there on the edge of the common we could look over a nearby stone wall into an enclosed field. The contrast between the open common and the stone-walled field was dramatic. There was not a bit of bracken in the field. It was carpeted with closely-grazed grass and dotted with a few gorse bushes and the occasional large sycamore tree. Three cattle stood there looking at us. They were the cause of the contrast between the common and the field and they represented the fourth great landscape-forming factor, living things.

Plants and animals are not just part of the landscape, they also play a part in forming it. Plants have a big influence on the soil and on the microclimate, by giving shade for example. But here in Britain overwhelmingly the biggest biotic influence on the landscape is grazing animals. The humble sheep and the stolid cow between them have transformed millions of hectares of land from woodland to grassland and moor. They will eat whatever they can get, but some plants are more able to survive constant nibbling than others. Supreme among the plants which can tolerate, even thrive on, constant nibbling are the grasses. All their buds are below ground, so they can easily regrow after being eaten. Plants with their buds above ground, like little trees, will eventually die if they're repeatedly bitten back. Old trees can survive intensive grazing, but they can't reproduce because their young ones get eaten. So grazing on its own is enough to turn woodland into pasture as long as it goes on constantly for the lifetime of the longest-lived trees.

Bracken, being poisonous, is hardly grazed by animals, but it can't stand being trampled. Sheep, with their nifty little feet, don't bother it but cattle tread on it and can eventually kill it out. I don't know how long it was since the common had been grazed, but it looked like a long time. Certainly no cattle had been there for a while and with such a dense stand of bracken there wasn't much there to eat for either sheep or cattle. Nor would those young trees have survived on the common if it had been grazed much during the past twenty years. The gorse has survived in the pasture because it defends itself with its spines. It likes more or less the same kind of soil as bracken so it often replaces bracken on land regularly grazed by cattle. The sycamores may well have been planted. Sycamore has often been planted for shelter in harsh upland areas because it's very tough, and also has the status of being a useful timber tree, which willow, rowan and so on have not. Or they may have got established at some time in the past when farming was depressed and grazing pressure was reduced.

The difference between common land and fields is all down to being able to control the pattern of grazing. Once you have a wall or a hedge round a piece of ground you can put a large number of animals on it for a short time and then move them off again. While they're there they will eat, browse or trample everything in reach and when they're gone the grasses will regrow vigorously. A few weeks later, when the grass has regrown, you can do the same again. Meanwhile on the open common the animals wander where they will and eat the plants they find most tasty, leaving the less palatable plants, like bracken and rushes, to thrive. So controlled grazing improves the pasture while uncontrolled grazing causes it to deteriorate.

Curiously, the way wild animals behave in natural grasslands is more like the controlled kind of grazing than the uncontrolled grazing on the common. Wild herbivores stick together in a tight herd, the better to defend themselves from the predators who are always on the lookout for an isolated individual. Wherever the herd goes it eats everything down to the ground, leaves a dressing of manure and doesn't return till the sward has regrown. Although it might at first sight seem more 'natural' to leave animals to wander around, in fact shutting them up in fields and moving them regularly from one field to another is much closer to their natural conditions.

All four landscape-forming factors, rocks, soil, climate and living things, are constantly interacting. The rocks affect the land form, the land form affects the climate, the rock and climate affect the soil, the soil and climate affect the vegetation, the vegetation affects the soil and climate, and all of these affect the agricultural potential of the land. This in turn affects how the land has been used by people, whether as woodland, rough grazing, grassland or arable.

Human beings are strictly speaking included in the fourth landscape-forming factor, the living things. But in a densely-settled country like ours the human influence is enormous. Without any human intervention at all, what is now Blawith Common would almost certainly be woodland. It was people who brought in grazing animals and kept them at high enough numbers to turn it into a bracken-dominated moor. Now that people have stopped putting their animals on it, it's starting the process of returning to woodland. The difference between the common and the adjacent fields is also a human artefact. The stone walls around the fields, painstakingly put together in some previous age, are the key to the difference in vegetation.

Even in such a relatively 'wild' part of Britain as the Lake District, the human influence is as great as all the other factors put together. So it can sometimes be useful to think in terms not of four factors which form the landscape but two: the natural and the human. The more you look at landscapes the more you see how every one is the result of an interplay between natural and human forces. At Blawith the decision to enclose the lower slopes of the hill and leave the upper part as common land was not arbitrary. The microclimate further up the hill is definitely worse and the soil probably is too. It wasn't thought to be worth the effort to enclose land that even after enclosure wouldn't be very productive. Although people have completely transformed the landscape, the way they've done it has been directed by nature.

Seeing the natural and the human as two complementary forces is a useful way of understanding what's happening in the landscape. But at the same time it's important to remember that we're part of the ecological community, with no more right to be here and thrive than any other species. Certainly we're much more powerful than any other and our effect on the world ecosystem is correspondingly great. But that's all the more reason to remind ourselves that we're part of the web of life, not superior to it or outside of it. When we forget this we destroy the very living systems on which we depend for survival.

The Natural and the Human

Even landscapes which appear to be entirely humanised or completely wild are in fact a result of this interplay of human and natural forces. However artificial a landscape may be, nature will find her way in there. The approach to Birmingham New Street station must be one of the most urbanised landscapes in Britain. The train creeps through a deep, dark cutting lined with high walls of sooty brick. The sun seems far away. But even here there are buddleia bushes growing in the cracks between the bricks.

At the other extreme it's hard to imagine a place on Earth where humans have had no influence at all. It used to be assumed that when Europeans first arrived in Australia and North America they were seeing wilderness virtually untouched by human hand. But now it's recognised that hunter-gatherers can have major impacts on ecosystems, especially where fire is used to improve the hunting. Today there's no part of the Earth unaffected by the industrial economy. DDT has been found in the body fat of both penguins and polar bears, and nowhere will escape the effects of global warming.

Most of the land here in Britain lies between these two extremes. Over two-thirds of it is farmland and most of the rest is moorland and other rough grazing.

Cities, towns and villages take up less than a tenth of the area. Virtually all of the land would be woodland if it weren't for us. The small area of woodland which does remain has been used as a resource by people down the ages and bears little resemblance to the wildwood of prehistoric times. Every bit of the island has been modified by us or our grazing animals, except for the peaks of the highest mountains which were always above the tree line.

So we have no wilderness in Britain. Even the Highlands of Scotland are the result of a blend of human and natural forces. We might like to think of the Highlands as the next thing to a wilderness. But what could be less natural than an open, treeless moor where naturally there would be woodland?

These Highland moors are an example of what's known as a semi-natural ecosystem. Semi-natural means that the ecosystem has been modified by human action but the vegetation has not actually been planted. In the case of moorland that human action is grazing. The grazing has been intense enough to prevent the reproduction of trees and thus in the long term to eliminate the woodland. The plants which are there now have either survived from the time it was woodland or have moved in of their own accord since then. They're all plants which can tolerate the present conditions of soil, climate and grazing. People haven't planted them but have been partly responsible for creating the conditions in which they thrive.

There are also semi-natural grasslands, composed of wild, self-selected grasses and wild flowers, but they're a tiny percentage of the total grassland area. Most of the grassland in Britain is cultivated, just like any other agricultural crop. There are also semi-natural woods. Many of these are coppiced woods, which have been felled and allowed to regenerate many times over hundreds or thousands of years. The structure of a coppice wood is quite unlike that of a wild wood. The trees are smaller and there's very little accumulation of dead wood. The regular coppicing also allows a range of sun-loving flowers, which must have been hard-pressed to survive in the wildwood, to thrive. No-one has ever planted a tree there, but it's not a natural wood, it's semi-natural.

You could say that a semi-natural ecosystem is one in which people have determined the structure and nature has provided the plants. Changing the structure from wildwood to grassland or from wildwood to coppice wood changes the range of species which can grow there. But these species are drawn from the pool of plants and animals which are able to colonise it of their own accord. That's what we mean by semi-natural.

Semi-natural ecosystems are jewels of biodiversity, the last refuge of wild plants and animals in an intensively humanised landscape. They're rare in the more fertile parts of the country, where almost all the land has been converted to productive farming. The ones which survive are now mostly nature reserves. In the uplands, where the productive potential of the land is lower, there's still a lot of moorland and rough grassland. It's the closest thing we have to wilderness, but it's not wild. It's the fruit of the long, long dance between people and nature.

The way this dance has been played out varies from place to place according to natural conditions and this has given us distinctive regional landscapes. A fine example of a regional landscape is the West Country, that curiously-shaped peninsula which snakes out from the south-west corner of England.

I live in central Somerset. The whole county, apart from its western tip, forms a great bowl, with a ring of hills round the edge, a fringe of clay vales and in the middle the Somerset Levels, flat marshland, now drained for farming. The rocks of Somerset are mostly young, soft and alkaline. Lime-loving plants, like the climber, old man's beard, are frequent. On the whole the villages are large, though there are some hamlets too.

The western rim of this bowl is the Quantock hills. You cross them and suddenly you're in the West Country, still in Somerset but in that western tip which has more in common with Devon and Cornwall than with the rest of the county. It has a very different feel. Everywhere there are hills, big ones and little ones, high ones and low ones, separated by narrow valleys with hardly a flat field anywhere. The rocks are old, hard and mostly acid. Plants of acid soils, such as foxgloves, are common. There are few big villages, most parishes consisting entirely of scattered hamlets and isolated farms. Perhaps the most evocative feature of the West County is that the hedges are raised up on great solid banks a metre or more high.

If you enter the West Country further south, from the chalk downs and clay vales of Dorset into hilly east Devon, there's a similar transition though it's not so sudden and obvious as the crossing of the Quantocks. Somerset and Dorset are both in the south west of England, but they're certainly not in the West Country.

There is variety within the West Country of course, but it's variations on a theme. One of these variations is to be found in the South Hams, that part of Devon which juts out into the sea to the south of Dartmoor. The land is a plateau of broad, flat-topped hills, all of much the same height, dissected by steep V-shaped valleys. The hilltops are just about the flattest land in Devon and are intensively farmed. Because the sea is not far away they're exposed to strong, salty winds, so trees are few and usually sculpted by the wind. The hedges are mostly trimmed short, hardly taller than the grasses and flowers which grow on top of the hedgebanks. The narrow valleys are a complete contrast. Here the trees have shelter from the sea winds and they can thrive and grow tall. The valley sides are often too steep for farming and so are densely wooded. These valley woods are often dense and jungly. Few sounds from the outside world reach you when you're walking in them and all you can see is woodland. Civilisation can feel far away. Even though you know there's a bare, ordered fieldscape above you on all sides, somehow it seems a bit improbable.

This distinctive pattern of alternating windswept plateaus and deep, wooded combes is repeated throughout the West Country peninsula wherever a dissected plateau lies beside the sea. This happens in quite a lot of places, especially in Cornwall, where everywhere is near the sea. It's also found across the Bristol Channel in the south-west peninsula of Wales, which is similar in many ways to the south-west peninsula of England. The country between Carmarthen and Newcastle Emlyn is an example. Within this general pattern each locality has its distinctive character. Some are less influenced by the sea than the South Hams.

In some places the valley sides are less consistently wooded than others, though whenever farming suffers a downturn these steep fields are the first to be abandoned and succeed to woodland. But this is a definite landscape type, one which can probably be recognised in many other parts of Britain. I've only mentioned the examples I know personally.

Near my home there's something similar on a miniature scale, though without the effect of the sea. It's a single hill which has three steep, wooded gullies cut into its side by small streams. The gullies are as little as twenty metres wide in places, but because you can't see out it can feel like being the middle of the wilderness, another world altogether from the pasture fields outside. My notebook records a visit to one of these miniature wooded valleys.

Withial Combe, 20th July 1995

Although quite clearly coppiced in the past it has a primeval feel. Sitting there in the deep green by the waterfall all my tension and worry slipped away. My mind relaxed as it rarely does.

The combe is at its most beautiful where the canopy is broken (usually by the death of a wych elm) and the sun dapples down on the horsetails, ferns and mossy logs over the little brook. It shines on gnats, dragonflies and butterflies in the columns of lit air between the tall trees.

A wood is a comparatively quiet place at this time of the year. Very few birds are singing or flowers blooming. The action has moved out to the fields, where a whole range of flowers are in bloom.

While the West Country landscape can be seen as variations on a theme, moving east to Wiltshire we find a county clearly divided into two distinct landscape types, the chalk downs and the clay vales. The downs are wide, open hills, formerly unfenced sheep-walk, now wire-fenced into big fields and mostly in corn. Trees are mainly in isolated clumps or occasional large woods. It's a dry landscape with no surface water. By contrast, the clay vales have smaller, thickly hedged fields with hedgerow trees, frequent small woods, streams and rivers. There are fewer long views. The heavy soil holds water well and grows lush grass which is ideal for dairy farming.

The old expression 'as different as chalk and cheese' refers to these two landscape types. They make such a contrast that they became a byword for how different two things can be from each other. The same contrast can be seen in other counties with chalk, but there the situation is complicated by the presence of other landscape types such as sandy heath, which there's very little of in Wiltshire.

It would be a mistake to think that variations in the landscape are always strictly governed by the physical conditions. We do have some free will in the matter. This was brought home to me once on a bus journey from England to the Netherlands. Soon after leaving the Channel Tunnel we crossed the border from France to Belgium. Suddenly, and precisely on the border, the landscape changed. On the French side was pure arable farming. On the Belgian side was grassland, grazed by beef and dairy cattle, with some maize for silage and just a little arable. The land is equally flat on either side and I find it hard to believe there's a dramatic change in soil type which exactly follows the political frontier. No, it's a cultural difference.

Individuals can also make decisions about the landscape which are not based on the physical realities of the land or strict economics. For example, I once knew a farmer who didn't like trees in his hedges so he took them all out. You could spend ages trying to find a logical reason why trees don't grow in the hedges on that farm while they do on neighbouring farms, but it would be in vain. It was a matter of personal whim. Many things in the landscape are a matter of personal whim. A field may have been abandoned at some time in the past and developed into woodland, not because it was particularly unsuited to agriculture but for some purely personal reason now long forgotten. If we try to understand that wood purely in rational terms we'll misinterpret it.

The interplay between human and natural influences can be seen in a page from my notebook. The scene is a cove on the north coast of Cornwall. A steep valley leads down to the cove itself, with a few flattish fields at the bottom and a small village.

Crackington Haven, August 1997

On the steep slopes there are many stages of vegetation: mature woodland; bracken with developing woodland; unused pastures with brambles encroaching, one of which has been planted with trees; grazed pastures; and climax scrub exposed to sea winds. The most recently abandoned pastures are being used as building sites, great terraces being cut into the steep slopes.

The woodland has the characteristic oaks of the West Country, small with wiggly trunks. There's some sycamore, perhaps more on the seaward edge of the wood. There's some ash, especially at the bottom of the slope, where it's regenerating well and is the main component of what appear to be areas of recent woodland. The understorey of the mature woodland is holly and hazel.

The 'stages of vegetation' I noted are stages in the process of succession. Succession is the way an ecosystem develops through time. The commonest example of succession you can see here in Britain is what happens when we stop grazing, mowing or cultivating the land. Bit by bit it gets colonised by woody plants, shrubs at first and then trees, till it develops into woodland. At Crackington I could see several stages of succession: grazed fields, abandoned fields being taken over by brambles or bracken, recently formed woodland and mature woodland. In one place people had speeded up the process of succession by planting trees. In others they were halting it by building houses.

There were also two examples of what is sometimes called climax vegetation, the end point of succession. One is the mature woodland. The other is the scrub, which I assumed could not develop to woodland because it's too exposed to the salt winds which so inhibit the growth of trees. The scrub was on a very steep slope directly facing the sea while the woodland was in a less exposed position.

The general picture is determined by nature: a steep landscape on hard, old rocks, more or less exposed to the influence of the sea. Given that context, the scene is set by humans. It's one of increasing woodland and decreasing farmland in a time when the kind of farming which is possible on this steep land is becoming less and less economically viable. The ebb and flow of human economics is the dominant influence. The details, such as the kinds of trees making up the woodland, are more determined by nature.

The mature woodland is a typical Cornish wood, composed mainly of oaks, 'small with wiggly trunks.' This tree form is largely a climatic effect, though soil and genetics probably play some part. The West Country is a narrow peninsula, thrust out into one of the windiest oceans in the world. You can find big, straight-trunked oaks in some sheltered inland parts, but not on the hills or near the coast, which amounts to most of the West Country. Hazel and holly in the woodland understorey are also typical of Cornwall. Like oak, they tolerate the relatively acid and infertile soils found there.

Sycamore is more resistant to salt winds than other broadleaved trees. I wonder whether I really did see more of them on the seaward side of the wood or whether I just assumed it was so because I expected it. I was more sure about the distribution of ash, which was growing mostly at the bottom of the slope. Ash needs a relatively rich soil and in areas where the soil is very poor it's often confined to the lower parts of woods, as plant nutrients are leached out of the soil higher up and accumulate at the bottom of the slope.

If landscape is the result of the relationship between human and natural forces, all too often this relationship has been seen as a fight. Throughout most of history there can have been little doubt which side we were all rooting for. Nature was relatively strong. Starvation was as real a prospect here in Europe as it still is in less fortunate parts of the world. In the 17th and 18th centuries moors and mountains were seen as ugly, unpleasant places. Good was seen as 'making two blades of grass grow where one grew before'. The fact that each extra blade of grass would displace another wild flower didn't figure as important in the consciousness of the times.

With the birth of the fossil fuel age our power to transform the landscape expanded exponentially and humans are now quite clearly winning the fight. Toady many people regret the losses we're inflicting on nature, both for nature's own sake and because it looks like we'll bring ourselves to ruin in the process. Starting with the Romantic movement of the early 19th century, a lot of us have changed sides. We've come to regard a 'good' landscape as one with plenty of biodiversity, little soil erosion, clean rivers and so on. Never mind how many blades of grass it will grow when we have overproduction of food.

Both these attitudes have some logic to them, but they see things as black or white. In reality there is not necessarily a conflict between productivity and biodiversity, especially when you look at the question in the long term. A monoculture of ryegrass fed with chemical fertiliser will produce more gross yield than a semi-natural hay meadow made up of perhaps a hundred species of grasses and wild flowers. But the ryegrass monoculture is liberally fed with fertiliser, which takes a lot of energy to produce and becomes a pollutant when it leaches into the groundwater we use for our drinking supply. The ryegrass field also needs ploughing up and reseeding periodically, which uses energy and exposes the soil to the possibility of erosion. If we deduct all the costs, including the contribution to global warming, the net output of the diverse meadow is likely to be higher.

The productivity of the semi-natural meadow is actually dependent on its diversity. The grasses and herbs which make it up include both short bushy species and tall thin ones, so it can form a denser sward and capture more sunlight than the monoculture. There are also deep-rooted and shallow-rooted species, so it can make more complete use of the water and mineral nutrients held in the soil. There are early-growing and late-growing ones, so it can make better use of sunlight, water and mineral nutrients through time. The diversity of plants and the fact that the soil is not disturbed by ploughing means that a healthy population of microbes, earthworms and other soil creatures can develop. These help to provide the soil fertility which would otherwise come out of the fertiliser bag. The diversity of herbs growing among the grasses gives the grazing animals a healthy, balanced diet and reduces the need for veterinary care.

Only diverse landscapes are truly sustainable. The idea that we can survive for long on this planet with just ourselves and a couple of dozen species we think are directly useful to us is unrealistic. Biodiversity is important for its own sake but it's also a matter of plain self-interest for us humans. The natural systems which support us are so complex that we can never fully understand them. We must preserve the whole in order to be sure we've still got all the important parts.

So we need to read the landscape as a whole. We're very much inclined to look at it from one point of view at a time: productivity, biodiversity, beauty or whatever. This habit is based on the assumption that you can't have all three together. But it's a false assumption. A hedgerow, for example, is both a functional part of a farm and a wildlife habitat. One group of creatures which lives in hedgebanks is the ground beetles. They spend the winter there and in spring they move out into the field, where they eat pests. Now is that a useful, productive function of hedges or one to delight nature conservationists? Of course it's both.

A Walk on Bindon Hill

Look at any landscape and what do you see? Trees, fields of grass and arable, roads, rivers, houses. It looks like a series of objects. But every one of them is changing all the time. Trees grow and a landscape which includes fast-growing conifers or poplars can look eerily unfamiliar if you revisit it after being away for only a few years. Grassland and arable would gradually succeed to woodland, rivers would change their course, and roads and buildings would gradually return to the Earth whence they came if people were not constantly tending them. What you see is not so much a series of objects but processes. Whether it's a process of growth, development or decay, everything in the landscape is constantly changing.

If in reality the fields look much the same from year to year and the river keeps the same course for generation after generation, it's because people do tend them. We slow down or stop the processes of change in order to keep things the way we want them. So the relationship between people and fields or people and river modifies the landscape and keeps it the way it is. If, on the other hand, the isolated cottage does turn into a ruin or the old road deteriorates to a track it's due to their relationship with the weather, woodworm and the wheels of the passing traffic.

So you can look at the landscape as a series of objects, but if you look at it as a series of processes and relationships it has much more meaning. You start to see not just the 'what' but also the 'why', which is the really fascinating aspect of the landscape. This page from my notebook illustrates the idea.

West Lulworth, Dorset, 3rd May 1999

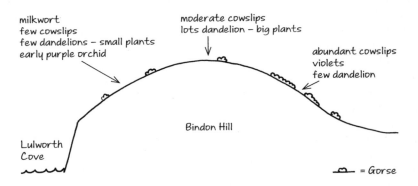

milkwort
few cowslips
few dandelions – small plants
early purple orchid

moderate cowslips
lots dandelion – big plants

abundant cowslips
violets
few dandelion

Bindon Hill

Lulworth
Cove

= Gorse

Skylarks, swallows, whitethroats.

This is the western end of the hill. It's quite tightly grazed. The rest of the hill is part of the military ranges, with much longer grass and different wild flowers, such as knapweed.

Lulworth Cove is famous for its rock formations. The sea has broken through the thin band of hard limestone which runs along the shore and carved out the almost perfectly circular cove from the softer rocks behind. But on this occasion I was more interested in what was going on Bindon Hill, the chalk down which forms the inland shore of the cove.

The most obvious process going on here is succession. Only the grazing of sheep and cattle prevents the grassland succeeding eventually to woodland. Despite the grazing, gorse has established itself and this change from herbaceous plants – grasses and herbs – to a woody shrub is a stage in the process.

The salty winds from the sea also act as a brake on succession here but in a different way from the grazing animals. The animals influence the kind of shrubs which can get established: spiny gorse can defend itself from browsing where softer shrubs couldn't. But the sea wind seems to me to influence the distribution of the gorse. It's thick on the sheltered northern side of the hill and sparse on the exposed top and southern, seaward side. So you could see succession here as the result of the relationship between vegetation, animals and microclimate.

The annual growth pattern of the plants is another process, quite distinct from succession. This again is influenced by the relationship between the plants and the grazing animals. The western end of the hill is fairly intensively grazed, so the grass is short. This enables the short-growing wild flowers, dandelion, cowslip, milkwort and violet, to grow. They couldn't survive the shade that tall grass would cast. The rest of the hill is in the military ranges and the grass is long. Perhaps this is because access to the ranges is limited so they can't always be grazed at the optimum time. Only the taller wild flowers, such as knapweed, are able to compete with the longer grass. In this way the pattern of grazing affects the range of herbaceous plants which is able to grow in each place.

On the western end of the hill, which I looked at in more detail than the military area, the wild flowers are not uniformly distributed. Violets are only found on the north slope of the hill, which is more shady, more sheltered and more thickly grown with gorse. Perhaps this is because they're plants of woodland and woodland edge and prefer slightly more sheltered and shady conditions.

The dandelions are more abundant and bigger on the flattish top of the hill than on the steeper sides. Since dandelions favour a soil which is rich in plant nutrients I reckon this is related to soil fertility. Four-legged animals are much more comfortable on flat ground than on a steep slope, so they spend as much of their time as they can on the flatter parts of the land. That's where they chew the cud and sleep and where they prefer to graze, so that's where most of their dung is deposited. This effect is reinforced by soil erosion, which is greater on the steeper land, leaving a thinner soil on those parts of the hill.

I can't say for sure why the cowslips favour the north side of the hill, nor why I found milkwort and early purple orchid only on the south side. There may have been milkwort and orchids in other places. I could easily have missed them. Not so the cowslips, which are abundant and highly visible.

The wild flowers I saw are to varying degrees characteristic of chalk downland. Dandelions are found on a wide range of soils, whereas milkwort and cowslip are much more typical of chalk or limestone grasslands. Apart from the knapweed they were all in flower when I was there. I'm sure if I'd been there a little later in the year, or if I were better at identifying wild plants when they're not in flower, I would have come up with a whole list of typical downland wild flowers. Gorse is only found on well-drained soils, which includes the chalk. Given the same conditions of grazing and climate but a different soil, you might see a landscape of very similar structure but with a different array of plants playing the different roles within it. On a deeper soil brambles might be taking the part of the gorse. On a more acid one the downland wild flowers might be replaced by tormentil, heath bedstraw and sheep's sorrel.

There were anthills on both parts of the hill. These only develop on pasture which is moderately grazed. They can't grow either where the grass is mown for hay or where grazing is so intense that they get trampled. Nor do they grow in woodland. Their growth is a process which tells us something of the recent history of the site. I also noted some birds, skylarks, swallows and whitethroats. These bring into focus the relationship of Bindon Hill with other places. The swallows need buildings to nest in and they probably find these in the village, a short distance to the north. They and the whitethroats are also migratory, so the connections reach as far as Africa. Neither do the grazing animals spend all the year on this one hill. They will be rotated around a series of fields of which this is only one and they probably overwinter in buildings.

So the landscape of Bindon Hill is composed of a whole web of processes and relationships. In fact the view of it you've just read is the product of another relationship, that between the landscape and the observer. My view of that landscape was influenced by my particular interests, the limitations of my knowledge, the time of year I visited it and the amount of time I spent there. Another person, another time, another purpose, all could have resulted in a different description.

This walk on Bindon Hill brings to the fore three important themes of landscape reading: everything changes, everything's connected, and everything has multiple causes.

Everything in the landscape is changing. Even the very hills, mountains and valleys, which look so solid and permanent, are changing shape on the incredibly slow timescale of the rocks. Changes in the vegetation are of course much faster than this. But even so there's a big difference between the slow, year-on-year process of succession and the much faster annual cycle of growth. These two cycles are going on at the same time but on very different timescales.

The annual cycle of deciduous trees is familiar to us all but herbaceous plants go through an annual cycle too. They start in spring with lush green leaves, which are actively turning the sun's energy into food. Everything looks fresh and young. In high summer comes the mature, reproductive phase when they flower and set seed, using the energy they've accumulated for reproduction. The plant is usually taller at this stage, to give the seeds more chance of moving away from the mother plant and colonising new ground. In many cases the leaves, now no longer active, go brown and tatty – hence the expressions 'gone to seed' or 'seedy-looking' for people who look past their best and uncared-for. The countryside loses the fresh, clean look it had in May and June. Because so many of the herbaceous plants have turned yellow or brown it can give an impression of dryness, even if the soil is wet from recent rain. Not all herbaceous plants follow exactly this annual cycle but it's common enough to dominate the feel of the landscape as the warmer months pass by.

If annual cycles are about the maturing of seeds and fruits, succession is about the maturity of whole ecosystems. Imagine a newly-ploughed field. Instead of sowing a crop, let's suppose that the field is left alone. It won't stay bare for long. First it will be colonised by mainly annual plants, then by herbaceous perennials, then by shrubs and young trees and then by woodland. This is not the end of the story because a young woodland like this will contain a very different range of plants and animals from a mature wood that has been in situ for hundreds of years or more. An ancient wood can be considered a mature ecosystem. The process can go from bare soil to the young woodland stage in a couple of decades or so. But from then on change is so slow that you can't see any difference in a human lifetime.

There's another kind of change in the landscape which falls between the short wavelength of the annual cycle and the very long one of succession. This kind of change is often caused by human action and it can be called rotational.

A classic example is the coppicing of an ancient woodland. This is a mature ecosystem, but over the centuries it's been repeatedly cut down and allowed to grow again, mainly through the trees resprouting from the stump. Just after felling it may look as though the ecosystem has suffered a major change, one which plunges it back to an earlier stage of succession. But no, this is just part of the medium-term cycle which the wood has experienced for hundreds if not thousands of years. The key difference between this kind of change and a successional change is that the mix of plants and animals which make up the ecosystem is not altered by it. Shade-loving creatures will have to move to another part of the wood which was not felled so recently while sun-lovers will move in the opposite direction. But the coppice wood will remain a coppice wood, made up of plants and animals which are adapted to this cycle of felling and regrowth.

Much the same happens when a timber plantation is felled and replanted. It can look like a devastating change has taken place. But if plantation is replaced by plantation it's a rotational change, not a change in succession. Foresters use the word rotation to mean the lifetime of one generation of trees. Farmers use the same word with a slightly different meaning. A rotation of crops on a farm means several years in which a number of different crops are grown one after the other, for example: wheat, followed by barley, followed by two years of temporary grass then back to wheat. All these rotational changes can go on year after year, century after century, without changing the nature of the ecosystem.

When you start out in reading the landscape it can be quite difficult to recognise what you see as a stage in succession rather than a permanent state of affairs. This was brought home to me once when I was giving some permaculture advice to a group of people who had just taken on a smallholding. We considered various options for different parts of the land – here a vegetable garden, there an orchard and so on. There was one area which I suggested might be left alone to succeed to woodland. It was mostly brambles, tall and dense, with a few young ash trees growing up through them. In other words it was in that stage of succession which comes between grassland and woodland – shrubs with young trees. One of the members of the group looked uneasily at the dense mass of bramble which towered over his head and said unsurely, "I suppose you've just got to have faith that it is going to turn into woodland." It can be equally hard to believe that cutting down a woodland is not destroying it. A woodland is only destroyed if something is done to prevent it regrowing. This either means grubbing out the stumps or grazing it sufficiently hard that none of the trees can regenerate.

In fact you can destroy the wood just as surely by grazing it without felling, though it will take longer because the existing trees will live out their lifespan.

The second theme which emerged from my walk on Bindon hill is the connectedness of the landscape. It was the birds which brought this to the fore. They, like many other mobile animals, make use of different kinds of habitat during different parts of the year or to fulfil different needs.

Greater horseshoe bats are an extreme example of this. They need a winter roost which stays at a constant cool temperature for hibernation and a summer roost which is warm, dark and undisturbed for daytime sleeping. Two important food items are dung beetles, which live in fresh cowpats, and cockchafers, which live in permanent pasture, that is pasture which is never ploughed up and reseeded. What's more the cattle can't be dosed with modern wormers containing Avermectin as this kills everything in the cowpat, including the bats' food. Tall, overgrown hedgerows or woodland edges are also needed. These provide a network of sheltered flight paths connecting the feeding areas with the summer roost. They also play a part in feeding, as one of the bats' hunting methods is to perch on a twig in wait for prey. The hedges must be thick at the bottom to provide good shelter, but unfortunately tall hedges are often neglected ones and these frequently become thin and gappy at the bottom.

What the greater horseshoe bat needs is a whole integrated landscape. No wonder they're nearly extinct. It's an extreme example but many wild animals need a mosaic of different ecosystems to provide them with all their needs. An intimate mix of woods, grassland, arable and water will support many species that even a rich and diverse ancient woodland could not. The simplification of the landscape, especially over the past half-century, has been one of many reasons for the decline in wildlife.

Migratory birds cast their net much wider, and who can say for sure whether a fall in the population of one of these is more due to changes here or in their winter home? The nightingale is a case in point. They're said to have been affected by the decrease in coppicing, which used to give them just the kind of bushy habitat they like for nesting. But this doesn't quite fit the facts since coppicing had almost completely died out by the middle of the twentieth century and the decline of nightingales has continued since then. It may be in part due to the dramatic rise in the deer population, which has thinned out the shrub layer in most woods. To add to the mystery it appears that at one time they used to breed in woods of tall trees with little undergrowth, but their behaviour has changed. Maybe the key lies in West Africa, where they overwinter, or somewhere on their migration route. Like so many things in the landscape, the decline of the nightingale is a mystery.

The third theme from Bindon Hill is that everything has multiple causes. This shows up most clearly in the distribution of the plants.

Take the gorse, for example. Its distribution on the hill is caused by soil, biotic and climatic factors, and the balance between them. It grows here because of the well-drained soil, its resistance to grazing and its partial tolerance of salt winds.

If there was no grazing on the hill it would probably be able to get established on the seaward side as thickly as it is on the landward. But it grows so slowly on the more exposed slope that even its thorns aren't enough to save it from the sheep.

We don't like complexity. We would always much rather put something down to a single cause. It's so much easier on our brains. It's never true but we fall into it all the time. For example, I might say "Rosebay willowherb is growing here but great hairy willowherb is growing over there because the soil is that bit wetter." What I really mean is that there are many different reasons why these two plants are growing in their respective places, but I reckon the only significant difference between the two places is soil moisture and great willowherb favours a wetter soil than rosebay. All the other factors are near enough the same and I leave them as understood. It's a useful shorthand. But the danger is that we tend to focus on one factor only and ignore the others. Then we can miss something significant.

In practice it's the observer who decides what's significant, although 'decides' is not quite right because we do it unconsciously. We often assume that the reason for something lies in the aspect which interests us most. Take a woodland with a wide diversity of trees in it. From a historical perspective we might say this is because it's an ancient wood rather than a recent one. From a botanical perspective we might say it's because it's on an alkaline soil, which supports a greater diversity of trees than an acid one. From a forestry perspective we might say it's because it hasn't been managed, as only a limited range of trees are worth growing as an economic crop. All these are true and we should remember that none of them is *the* reason.

So if I sometimes attribute something to a single reason, please remember that I'm simplifying. Most of the time we have to simplify in order to make any sense of what's before us. Nature is that complex.

How to Read the Landscape

Landscape reading can be done at various different levels. You can do it from the window of a train or a car and get a general idea of the country you're passing through. This is great fun. I often spend a train journey with a book lying unopened in my lap because I can't take my eyes off the fields, woods and rivers as they fly by. Although this is a very superficial way of reading the landscape it has the advantage of great breadth. You can see many different places in a short time so comparisons between different landscapes come into sharp focus.

Obviously you see much more of an individual landscape as soon as you swap wheels for feet. My notebook is full of insights I've had on a first visit to a place. If you work as a permaculture consultant a single visit is often all you get. This is usually enough for the purpose of giving some basic advice but no-one would pretend you can prepare a full, detailed design based on one day's observation. For that you really need to live with the land for at least a year, observing it carefully as it changes through the seasons. The same applies if you're just reading the landscape for pleasure: the more intimately you know a place the more of its secrets are revealed to you.

Even a second visit to the same place can be very valuable. I find I often notice as many new things the second time round as I did the first. The mind can only take in so much at once. Sometimes a change will have occurred since you were last there which reveals something you couldn't see before. This can also happen when you've already known the landscape in question for years.

An example comes from Ragmans Lane Farm, a place I know well because my work takes me there for about a quarter of each year. At the top of a small rise there are two maple trees standing by a gateway. Although they look to be the same age, one is more than twice the size of the other. The smaller one is visibly shrinking as branches die back while the other grows bigger each year. I'd never even wondered why until one year the land was let out to a neighbour who rotated his cattle round the farm on a different system from the one which had been used before. The new system meant that the cattle passed back and forth through this gateway for a longer period of the year and two prominent tracks of bare soil appeared where they habitually walked.

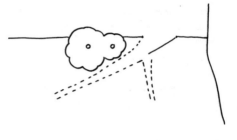

At once it was obvious that the smaller tree was suffering from soil compaction in the root zone. The most heavily used track and the gateway are both much closer to the small tree. In previous years, when the impact of the cattle wasn't enough to make a clear bare-soil path, it was still enough to harm the tree. But it was only when the cattle paths became visible that I realised what was happening.

Seeing the same landscape in all four seasons of the year is important. The vegetation can change quite dramatically. If you visit a woodland in winter, when the trees and shrubs are leafless and the herbaceous plants have died down, it can feel like a light and airy place. Visit the same spot in high summer and you may find the sunlight cut off by a closed canopy and your vision blocked in every direction by thick shrubs and tall nettles. The character of the place can be so different that you feel sure you've missed your way and come to a different wood.

Although seasonal differences are usually most dramatic in woodland, they're there all the same in hedgerows, grassland and other kinds of vegetation. It's not just the abundance of leaves and herbage that changes, but some plants are particularly prominent when they're in flower. This can give the impression that a particular species is more abundant than it really is. For example, if you see a landscape of woodland edges and untrimmed hedges in late winter you may be impressed by the number of hazel trees. Domes of golden catkins catch the eye on all sides and they seem to outnumber all the other shrubs combined. See the same scene at Easter time and you could be forgiven for thinking that almost all the shrubs in the area are blackthorn. The gentle frosting of its white blossom dusts every hedge and woodland edge with silver. And so on through the year. Whatever's in flower at the time seems to be more common than it really is.

The same is true of grassland. See a field at buttercup time and you may think every wildflower there is a buttercup. It may almost seem as though there's no grass in the field at all. Come back in a month's time and you won't notice any buttercups unless you look for their leaves. If the field is rich in wild flowers you may see blooms of other colours, if not it may seem as though there's nothing there but grass. So if you want to have a complete picture of the plants which are there you need to visit regularly throughout the spring and summer. Even a botanist who can identify wild flowers when they're not in bloom will need to come more than once because some wild flowers will not even be showing above the ground when others are in full flower.

Landscapes can look different from one year to another even if they haven't fundamentally changed. It's always a 'good year' for some plant or other, but this usually means that they're more visible rather than more abundant. The flowering hazels I mentioned above, for example, will be much more prominent in a good catkin year than in an average year. Common daisies are more visible in a year when the grass has been grazed down short at flowering time than in a year when it's taller and hides the little plants. The daisy population may be the same whether you can see them or not.

The time of day can make a difference, especially for observing animals in the landscape, as this extract from my notebook illustrates.

The Field, 19th June 2000

> I was up early in the morning and stood in the middle of the field. No human sounds at all. Under the bird song there was a steady buzz of bees working the flowers. I looked around and they were everywhere, at least one per square metre, probably more. Later in the day, when it got warmer, they were not in evidence.

We'd camped in the field that night, the first time I'd spent a night there for years. The early morning is a good time for observing all sorts of animals. Rabbits, for example, are much more active then. But rabbits and other mammals are usually detectable by the signs they leave behind them, whereas insects are usually not.

The time of day doesn't just affect what's there to see but also how we see it. There's a big difference between a scene lit up by the low-angled sunshine of early morning and the same scene cast into shadow by the low-angled sun of late afternoon. Suppose you visit a field with tall hedges on either side just as the magical glow of pre-sunset lights up the eastern hedge. You'll see masses of detail in that hedge: what kind of shrubs it contains, whether any are in bloom or, if it's autumn, whether any are changing colour. By contrast the western hedge will seem dark and homogenous and any sunlight which does filter through it will hinder your vision rather than help it. In hilly country you can get the same effect with whole landscapes. It's very easy to jump to wrong conclusions when you see a scene at just one time of day.

Jumping to conclusions is always something to be on your guard against. On the other hand it's important to trust your eyes and not to dismiss something you've seen because your brain tells you haven't seen it. That's just the mistake I made on the day I noted the following:

Brean Down, 14th August 2003

Trust yourself! Looking at the wind-flagged hawthorns I thought "Those look like they have a browse line. But surely not – this place is only grazed by rabbits." Later on we saw both goats and cattle. (See photo 4.)

A browse line is a tell-tale sign that grazing animals are around. They eat the leaves off the trees and shrubs up to the height they can reach, leaving a distinct 'tide-mark' with bare stems below and leaves above. I would have done better to take the browse line as an indication that the down is grazed by animals larger than rabbits. After all, the goats and cattle might have stayed out of sight the whole time I was there. Then I would have gone away with a mistaken impression of Brean Down.

There aren't many tools of the trade of landscape reading. I always try to remember to carry a notebook and camera. It's hard to remember everything you've seen, and it can be interesting to make comparisons both between different places and between the same place through time. The only really essential equipment is a pair of eyes. But there are three important aids which can add to what your eyes can tell you: plant identification guides, maps, and talking to people.

You can start out reading the landscape without knowing how to identify many plant species. For example, if you see young trees growing in a field of grass it's reasonable to deduce that you're looking at a process of grassland turning into woodland. The young trees will grow and as they get bigger they'll become the dominant life form and shade out the grass. You don't need to know what kind of trees they are, let alone which grasses and herbs are present in the grassland.

A wood consisting of a few big old trees with wide-spreading branches surrounded by lots of young trees growing close together suggests a later stage in the same process of wood formation. A tree with spreading branches can't have grown up in a woodland surrounded by other trees. In that situation it would have grown tall and narrow, shaded by its neighbours on all sides and reaching for the light above. So what we have here is a new wood growing up around trees which formerly stood far apart on grassland or heath. You don't have to know what kind of trees they are to work that out.

Having said that, you can discover more about the landscape if you can identify plants. Some indicate specific things about the soil, and thus also about the rocks below. Others tell about the history of the ecosystem. For example, certain plants are characteristic of ancient woodland rather than more recently formed woods. A whole new depth of meaning opens up as you become more and more familiar with plants. Nor is it difficult to learn. At the age of thirty I probably knew some half dozen each of trees, wild flowers and birds, even though I'd lived in the country all my life. Then I got interested and I started learning fast. It wasn't an effort, it was fun, and I'm still learning and enjoying it.

The first thing to do is to get yourself one or more guides, as identification books are known. You can get general ones, which cover all the wildlife of

Britain, or specialist ones covering either trees, birds, wild flowers, grasses and so on. A general guide may seem attractive because you only need to take one book with you when you go out. But because they cover such a wide range there's not much room for each species and often the descriptions are too brief for a really certain identification. For the same reason they can't include every species, though they probably will have most of the important ones.

For the purposes of landscape reading the most important groups are trees and wild flowers. Although animals are just as much a part of the landscape as plants they're less useful in landscape reading because they move around and hide while plants stay put. Soil indicator plants are very useful in practical landscape reading. These are plants which only grow in one particular kind of soil, so they can suggest which crops would be suitable or how the soil should be treated in the place where they grow. Other plants give a less clear message, but all have their preferences and much of the joy of landscape reading lies in watching the dance between soil, climate, plants and humans which has led to the landscape as it is today.

Grasses, ferns and mosses are on the whole difficult to tell apart, even with the aid of a guide. But fortunately there are only two species which are really important in landscape reading, bracken and rushes, both of which are common and easily identified. Other ferns and grasses do come into it, but you really don't need to know them in order to get started. It's probably not worth getting a guide to these plants.

It can be more difficult learning to identify both wild flowers and trees if you live in a town rather than in the country. Many of the apparently wild flowers you see will be garden escapes and many of the trees will be unusual ornamentals. You could waste a lot of effort learning to identify trees which have no significance in the landscape other than to indicate that someone has planted an ornamental tree. The best solution is to get a guide which doesn't include many exotic ornamentals and then regard any tree which isn't in the book simply as an 'unidentified ornamental'. For more detailed advice on choosing guides see Appendix A, Further Reading, on pages 320-321.

At first, with both trees and wild flowers, you need to look at specific features, like leaves, petals and buds, and study them carefully before you can be sure of the plant. But bit by bit you become more familiar with the plants and don't need to look so carefully at details. After a while you get to know the general character of the plant, what birdwatchers call its 'jizz', and recognise it without necessarily being able to say why – though you can always look for a telling detail if in doubt.

In winter you can sometimes identify a tree by the dead leaves at its foot. But this can be misleading. Leaves get blown around by the wind and there's no guarantee that the leaves on the ground come from the tree above them. Some leaves last longer than others. Ash, lime and willow leaves disappear quickly. They're the most palatable to earthworms, who rapidly pull them below ground and consume them in their burrows. Beech, oak, sweet chestnut, sycamore and maple are much tougher and can last all winter or longer. So, as leaves often get blown around by the wind, a tree with a few oak leaves under it could easily be an ash.

In summer, when every tree and shrub is in leaf, of course it's easier. But do be sure that the leaf you're looking at really is attached to the tree you're trying to identify. The branches often intermix and it's a common mistake to spend time examining a leaf that turns out to belong to a neighbouring tree.

Tree guides can teach you to identify individual trees, but when looking at the wider landscape you may want to know the broad composition of whole woods and hedges you can see at a distance. This is a much less precise art. Obviously you can't rely on those little details of leaf or twig which are diagnostic when looking at an individual tree. It's much more about colour, and the relative colours of different species change through the seasons. Appendix B, Identifying Trees at a Distance, sums up the notes I've made on this over the years. My observations are from the south west of England and South Wales, so the dates may need modifying for other parts of the country, and for global warming.

Maps can be useful in landscape reading, especially for the historical aspect. The patterns made by the fields tell us a great deal about the history of the land, as we shall see in Chapter 5, The History of the Landscape. These patterns usually show up better on a map than they do from the ground. A map gives you a view which is both vertical and much wider than you can see from the ground. Having said that, I have known the view from the top of a steep hill to give me an insight into field patterns which I hadn't had from the map.

Rodney Stoke, 22nd April 2005

Sitting on top of the hill, the pattern of fields below is much more eloquent than it is on the map.

These are quite clearly fields enclosed from the furlongs of the old open fields.* From up here they jump out at me, but on the map I didn't recognise them.

* See pages 93-94.

In that case I was looking at medieval field patterns. But maps can also help with more recent history if there have been changes since the map was made. Maps can show hedges which have been removed, ponds that have been filled in or old cottages which are now low ruins hidden by the vegetation. Sometimes an old map can confirm something you've tentatively read from the landscape itself. I have an example in my notebook from my early days as a landscape reader.

Mid Devon, Easter 1991

Basil's wood:

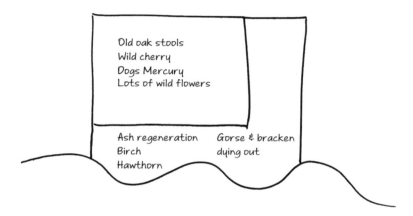

Old oak stools
Wild cherry
Dogs Mercury
Lots of wild flowers

Ash regeneration   Gorse & bracken
Birch   dying out
Hawthorn

Everything I noted in the north-west part of the wood indicates ancient woodland, everything in the south and east, recent woodland. I deduced that the north-west had long been woodland and the south and east had till recently been grassland. Later I checked the Ordnance Survey map. It showed the north-west corner as woodland and the south and east as grassland with scattered trees. This confirmed my appraisal and also told me that the grassland had had scattered trees on it at the time the map was made.

As well as showing you the history of the landscape maps can show things which are still there but which you might not have seen. Springs, for example, are important features but sometimes not very conspicuous. And, of course, maps help you to find your way about. But remember they're not always a hundred percent accurate.

The Explorer range of maps made by the Ordnance Survey are the best available for reading the landscape. They're at a scale of 1:25,000 and show a great deal of detail, including all the field boundaries. The smaller-scale Landranger, at 1:50,000, really doesn't show enough detail to be of any use. All the larger scale maps are now computerised. This has its advantages. For example you can ask for a print-out of the area you're interested in, no more and no less. In the days of paper maps you often had to buy two or more sheets if the area you were interested in crossed the boundary between one and another. But they're expensive, and neither the quality of the cartography nor the level of detail are a patch on the old paper maps. The 1:2,500 paper maps are like gold-dust if you can find them second-hand. Needless to say the map which showed me the history of Basil's wood was one of the old paper ones.

Since the underlying rock is such a big influence on the landscape, geological maps can be useful. They're quite easy to understand once you know the names of the different kinds of rocks. These are explained at the end of the next chapter. Soil maps, on the other hand, are so full of jargon that they're not much use to a lay person.

Plant guides and maps are basic tools which you can equip yourself with in advance and keep by you all the time. But local people who are willing to share their knowledge with you are a much less predictable resource, although often more valuable. They can sometimes tell you things which you couldn't have worked out for yourself, as this extract from my notebook illustrates.

Harbourneford, 15th May 1992

> Phillip told me the story of the patch of docks in the little field across the stream by the quarry. A neighbouring farm across the valley had their slurry pit overflow during a flash flood. The slurry came down the lane, in the gate and all along the bottom of the valley. The cows who produced the slurry had been eating silage full of docks, so it was full of dock seeds. They germinated in a swathe where the flood had been, right along the valley. But they only survived in the little field because it's not grazed by sheep, because the fencing's too bad to hold sheep.

Sheep will eat docks and cattle won't, at least not when they can avoid it. Sheep also need much better fencing to keep them in than cattle do. The little field by the quarry was the only one in the valley with fencing too bad to take sheep and that's why the docks survived there. I might have worked out that part of the story on my own but not how the docks came to be there in the first place.

Over the years I've learned a great deal about the landscape from talking to other people who live or work in it. It's not just the information that's useful but the different perspective each of us has according to our relationship with the land. A farmer or forester may see things which pass a nature conservationist by and vice versa. We all see the landscape through the filter of our personal experience.

Once I was designing a garden for a young farmer and his wife. He's generally a person of radical views with an open mind. At the time he was converting his farm to organic and looking at installing a wind farm on part of his land. But his reaction to my design revealed how dyed-in-the-wool a farmer he is.

I'd placed the things they'd asked for – vegetable garden, fruit trees, chicken run, flower beds and lawn – on the design but there was still plenty of land left over. Rather than give him a large area of lawn that would shackle him to the lawnmower for the rest of his life, I proposed a shrubbery for the surplus space, mainly of native shrubs. It would look attractive throughout the year, provide wildlife habitat and need virtually no maintenance. He looked very serious and said, "Well, I like most of the design, but not the shrubbery. I don't want to lose that much land."

It was almost as though he believed that once you plant trees or shrubs on a piece of land it ceases to exist as land. On a farm there is a grain of truth in this. Once you plant a field up with trees you can't farm it, at least not in the sense of growing arable crops or grass, which is what most farmers think of as farming. But in a garden?

He's by no means the only farmer I've come across with this attitude to trees. But it was really brought into contrast for me one day when I heard a forester refer to 'bare ground'. What he meant was not soil with no plants on it at all, but land without trees. It's a normal forestry term and this wasn't the first time I'd heard it, but it reminded me of the farmer and the shrubbery. The forester's perspective is exactly the opposite: if a piece of land doesn't have trees on it there's nothing going on there at all.

I don't mean to put people in compartments. After all, sometimes the farmer and the nature conservationist are one and the same person. But we do all have our own approach to the land and the challenge for all of us is to see it in the round, from every point of view. On the other hand we can never know everything about a landscape. Living processes are just too complex for that. A sense of mystery and wonder will always remain.

Rocks,
The Bones
of the Landscape

The town of Callendar lies on the boundary between the Highlands and the Central Lowlands of Scotland. Approaching it from the north you drive through a landscape of high rocky hills, deep glens and long narrow lochs. The hillsides are covered in bracken and heather with a scattering of birch trees. There's no cultivated land and only the occasional field of grass. Houses are few and far between and there's hardly a town to be found between here and the north coast, some 200 miles behind you. The road, like most roads in the Highlands, runs along the bottom of the glen. Suddenly you turn a corner. The hills on either side fall away and you're in a different landscape, one of gently undulating fields of crops and grass, neat wire fences, sycamore trees and farmsteads. In a few moments you're driving down the main street of Callendar. The contrast could hardly be more complete.

You've just crossed the Highland Boundary Fault. A fault is a split in the rocks and this is a huge one. It stretches from one side of the country to the other. The hard, old rocks of the Highlands have slipped down hundreds of metres on the southern side of the fault and stayed still on the northern side. On the southern side they're now covered by the younger, more fertile rocks which make up the Central Lowlands.

The clear difference between the two landscapes which meet at Callendar gives a dramatic display of how the underlying rocks affect the land. The effect may not be so obvious in other places but it's there all the same. The rocks influence the landscape in three main ways. Firstly, they affect the landform: harder rocks make hills and softer rocks make valleys. Secondly, they affect the presence of surface water such as streams and ponds, as some rocks hold water and others don't. Thirdly, they affect the soil and thus the vegetation. This in turn affects how people use the land, with woodland and extensive grazing on the least fertile land and intensive cultivation on the most fertile.

Kinds of Rocks

An important theme which runs right through this chapter and the next one, which is about the soil, is the difference between acidic and basic minerals. In fact it's a theme which crops up throughout the book. Basic minerals are rich in the elements which chemists call 'bases'. Many of these are elements which plants need for their nutrition and they make a rock or soil which contains them more alkaline.

Acidic minerals contain few plant nutrients and are, as you'd expect, acid. So rocks which are made predominantly from basic minerals give rise to soils which are rich in plant nutrients and alkaline, while rocks made mainly from acidic minerals give soils which are poor and acid. The effect of this difference on the vegetation is easy to imagine. Rich, alkaline soils can support deep, lush grass and waving fields of corn, while poor, acid soils are the home of heather and birch. Of course these are the extremes and there are a whole range of gradations between them but the balance between acidic and basic is one of the main ways in which the hidden rocks beneath our feet influence the landscape we see around us.

The most primitive rocks are those which have come straight up from the molten magma below and solidified in the Earth's solid crust. They're called igneous rocks, which means rocks from the fire. There's a whole range of igneous rocks, from the most acidic to the most basic. Towards the acid end of the spectrum is granite. In its molten form granite is viscous and rises very slowly from the magma, cooling gradually as it comes up through the solid rocks of the crust. Because it cools slowly it forms big crystals, some of them big enough to be seen with the naked eye. It's made up of a wide range of minerals but includes a high proportion of silica. Silica is a simple compound of silicon and oxygen. It contains nothing else and is very acidic. It's also exceptionally tough and slow to break down. When it does break down it rarely gets smaller than a grain of sand. In its pure form silica is known as quartz.

All igneous rocks are hard and resistant to erosion but granite is tougher than most. The highest mountains in Britain, Ben Nevis and Cairn Gorm, are made mostly or entirely from granite. Because it rises slowly through the Earth's crust granite usually takes on the form of great rounded masses. In the English West Country a single mass of granite with a string of domes on its back gives rise to a chain of moors from Dartmoor to the Isles of Scilly. These domes were formed below the surface and have gradually emerged as the surrounding rocks have been eroded away over millions of years. Granite is impervious to water and where the land surface is flat drainage is poor enough to form peat bogs.

The West Country Moors

Basalt lies at the other end of the scale from granite: rich in bases and very liquid when molten. It's so liquid that it floods out of the ground and forms horizontal sheets of lava. In places it has poured out of the Earth repeatedly to form layer upon layer of rock. In some cases these layers have built up to a great height and now stand up as mountains. The north-western part of the Isle of Skye and much of Mull have been formed out of repeated flows of basalt. Like granite, basalt contains many different minerals but the proportions are quite different. There's much less silica and much more of the basic minerals. So a soil derived from basalt is richer in plant nutrients and more alkaline than one derived from granite. In the drier parts of the Highlands, away from the west coast, you can pick out the granite mountains from ones made of more basic rocks because they're covered in heather while their neighbours support more demanding plants such as bracken and grasses.

Although igneous rocks are hard they don't last forever. Even the hardest rocks eventually get worn away by the forces of erosion. But as they erode another distinction between acidic and basic minerals is revealed: they break down in different ways. The tough quartz doesn't decompose chemically and is hard to break down physically. So those large crystals of quartz which originally formed as the granite slowly cooled stay much as they are, inert little lumps about the size of a grain of sand. Basic minerals, on the other hand, can be physically broken down and chemically changed. The result of this chemical change is the range of soft, plastic minerals we know as clay. A single particle of clay is microscopic, tiny compared to a grain of sand.

The eroded material is carried away, mostly by streams and rivers, and deposited lower down, in the sea, in lakes and by riversides. These deposits are laid down over the ages, layer upon layer, the weight of the upper layers compressing the lower ones until they're transformed into new rocks. These are called sedimentary rocks. While igneous rocks are mainly found in the mountainous north and west of Britain, sedimentary rocks predominate in the lowlands of the south and east.

Water does more than move the eroded material, it also sorts it according to the size of grain. This happens because it takes more energy to move a heavy object than a light one. It takes more energy to move a sand grain than a clay particle and more again to move a stone. A young river rushing down a mountainside is full of energy. It can move stones, at least when it's in spate, but when it reaches flatter ground it slows down and drops them. It's no coincidence that gravel beds are usually found on the banks of rivers, such as the Thames. It was the rivers themselves which made the gravel beds. Sand is often carried as far as the sea, while clay particles can be held in suspension in the water and will only settle out when the water is still. This means clay can be transported further than sand and is deposited separately. Something of this sorting effect can be seen round the coast today. Sandy or pebbly beaches form where the waves beat against the coast, while the still estuaries fill with mud.

These different deposits turn into different kinds of sedimentary rocks. The sand forms sandstones and the clay forms what geologists call mudstones but which most of us simply refer to as clays. With some more pressure from above clays turn into shales, which are a bit harder and somewhat laminated. Sandstones tend to be harder than clays and when the two are found side by side the sandstone makes the hills and the clay makes the vales. Clays are impervious to water and a clay landscape is full of rivers, streams and marshes. Most of the marshes have long since been drained but place names with 'marsh' in them, such as Middlemarsh or Dilton Marsh, can sometimes point to a village sited on or near the clay.

Sandstones may or may not be permeable to water, depending on whether or not the individual sand grains are cemented together by some other component in the rock. The millstone grit which forms much of the Pennines is a strongly cemented sandstone and peat bogs are characteristic of it. The greensand of southern England is uncemented and freely-draining. Ponds and brooks are rare on it. But the sandy soils which are formed from sandstones of all kinds are usually well-drained unless an impervious layer below interferes with the drainage. Very sandy soils are excessively well-drained and dry out quickly after only a short time without rain.

Clays and sands are also quite different from each other in terms of acidity and plant nutrients. The sorting by size which takes place during the process of erosion also has the effect of sorting the particles by base content. Most of the sand-sized particles are made of the tough, resistant silica, which is nothing more than silicon and oxygen. The more complex minerals which contain bases have been decomposed and converted into clay particles. So, regardless of what mix of rocks the erosion process started with, most of the bases have ended up in the clay and most of the silica in the sandstone.

This means that soils derived from sandstone are usually acid and poor in nutrients, while those derived from clay are usually alkaline and rich in nutrients. In practice all soils contain both sand and clay but the proportions vary from extreme sandy and clay soils at either end of the spectrum to soils with a balanced mixture of the two in the middle. The source rock is the main influence on the proportions of sand and clay in a soil, and where sandstones and clays are found close together there can be some sharp contrasts in landscape.

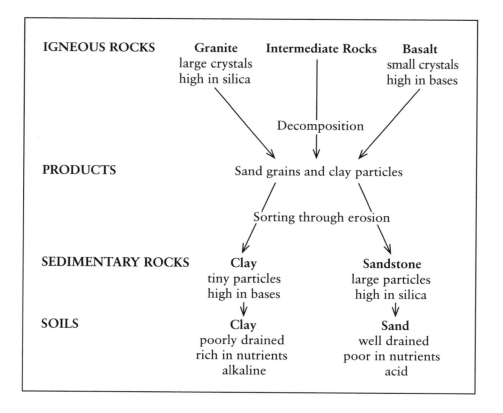

Very sandy soils often underlie lowland heath, with its characteristic vegetation of heather, gorse and bracken. Ashdown Forest in the middle of the Weald of south-east England is an example. It makes a very obvious contrast with the green, hedged fields of the Vales of Kent and Sussex on either side of it, which are both predominantly clay. Sandy soils have often been left as common land for rough grazing while the clay land was farmed intensively and individually owned. Without chemical fertilisers it was just not possible to get a sustained crop from the sand. Even now there's more common land remaining in areas with a lot of sandy soils. Surrey, for example, has more commons than any other county in southern England.

This pattern can be carried over from an agricultural landscape to an urban one. Hampstead Heath in London is an island of open land in a sea of bricks and mortar. It's also an island of sandy soil surrounded by clay, so it was common grazing while the surrounding clay was divided up into privately owned farms. As the town expanded, individuals could sell their land for building but the heath remained common property which no one person could sell. Hampstead Heath owes its survival to the difficulty of getting a group of commoners to agree on disposing of their common property.

Sandstone and clay come from purely mineral origins, but there are two other important sedimentary rocks which come from biological sources: limestone and coal. Limestones are formed in relatively clear seas, where there's little or no debris from erosion. The only mineral present in large enough quantities to accumulate on the bottom and form rock is calcium carbonate, or lime. Most limestones are made from the accumulated skeletons of billions of little sea creatures which garner every speck of lime they can from the water to make their skeletons. Some are reef-forming, like corals, others free-swimming. When they die everything but their skeletons rots away and disappears. Over millions of years the accumulation of lime on the sea floor is consolidated into layers of limestone which can reach thousands of metres thick. Limestone has two important characteristics which lead it to form distinctive landscapes: it's permeable to water and it's very alkaline.

Limestones almost always make hills. Some are hard, like the Carboniferous limestone of the Pennines and Mendips, while others, like chalk, are soft. But even the chalk is resistant to erosion because it's so permeable. Rainwater sinks straight through the rock, leaving none on the surface, and surface water is the main agent of erosion. After exceptionally heavy rainfall there can be some surface flow but this is usually just enough to erode the soil, not the rock.

Carboniferous limestone, so called because it dates from the same geological age as the coal, makes a mildly rugged landscape. The grey rock is often exposed on the surface and it can make spectacular cliffs. People have gathered loose stones from the surface and made them into the characteristic field walls. By contrast, the softness of chalk has given its hills a rounded, pillowy profile. Chalk is generally too soft to make walls and the flints which are found in it too small for dry-stone work, so there are no field walls in chalk country. But despite its softness the chalk stands up above clay vales and sandstone hills alike.

The Cotswold limestone is intermediate, less rugged than the Carboniferous and less rounded than the chalk. It's hard enough for stone walls and the Cotswold country is famous not so much for its field walls as for picture-book villages made of the honey-coloured stone.

As well as lacking surface water, limestones are very alkaline. Lime, the very stuff they're made from, is calcium carbonate and calcium is one of those bases that make clays more alkaline than sands. Limestone soils are much more alkaline than the average clay soil. Generally speaking there's a spectrum with sandy soils at the acid end, clay in the middle and limestone soils at the alkaline end. On the other hand, limestone soils aren't necessarily rich in the other bases which plants need for their nutrition, so they're not necessarily fertile.

One characteristic of alkaline soils is that they can potentially support a much greater diversity of plants than more acid soils. Of course there are other factors which affect diversity, especially human activity. But given a semi-natural ecosystem, like a woodland or an ancient hedge, a high diversity of plants is a clue that the underlying rock is limestone, or at least a very alkaline clay. Another clue is the presence of lime kilns. These were used in the past to burn limestone for making whitewash and agricultural lime. They're usually square structures, four or five metres wide, with a characteristic triangular hearth recessed into the base, though some of the later ones have round-arched hearths. They can also sometimes be found on the coast, on a creek or inlet where limestone was brought in raw and processed near the point of use. This, of course, is an indicator that the local rock is not limestone, otherwise why bring it in by sea? (See photo 7.)

Coal is the other sedimentary rock of biological origin. It comes from the plant remains of swamp forests. In the waterlogged conditions of a swamp dead plants don't decay completely but turn into peat and if peat is subjected to great pressure it turns into coal. The swamps which formed the coal probably grew in great river deltas. A delta is a very unstable landscape, with channels constantly shifting. Any particular place could spend some time as a swamp and then some time as a sand bar or mud flat. These mineral deposits eventually became sandstone and shale and their weight compressed the peat into coal. This is why coal is found in thin seams between thicker layers of sandstone and shale, never in pure deposits hundreds of metres thick as limestones often are. This complex of rocks is known as the coal measures though only a small proportion of the rocks in them is coal.

Coal has had a major impact on the landscape of Britain. More than any other factor it has determined which parts of the country are urban and which are rural. Most industrial processes required a greater bulk and weight of coal than of raw materials, so it was easier to transport the cotton, wool or china clay to the coal than vice versa. Not every coalfield became urban but the biggest and most productive ones did.

These four sedimentary rocks, sandstone, clay, limestone and the coal measures, make up most of the solid rocks of lowland Britain. In some places it's easy to see the contrast between one and another, as with the chalk and clay of Wiltshire which I mentioned in the last chapter. But in other places the change from one sedimentary rock to another makes itself felt more subtly. A page from my notebook gives an example. It's about a walk I took from Ragmans Lane Farm into the nearby Forest of Dean. The farmland which lies outside the Forest overlies various limestones, while the Forest itself lies on coal measures, mostly made up of infertile sandstones. It's a steep climb up from the farmland into the Forest.

Ragmans Lane, January 2001

There's a very clear boundary between the limestones below and the sandstone of the Forest above. It lies exactly on the Ruardean to Lydbrook road. The geological map shows the first coal seam here. On the uphill side of the road there's a strip of gruffy ground,* now covered by a beech plantation. The course of the road, just downhill from the coal pits, was presumably determined by the need to get the coal out.

The coal seam actually separates the limestone from the sandstone. Below is a network of small fields and lanes with scattered cottages, farms and smallholdings – a typical rural-industrial settlement pattern. The trees include oak, ash, lime, cherry, beech, wych elm, birch and maple with old man's beard. Above, beyond the beech plantation, the deciduous trees are restricted mainly to oak, beech and birch. The soil texture, which was silty below, is sandy here. The pattern of lanes, cottages and small fields is the same except that suddenly there's a lot of unfenced land, presumably common – typical of less fertile soils. Ahead is the edge of the forest plantation.

* Somerset phrase meaning the humps and hollows of old mine workings.

As well as the igneous and sedimentary rocks there's a third major group, the metamorphic rocks. These started life as igneous or sedimentary and have been transformed by extreme heat, pressure or both into something which is distinctly different and usually much harder. Clay is turned into shale by moderate pressure but more pressure turns it into slate, which is considered a metamorphic rock. Sandstone metamorphoses into quartzite, in which the sand grains are fused together making one of the hardest of all rocks. Gneiss and schist are coarse-grained rocks made from a variety of sedimentary and igneous originals. They're the commonest rocks in the Scottish Highlands and much of Mid Wales. Metamorphic rocks are uniformly hard and impermeable. Regardless of what material they were originally made of they tend to form hills or mountains with thin, poorly-drained soils. There's not the contrast between one and another that there is between, say, a clay vale and a limestone ridge. So individual metamorphic rocks have less significance in landscape reading than individual sedimentaries.

All the rocks I've mentioned so far are solid rocks, but mineral material doesn't need to be consolidated in order to be considered a rock. Surface deposits which have been laid down recently and haven't yet been compressed into solid rock are known as drift deposits. Wind-blown sand in coastal dunes is an example. Fine material deposited by rivers is called alluvium, and almost every river which is moving slowly enough to shed some of its load is bordered by a band of it. Peat is formed under waterlogged conditions, both in hill areas where the rainfall is high and in low-lying basins where drainage is poor. Boulder clay is a mixed bag of materials left behind by the ice sheets of the Ice Ages. As its name suggests, it's mostly made up of clay and stones. In East Anglia it overlies chalk and here it has made some of the most fertile soils in the country. Clay on its own can be badly drained due to lack of structure and chalk on its own can be poor in plant nutrients, but in combination the clay provides the nutrients and the chalk improves the structure.

Although solid rocks largely determine the landform, drift can be much more important in terms of vegetation and land use. For example, a geological map which shows only the solid rocks shows a wide band of chalk down the middle of East Anglia. But when you're there you have to search hard to find anything resembling chalk downland as most of it is covered in a swathe of boulder clay.

Hills and Valleys

The growth of sedimentary rocks tends to be a quiet business. Sand, clay or lime accumulates slowly, mostly at the bottom of the sea, in flat, horizontal beds, known as strata. It takes a great deal of force to raise these flat strata up above sea level and turn them into the hills and valleys we see around us. This force comes from the collision of the Earth's great plates. The Earth's crust is not a seamless skin of rock but a series of plates, fitting together rather like the pieces of a tortoise's shell. But unlike a tortoise's shell the plates of the Earth's crust are constantly moving relative to each other, very slowly, but inexorably. When two plates collide, great mountains are thrown up. The Alps are caused by the African plate colliding with the one which includes Europe, the Himalayas by India striking the great mass of Asia, and similar collisions have happened many times before in the history of the Earth.

The crunching of the Earth's surface allows igneous rocks to well up from the heart of such collisions and metamorphic rocks are formed in the enormous heat and pressure they release. Sedimentary rocks can be thrown up to form the highest mountains and further away from the site of the collision, where the shock waves are gentler, they can be pushed up into gentler hills. This happens either by faulting or by folding. The Highland Boundary Fault which I mentioned at the beginning of this chapter is an example of faulting. Folding turns flat strata into wave-like folds. Folds come in all sizes: a single fold can make a whole range of hills or you can hold a piece of rock containing a complete fold in the palm of your hand.

When folding happens you might assume that the upfolds would be the hills and the downfolds valleys. This does happen sometimes but it's just as often

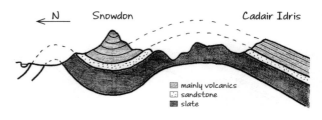

the other way round, with the downfold left as high ground after the upfold has worn away. This happens because the rocks in the downfold are compressed and hardened by the folding process, while those in the upfold are stretched and weakened. These weakened rocks erode away more quickly than the strengthened rocks of the downfold. Snowdon, the tallest mountain in Britain south of the Highland Line, is a downfold. Made up of hard igneous rocks, further hardened by being in a downfold, it has survived, while the loosened upfolds on either side were worn away by the relentless forces of erosion. Even soft rocks can make hills if they're in a downfold. The little hill of Glastonbury, on which I sit as I write this, is mainly made of clay. It's a hill because it's a downfold and in its own modest way it withstood the forces of erosion while the land around was worn down to sea level and became marsh.

The Forest of Dean is another downfold. Because it's tilted towards the south, you don't appreciate that it's a range of hills when you approach it from that direction. But from the east, west and north you climb a steep escarpment to enter the area. The limestones and sandstones which make up the Forest are hard rocks. Nevertheless they've been worn away from the surrounding country where once they formed upfolds, revealing the clay that lies beneath. Only in the downfold, which includes the Forest itself, have they survived to form hills.

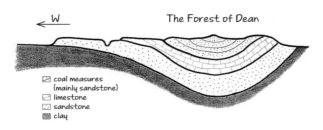

The difference between limestone and sandstone has had little effect on the shape of the land but it's had a big effect on the way the land is used. The alkaline limestone is much more fertile than the poor, acid sandstone of the coal measures. Although much of the limestone lies within the legal boundary of the Forest it has been farmed for many centuries, while the coal measures have always been wooded.

The South Wales coalfield, to the west of the Forest, is a larger downfold, made of the same rocks, also tilted to the south. But it's deeply dissected by the rivers which have carved out the famous Valleys, so its structure is much less clear. Between these two downfolds lies the Vale of Usk, a broad clay plain where formerly there was an upfold.

The Weald and Downs of Kent and Sussex form a series of ridges and valleys lying approximately east-west. This is not a sequence of separate folds but a single large upfold which has eroded away, exposing a series of rocks which erode at different rates. The overall landform is the result of the whole

The Weald and its flanking downs

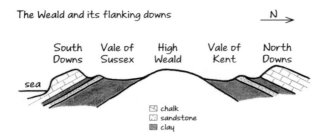

region being an easily eroded upfold but the detailed topography is the result of differences in erodibility. Chalk and sandstones stand up as hills while the vales are made of clay.

Nevertheless upfolds can form hills, especially when the rocks which were laid down first, and are thus at the heart of the upfold, are the hardest. The Mendip Hills, near my home in Somerset, are an example. Most of the range is made of hard limestone but the sandstone which lies under it is harder and stands up above the plateau formed by the limestone. The clay, which was laid down later, is much softer than either and has eroded away to leave the Mendips as a clear range of hills, visible from far off.

The Mendip plateau was mostly rough common grazing till the end of the eighteenth century. Then all the limestone area was enclosed with neat stone walls and divided into farms. Only the sandstone Black Down has remained as an unenclosed common of heather and bracken. Its soil is acid and poorly drained compared to the limestone and being higher it gets even more cloud and rain than the rest of Mendip, which is notorious for its wet and misty weather. The clay vales at the foot of the hills had been intensive farmland for many centuries before this. This is where most of the villages lie. The warm, sheltered microclimate on the southern side of the hills has made it a centre of strawberry growing.

Where strata have been tilted rather than folded the harder rocks stand out as a ridge with one steep side and one gently sloping. The gentle slope more or less follows the angle of the strata and is known as the dip slope. The steep slope, known as the scarp, is formed where the exposed strata have been eroded away. The whole ridge is sometimes referred to as a scarp, though strictly speaking this term only applies to the steeper side where the slope cuts through the strata. The Cotswolds are a typical example, with a steep scarp slope and a dip slope that's so gentle it's hard to say where the hills end and the

vale begins. My notebook records a train journey through the Cotswolds, from the scarp slope on the north-west side of the range to the dip slope on the south-east.

From Gloucester to Swindon, 10th March 2003

In the Vale of Gloucester the fields are mostly grassland, some with a trace of ridge and furrow.* On the left lies the Cotswold scarp, an irregular line of steep hills, partly wooded and partly grassland. There's a clear contrast between the dark green of the intensively farmed fields at the foot of the scarp and the pale buff-green of the permanent pasture on the steep slopes, some with scrub on them.

At Stonehouse the railway line makes a broad sweep to the left into a gap in the hills and soon the train stops at Stroud. The town is shaped like the fingers of a hand, spreading up through five valleys that snake back into the hills from their meeting place at the town centre. The railway follows one of the valleys out of the town.

River, canal, road and railway are crowded into the narrow valley, along with buildings old and new. On the steep hillsides small fields, mostly used for pony grazing, are interspersed with woods of ash and hazel with old man's beard. The train climbs all the time, winding through the steep valley, hooting at each bend. Now the slopes are covered in woods of ash, beech and birch and the houses are left behind.

Where the going gets too steep the train goes into a tunnel and comes out in a cutting of Cotswold stone. The strata are horizontal as near as the eye can tell. They only appear to dip away from us because the railway line is still climbing. We come out in a woodland which has mostly been converted to conifer plantation.

Soon we're in wide open country of big arable fields bounded by low stone walls. The land is undulating, with wide, shallow dry valleys.** Often there's not a house or a barn in sight. It has the feeling of hill country, if only because you can't see any hills so you must be on top. Here we're well back from the scarp slope, but still high on the Cotswold plateau.

As we descend the dip slope there's no sensation of going down, but the landscape subtly changes. Occasionally a field is bounded by a hedge instead of a stone wall. The pure arable gives way to a mixture of arable and grass. At Kemble there's a short tunnel and a cutting, still through the almost horizontal limestone strata. But beyond it there are no more field walls, only hedges, though the houses are still of Cotswold stone. The land is less undulating, all grassland now.

The fields get smaller, the hedges thicker, with more oaks in them, and occasional ditches beside them. Then there's elm in the hedges, some ridge and furrow* in the grassland, and houses of brick. Imperceptibly the limestone of the Cotswolds has merged

into the clay of the Vale of White Horse.
As the train enters Swindon you can see the next scarp, the chalk of the Marlborough Downs, rising on the horizon ahead.

* Medieval cultivation ridges. See page 91 and 94-95.
** A sign of limestone country. See page 41.

All these landforms are made by the simultaneous action of two forces: folding and erosion. Even as the upfolds begin to rise they start to erode. There have been two major forces of erosion in Britain and these have given us two very different kinds of landscape. To get an idea of this I'd like to look at two contrasting valleys: Glen Coe in the Highlands and the upper Dart valley in the West Country, especially the section between Dartmeet and Holne Chase. They're both upland valleys draining an area of moorland but they could hardly be more different in appearance.

Glen Coe is spectacular. It's not just the height of the mountains on either side of it but the way the sides of the glen climb up almost vertically from its wide, rounded bottom, reaching up and up, leaving the glen itself as a great space between towering walls. In cross section it's like the letter U. Curves in its course are gradual, so you have a long view up or down the glen. A small river runs along the bottom and feeds a lochan (a small lake) at the lower end. Surely it's far too small a river to have carved out this great void?

The Dart, as it cuts its way though the granite of Dartmoor is a secret river, hidden deep in the bottom of its narrow valley. The valley sides rise straight from the banks of the rushing river, not vertically but steeply, till they gradually round out onto the plateau. In cross section the valley's like the letter V. From the riverbank you can't see far in any direction, not just because most of the slopes are wooded, but because the narrow valley constantly changes direction, so whichever way you look a protruding spur blocks the view.

The reason why these two valleys are so different is, in a word, ice. An Ice Age glacier made Glen Coe what it is, but the ice never got as far south as Dartmoor, and the upper Dart valley is the result of water erosion alone. Water and ice erode the rock in two very different ways.

Water always follows the easiest way down a hill. This may be straight down, but it's just as likely to be off to one side, deflected by a hump or attracted by a dip. Sooner or later some other irregularity pulls it

Interlocking spurs on the upper Dart

back the other way and the young stream develops a zig-zag course down the hillside. Once a crooked course like this is started it gets entrenched as the stream bites down into the rock. The spurs of land on either side of the river fit into each other like the cogs in two cogwheels. This pattern of interlocking spurs is almost universal among young rivers and streams unless their course has been modified by human action. It's the reason why you can't see far up and down the course of the upper Dart.

The Dart is a young river, rushing down the side of the moor, cutting into the rock and carrying the eroded material away to deposit it in the lowlands. The narrow bed of the river is the main focus of erosion. That's where most of the water is and it's the water which erodes. It wears down its bed by drilling holes in it with stones. When the river's in spate any irregularity in the riverbed causes an eddy. Stones are whirled around in the eddy and eventually wear away a rounded hole. Eventually the holes coalesce and the course of the river is lowered. (See photo 6.) The powers of erosion are concentrated in the river bed and this gives the valley a v-shaped profile.

If a river cuts through rock like a series of drills, a glacier wears it away like sandpaper wrapped around a block of wood. The ice picks up stones which become frozen to its bottom and sides and abrade the solid rock as the glacier slowly moves forward. The weight is enormous and though glaciers move at less than a snail's pace they can erode at least as much as the frenzied fury of a river in spate. Because they move so slowly glaciers have to be huge in order to shift the same volume of H_2O in a given time. So, when a young river up in the hills becomes a glacier the valley gets much bigger. The interlocking spurs are worn away and the base of the valley broadens out. Ice is less flexible than water so any curves in the valley are smoothed out and become more gradual.

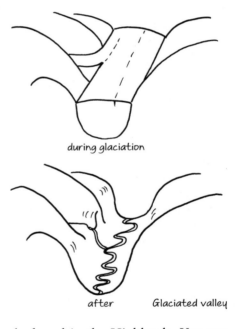

during glaciation

after   Glaciated valley

U-shaped valleys like this are not only found in the Highlands. You can see them in any part of upland Britain which was reached by the ice during the Ice Ages. (See the map overleaf.) Many of these areas are much less harsh than the Highlands and are farmed more intensively than the typical Highland glen. This gives rise to a distinctive landscape type. The flat valley floor is filled with hedged or stone-walled fields and dotted with farmsteads or the occasional village. The steep sides are usually in larger fields of less productive pasture, with occasional woodland, while the fell tops are open moorland. This pattern is particularly characteristic of the Pennines and the

The limit of glaciation

limit of last Ice Age

maximum limit of Ice Ages

Lake District but can be seen in parts of Scotland and Wales too. We could call it the dale and fell pattern.

This is a pattern repeated again and again over a wide area, on a par with the combe and plateau pattern. (See page 6.) In the combe and plateau landscape all the cultivation and most of the grazing are on the flat plateau above because the valleys consist of nothing but steep sides. Dale and fell country is the other way up, with most of the cultivation on the flat valley bottoms. As the fells are usually much higher than the gentle plateaux of the West Country the microclimate on top is usually too cold, wet and windy for anything but rough grazing and moor. In origin, one of the two is an unglaciated landscape with V-shaped valleys, the other a glaciated landscape with U-shaped valleys. (See photo 5.)

In the later, lowland part of their course there's little difference between the rivers of glaciated and non-glaciated parts of the country. The mark of a river's mature phase is a rhythmic, snake-like course made up of those sweeping curves known as meanders. Meanders both erode and deposit material. They don't erode downwards so much as sideways and thus widen and flatten the valley floor. They do this because the current is stronger on the outside of a curve and slacker on the inside. So they erode from the hills on either side of the valley, which are often no more than low bluffs, and deposit the eroded material on the valley floor. The current is also stronger on the downstream side of a curve than the upstream side so the river eats away on the downstream side and leaves material on the upstream side. Thus meanders slowly work their way down the valley, flattening the valley floor as they go. In this way they gradually fashion a flat flood plain. Meanders can be seen on mature rivers and streams all over lowland Britain, and sometimes in upland areas where the land levels out for a bit. It's not uncommon for a stream to go through a meandering phase with a little flood plain where the slope's gentle and then back to a V-shaped valley through a defile of hard rocks.

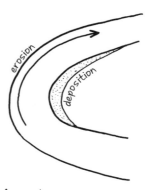

A meander

Erosion on all rivers happens in occasional bursts when the river is in spate. Sometimes you can see how fast a meander is eroding when some fixed structure, such as a fence, gets left behind in the river as the bank disappears beneath it. In the Highlands I once saw a small meandering stream with a line of living alder trees in the middle of its course with a corresponding line of alders left high and dry on the inside of the bend.

Any river or stream with a dead straight course has probably been straightened by human hand. Sometimes the original meandering course can be seen as a dip in the fields of the flood plain. In other places it may be too slight to see under normal conditions and only reveal itself after heavy rain, when it fills with water. Or it may survive only as the meandering dotted line of a parish boundary on the Ordnance Survey map. Rivers can also change their course of their own accord and then you can see similar ghost meanders beside a river which still has its natural curving course.

Getting to Know the Rocks

One of the most fascinating things about the British landscape is how varied it is. From the front window of my house I can see four distinctive landscapes. On the right are the Mendips, a plateau of stone-walled fields and ash woods. Straight in front of me, stretching away to the sea, are the Somerset Levels, former marshes where drainage ditches take the place of hedges. Half-left in the distance lie the Quantocks, old sandstone hills topped with heather, bracken and oak woods. Further left and closer to hand is a belt of hedged fields and stone villages on a mixture of clay and limestone which rises from the edge of the Levels to form the low ridge of the Polden Hills.

If you travel from here to the south coast, some thirty miles away, you pass through yet more kinds of country: chalk downs, sandy heaths, deep clay vales and hills of the golden yellow Ham stone that makes the houses in those parts glow warmly against the green fields and trees. To the north, just over the Mendips, there's a jumble of hills and narrow valleys which harbour a former coalfield, and patches of land with a soil so red it was till recently quarried for pigment.

This variety of landscape is mostly down to the variety of rocks. Climate certainly does vary from one part of the country to another, but on the whole it changes gradually and doesn't produce contrasting landscapes side by side. Where it does change abruptly it's usually because of abrupt changes in the landform and these are caused by changes in the rocks. As for the human influence on the landscape, although we've always tried to get the same things from the land – food, shelter, fuel and so on – the way we've gone about it has varied from place to place according to natural conditions. Where there's an abrupt change in the human element in the landscape over a short distance it's invariably because of a change in the rocks below.

Some parts of the country have less variety in them than Somerset but as a whole Britain is remarkable for the variety of its rocks. In most countries of the world you'd have to travel hundreds of miles to see the varieties of landscape that here you can see in dozens of miles. Over the long course of geological time the land which is now Britain has drifted over different parts of the globe. At every stage, in situations varying from desert to deep sea, it has acquired deposits of rock. It's not just the number of different rocks which gives us the variety of landscapes but also the fact that so many of them crop up on the surface. Despite the details of folding and faults, if the land had stayed level on the large scale only a few of the most recent rocks would cover the surface and all the older ones would be hidden below. The most important single event in our geological history was the one which brought them to the surface.

In the magma below the Earth's crust there are convection currents, great movements of molten rock flowing from the interior towards the surface of the planet. About sixty million years ago one of these raised up the north-west of Britain, making it higher than the south-east. Whenever rocks are lifted up they start to erode, and the soft young rocks overlying the north-west were worn away to reveal the hard old rocks below. The reason why the north-west of Britain has mountains and the south-east doesn't is simply that the older rocks are exposed in the north-west. It's not so much the effect of the original uplift as the difference in the rate of erosion between the old rocks and the younger ones. If all the rocks eroded at the same rate Britain would now be level again. The Highlands are high because the rocks there are hard.

In the south-east the layers of younger, sedimentary rocks were tilted by the uplift. As the raised part of each layer eroded away a series of dip-and-scarp ridges was formed, including the Chilterns, Cotswolds, Yorkshire Wolds and many other ranges of hills. There they lie like the edge of an open carpet sample book, each sample different from the next and each leaving a little bit of the one below it showing. The scarp slope of these ridges almost always faces somewhere between north and west.

There's a steady progression in the age of the rocks right across Britain from the Isle of Lewis to the Suffolk coast. A journey from one to the other is a journey through time as well as through space. If you cut across the line of this journey at right angles, you leave the older, harder rocks to the north-west and the younger, softer ones to the south-east. It's often said that if you draw a line from the mouth of the river Tees to the mouth of the river Exe – approximately from Middlesborough to Exeter – you divide highland Britain from lowland Britain. It's a very wiggly line but it does indeed divide the majority of the high hills and mountains from the broad vales and low hills. This is why the welling up of that convection current in the magma was such an important event in the formation of the British land-scape. It formed its biggest single feature: the division between the rugged, infertile lands of the north-west and the soft, fertile south-east.

Highland and lowland Britain

In getting to know the rocks of your area you may be lucky enough to get a glimpse of them directly. They may be exposed naturally in mountainous areas and on sea cliffs or revealed by human action in quarries and gravel pits. You can also see them in stone walls, but field walls are much more reliable than buildings because people have often transported good building stone. The more prestigious the building the further they were prepared to bring the stone.

Some English cathedrals are even built from Caen stone, imported from Normandy. Norwich Cathedral, conveniently situated on a navigable river, is of Caen stone. On the other hand the round towers of some East Anglian parish churches are a sign that the only local building stone is flint. Flints come in small sizes and with small stones you can't make a corner strong enough to support a tall tower. But any building, even a cottage may be made of stone from a mile or two away. On my trip over the Cotswolds I couldn't mark the exact boundary between limestone and clay by the change from stone houses to brick.

More often you'll be dependent on indirect clues to give you an insight into the local rocks. The most useful of these to the landscape reader is the soil. The soil can't tell you exactly what the rock is but it can tell you whether it's sandy or clay, acid or alkaline and give a good idea of how permeable it is to water. The easiest way to read the soil is by looking out for soil indicator plants, and this is dealt with in some detail in the next chapter. But remember that the soil isn't always a reliable guide to the solid rocks below. Some soils are formed from drift deposits which may be too thin to appear on a geological map.

The landform will tell you which are the more erodible rocks and which the least. But hardness is only relative. The Brecon Beacons, which are the highest mountains in South Wales, are made from the same Old Red Sandstone rock as Strathmore, the broad valley that lies between the Grampians and the Sidlaw Hills in eastern Scotland. In South Wales the Old Red is a relatively hard rock, but the Grampians and Sidlaws are made of even harder rocks so here the Old Red is relatively soft.

Another clue is the absence of surface water where you would expect it. A small valley without so much as a stream or a ditch at the bottom is a sure sign of a permeable rock, probably some kind of limestone but possibly a porous sandstone. This is known as a dry valley. Conversely, if you know you're in limestone country and come upon a spring, a stream or a pond you can be pretty sure that you're on an outcrop of some other rock, unless it's an artificial pond which has been lined with clay or plastic. Rivers can flow through limestone country but not small streams. This is most important if you're reading the landscape with a view to siting a pond. If there's no sign of surface water now it will be difficult to introduce it later. The pond will have to have an artificial liner and filling it by natural means may be difficult.

This extract from my notebook shows how the behaviour of the surface water gave me a clue about the rocks beneath my feet. I was on the edge of the Forest of Dean, where the two main rocks are an impervious sandstone and Carboniferous limestone.

Simmonds Yat, 10th June 2006

Beside the woodland path there's a little brook that disappears down a sink hole. All around there are rocks on the surface which are quite plainly sandstone. They glisten in the sunlight that filters down through the trees. But the disappearing stream is a sure sun that there's limestone below. The gully carved through the sandstone by the stream must be just deep enough at this point to reach a layer of limestone below. (It continues down the slope as a dry gully.)

A little further along the path, in a felled area full of foxgloves and scattered with the same sandstone blocks, a small patch of limestone showed through from below. Its smooth, jointed surface and pale grey colour were quite unmistakable among the brown boulders with the slight sparkle. This confirmed that there's a thin layer of sandstone over limestone here.

Putting all these clues together – walls, buildings, soil, landform and surface water – you can often get a good idea of the kind of rock you're on. Carboniferous limestone, with its grey field walls, occasional exposures of grey rock, ash woods and absence of surface water is quite distinctive. On the Pennines it makes a strong contrast with the dark millstone grit, which is often covered in heather or purple moor grass and tends towards bogs (or mosses as they're known) wherever the land is flat. Heavy clay vales, where elm hedges are common and the ridge-and-furrow pattern of medieval cultivation often survives in the grassland, are hard to confuse with light sandy country with its birch trees and bracken.

To confirm your diagnosis you need a geological map. You can get a broad idea of the local geology from a small-scale geological map, even one which covers the whole country. But since the nature of the rock can change completely over short distances a large-scale local map is much more useful. It will also add a fascinating dimension to your landscape reading. They're usually at a scale of 1:50,000 with the colours for each rock superimposed on a faint outline version of the Ordnance Survey (OS) map. Sometimes the outline of the OS map is somewhat obscured by the geological information and this can make it hard to locate where you are with certainty. It can be a great help to have an ordinary OS map with you to help with this. Comparing the two maps usually clears up any doubts.

Some geological maps show the solid rocks only, leaving out drift deposits such as peat and wind-blown sand. Others show both drift and solid, which is much more use for landscape reading. You can look at them in the reference section of some public libraries. Each county usually has one library where all the geological maps are held. Or you can buy them from the British Geological Survey. (See Appendix A.) They're easy to use as long as you're familiar with the names used for different kinds of rock. The list below gives you all the most common ones.

NAMES OF COMMON ROCKS		
D = drift I = igneous M = metamorphic S = sedimentary		
Alluvium	D	deposited by rivers
Andesite	I	fine-grained, intermediate
Arkose	S	a sandstone
Basalt	I	fine-grained, basic
Boulder clay	D	various materials deposited by ice sheets
Breccia	S	angular stones cemented together
Brownstone	S	usually sandstone
Chalk	S	a pure, soft limestone
Chert	S	a form of silica, called flint when found dispersed in chalk
Clay	S	soft and plastic
Coal measures	S	a mix of sandstones, shales and coal seams
Conglomerate	S	rounded stones cemented together
Crag	S	mostly unconsolidated sand
Diorite	I	coarse-grained, intermediate
Dolerite	I	medium-grained, basic
Dolomite	S	limestone containing magnesium carbonate as well as lime
Flagstone	S	thinly bedded sandstone
Fuller's earth	S	clay or harder rock made from fine volcanic ash
Gabbro	I	coarse-grained, basic
Gault	S	a clay
Gneiss	M	coarse-grained, extremely metamorphosed
Granite	I	coarse-grained, acidic

Greensand	S	a sandstone
Greywacke	S	sandstone cemented with mud
Grit	S	coarse sandstone with angular grains
Head	D	unsorted stones and clay
Hornfels	M	clay metamorphosed by heat alone
Limestone	S	made of lime (calcium carbonate)
Marble	M	(not found in Britain; 'Purbeck marble' and 'forest marble' are limestones)
Marl	S	lime-rich clay
Moraine	D	unsorted material deposited by glaciers, usually in ridges
Mudstone	S	similar to clay
Oolite	S	limestone composed of tiny spheres
Peat	D	partly decomposed organic matter
Quartzite	S/M	sandstone cemented with quartz, very hard
Rhyolite	I	fine-grained, acidic
Sandstone	S	coarse-grained, mostly quartz
Schist	M	medium- to coarse-grained
Shale	S	laminated clay
Siltstone	S	similar to clay
Slate	M	clay metamorphosed by pressure alone, highly laminated
Tuff	I	compacted volcanic ash, acidic

The Soil
and What Plants
Can Tell Us About It

The big sand grains and tiny clay particles which I introduced in the last chapter play a leading role in this one. The different kinds of soil produced by sand and clay give the landscape much of its character and variety. Clay is rich in bases, usually somewhat alkaline, and has the quality of plasticity. Water passes through it slowly and poor drainage can be a problem. Sand is poor in bases, usually somewhat acid, and lets water run through it freely. So sandy soils tend to be low in nutrients and prone to drought. Very few soils are either pure clay or pure sand. Most are a mixture, which is known as a loam, and loams tend to avoid the problems of either extreme. Clay loams are predominantly clay and sandy loams predominantly sand. When we talk casually about 'clay' and 'sandy' soils we usually mean these loams. A medium loam is a soil with a balanced mix of both.

Clay loams used to be called 'wheat and beans land' or 'four horse land'. They were rich enough in nutrients to grow the more demanding crops before the days of chemical fertilisers, but the heavy, plastic soil was hard to work and needed twice the number of horses for ploughing than the lighter sandy loams. These two terms, 'heavy' and 'light', are often used to describe clay and sandy soils respectively. While clay and sand are the extremes in particle size there are also particles of intermediate size, known as silt. Silt is usually not as rich in nutrients as clay but it can be equally difficult to drain, so soils which are rich in silt are also considered heavy.

Light		Heavy
Sand	Silt	Clay
freely draining	slowly draining	slowly draining, plastic
low in nutrients usually acid	intermediate	high in nutrients usually alkaline

Sandy loams are often sought after for market gardening. As they're quick to drain you can work a light soil sooner after rain than you can a heavy one, and vegetable growing involves lots of different tasks throughout the growing season. You do need to add a lot of manure to a sandy soil, both to make up for the lack of nutrients and to improve its ability to hold water. But the trade-off is well worth it. Rabbits also appreciate sandy soils because it's much easier to dig a burrow in a light soil than a heavy one. If you see signs of a lot of rabbits it's a clue that you're on a sandy soil, and likewise bracken, which is one of the most characteristic plants of light land. These are no more than clues because both rabbits and bracken can be found on heavy soils too, but the more of them there are the stronger the clue.

As the rainwater washes down through a sandy soil it tends to leach out what bases there are in it. This leaching goes on in all soils but it's much faster in sandy ones. The bases may be redeposited lower down in the landscape, where the water re-emerges at a spring or in a stream. This is often the pattern on heaths and moors, where the sweet grasses growing by a spring or on the banks of a stream may provide the grazing animals with a disproportionate amount of their nutrition compared to the heather and moor-grasses of the open moor.

The problem with heavy soils is that water moves through them too slowly. Excess water damages plants by denying air to their roots. Normally air enters the soil through a network of pores and channels which are partly filled with air and partly with water. When the pores are entirely filled with water the soil is said to be waterlogged and the plant roots can't respire. Other forms of soil life such as earthworms suffer in the same way. The lack of oxygen also converts nitrogen from a solid to a gaseous form, so it's lost from the soil. Plants need a lot of nitrogen and when they're short of it their growth slows down and their leaves turn yellow. The most common cause for a patch of yellow, stunted plants in an arable field is poor drainage. In a pasture a patch of rushes often indicates an area which is frequently waterlogged.

One thing which causes waterlogging is soil compaction. Compaction is caused by driving tractors or pasturing cattle on a heavy soil when it's too wet. In a garden, walking on the soil or working it too soon after rain will have the same effect. Compaction squeezes the pores out of the soil and in extreme cases it becomes a solid mass. But even mild compaction makes the soil drain badly and reduces the vigour of plant growth.

Compaction can also make the soil more vulnerable to erosion. Rainfall can't percolate into a compacted soil, so it starts to run along the surface. As it runs it takes soil with it and erosion begins. The steeper the slope the faster it runs and the more soil it can carry, so slopes are much more vulnerable to erosion than flat ground. If the soil surface is covered with plants, as in a permanent pasture, the flow of water is slowed down and it doesn't erode very much. But if the soil is bare after ploughing, or only partly covered by the immature plants of a young arable crop, it erodes more. Maize is one of the worst crops for soil erosion because the plants are so far apart that when they're small they hardly cover the ground at all. Outdoor pigs, although great for animal welfare, can be even worse. Not only do they bare the soil they also compact it.

Hedges which run across the slope help to reduce erosion by checking the downhill flow of water. It's well known how the destruction of hedges over the past few decades has increased erosion but hedges also store a record of the erosion which has gone on in the past. Soil banks up on the uphill side of a hedge and is eroded away from the downhill side, creating a terrace which is known as a lynchett. Lynchetts often remain long after the hedge has been removed. A distinct step running across the slope of a grassy field is a sign that there was once a hedge there. Where the soil on one side of a hedge has never been ploughed, such as on the

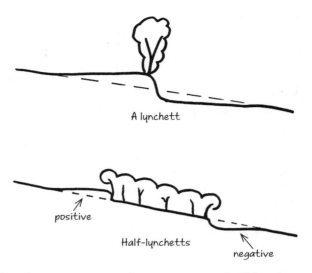

A lynchett

positive

Half-lynchetts

negative

boundary between a field and an ancient wood, you can get a half-lynchett. On the uphill boundary of the wood it will only be formed by soil washed down from the field and on the downhill side only by erosion from the lower side.

You can sometimes see little half-lynchetts forming where there's a new fence with ploughing on one side and undisturbed grass on the other. Outdoor pigs will rapidly form small half-lynchetts at the top and bottom of their run if it's on a slope, and this of course will persist after they've gone unless it's ploughed out.

Where you see deep lynchetts in hilly country which is all down to permanent grass you can be sure there has been arable cultivation at some time in the past. I know a steep field of a couple of acres in just such country on the edge of the Forest of Dean. It has a lynchett nearly two metres deep at the bottom while the soil at the top is so thin that the bedrock shows through in places. It may have been cultivated in times of desperate overpopulation, such as the thirteenth century, or during the U-boat siege of the Second World War. But cultivating ridiculously steep slopes can just as easily happen in times of prosperity and peace. In the 1970s I saw people ploughing a bank on the chalk downs so steep that they could only plough downhill and had to drive back up empty. There can't be much soil left on that bank now.

Soil erosion is most serious where the soil is very thin in the first place. Once when travelling through the chalk country by train I thought for a moment that I'd seen a snow-capped hill. Another time, driving at dusk on an unfamiliar road, I thought I was surrounded by snow-covered fields. Both times I quickly realised it was not snow but soil erosion. Chalk soils can be extremely thin and if only a little of the soil is lost it becomes so thin that the plough brings up the white rock from below. The lumps of chalk lie on top of the soil and from a distance you can't see the soil at all. Even when the corn comes up the chalk can hide the young plants for several weeks till they grow tall enough to overtop it. (See photo 22.) Often you can see a recently ploughed chalk field which is partly white and partly brown, or one with a young cereal crop in it which is partly white and partly green. The white parts of the field are the steeper parts, where erosion is worse. The colour difference can be used to map out the steeper and flatter parts of the field quite accurately.

Although loughing steep slopes is the main cause of severe soil erosion it can also happen less dramatically on flatter land. Here the main cause is the loss of soil structure. Structure means the way the soil is divided up into discrete pieces with pore space between them, as compared to the solid mass of a severely compacted soil. Crumb structure, which you can see on the surface of a fertile garden or farm soil is one kind of soil structure. These little crumbs, just a few millimetres in diameter, give the ideal conditions for plant growth. They allow excess water to drain away yet store an abundant supply of the precious liquid in the pore space between them. They let the oxygen-rich air enter from above and allow plant roots to penetrate the body of the soil in their quest for the nutrients it contains. If you pick up a handful of this crumbly soil and let it run through your fingers you can feel the mellowness of it. This is soil to put abundant, wholesome food on your plate. A soil with a flat, solid surface is quite the opposite.

Take a sample of these two kinds of soil, one well-structured and the other more of a solid mass. Put them side by side and you'll immediately see a contrast in colour. The well-structured one will almost certainly be darker than the structureless one. This is because of its humus content. Humus is a black, jelly-like substance, manufactured by earthworms and micro-organisms from the dead plant material they find in the soil. By and large, the darker the soil the more humus it contains. Humus is the very key to soil fertility, not least because it forms and maintains good soil structure. It builds up slowly in an undisturbed soil, such as grassland or woodland, and is lost quickly when the soil is ploughed. Ploughing lets unnatural volumes of air flood into the soil and the resulting excess of oxygen stimulates the microbes into a state of hyperactivity in which they rapidly consume the humus.

Although good structure is the basis of soil fertility other factors are important too, particularly the supply of mineral nutrients and the level of acidity. But although soil fertility is vital to our existence, it's not always and everywhere desirable. We'd have a very boring landscape if all soils were a model of perfection because it would be inhabited by a very small number of plant species. In fact the vast majority wild plants can only grow successfully on soils which are hopelessly infertile from a farmer's or a gardener's point of view. Some grow in wet soils, others in dry, some in acid ones and many in soils which are short of nutrients. So infertile soils are essential to biodiversity. I believe that in the long term only a biodiverse landscape can be a truly productive one. If that's so, then, paradoxically, some patches of infertile soil are also vital to our existence.

Most wild plants don't actually prefer infertile soils. They tolerate them. The reason they don't grow in fertile soils is that there's a small number of very competitive plants which thrive where soil conditions are ideal and can out-compete all comers. Amongst herbaceous plants the classic competitors are stinging nettles, docks, hedge bindweed and chickweed. They can make such good use of high fertility, especially a high level of nutrients, that bit by bit they crowd out any less vigorous plants trying to grow in a fertile soil. But they don't like infertile soils and here all the other plants have the advantage.

Most plants actually grow better in a more fertile soil than the one they normally grow in [sic] the wild. We could take a range of plants from soils which are acid, poorly drained, deficient in nutrients and so on and plant them in the nice fertile soil we provide for our vegetables in the garden. As long as we kept them weeded they'd grow well, probably better than in the soil where they're usually found. But if we were to stop weeding and leave them to their own devices, sooner or later the competitor plants would move in and crowd them out.

Alder trees are an example of a highly specialised tree species. Unless they've been planted by human hand they almost always grow in wet soil. In fact they're so faithful to wet soil that if you see a clump of alder trees you can usually be pretty sure that there's a river or stream at their feet. One reason why they're adapted to wet conditions is that they can fix nitrogen. That's to say they have micro-organisms living in their roots which can take nitrogen from the atmosphere and turn it into a mineral form which plants can use. This is very useful in waterlogged conditions, where nitrogen is easily lost from the soil. It gives alders an edge over trees which can't do it. But when people plant them in a moist but well-drained soil they actually grow better than they do in their usual habitat. So why are wild alders confined to wet soils? It's partly because their seedlings are vulnerable to drought in their first summer, which makes it hard for them to get going away from a constant source of water. But if an exceptionally wet summer was to remove this obstacle they still wouldn't thrive in a well-drained soil because they'd be crowded out by competitor trees which are more vigorous under ideal conditions.

The prime example of a competitor tree is the beech. It can probably out-compete all other native British trees on a fertile soil but it can't stand poor drainage, let alone the wet riverside conditions which the alder thrives in. Although beech has the potential to become dominant over large areas it will never oust the alder from its perch on the river bank.

Soil Types

So far I've looked at the differences between soils mainly in terms of the proportions of sand, silt and clay they contain. If you ask a farmer or gardener what kind of soil they have they will most likely reply that it's a sandy loam, a clay or some such. This is a good way to classify cultivated soils because there's not much else to distinguish them. But the soils of permanent pasture, moorland, heath and woodland have never been ploughed, or were ploughed so long ago that natural conditions have reasserted themselves. In these semi-natural ecosystems the difference between one soil and another is much more than simply a matter of how much sand, silt and clay they contain.

The difference lies in the soil profile, the vertical cross-section of the soil. Cultivated soils usually show a clear distinction between the topsoil, the upper layer which is regularly ploughed or dug, and the subsoil below. The topsoil is where most of the biological activity takes place. Most of the plant roots, soil creatures, microbes and humus are concentrated there. The humus gives it a darker colour than the subsoil. But in wild soils there's a lot more to the soil profile than a simple division between topsoil and subsoil. Most of the variation is due to the soil's relationship with water. At one end of the spectrum are soils which drain excessively and at the other are soils which hardly drain at all. Somewhere in the middle lies the happy medium, the naturally fertile soil.

Podsol

pure humus

pale, leached

hard pan

reddish-brown

At the excessively drained end of the scale lies the podsol. A podsol is always very acid. Rainwater flows down through it, leaching out minerals, humus and clay particles and redepositing them lower down. The upper part of the soil is pale, in extreme cases an ashy white, because the iron and other minerals which give soil its colour have been leached out of it. Lower down there's a thin, hard layer where the leached materials are redeposited, known as a pan or a hardpan. This can be a complete barrier to roots and water and if it's high in iron it can be extremely hard. At the surface, lying on top of the pale upper layer, is a layer of pure humus. In other soils surface organic matter is taken down and mixed with the mineral matter by earthworms but a podsol is too acid for earthworms.

This is an extreme kind of soil and it forms where conditions are extreme. Usually the starting point is an acid parent material such as granite or sandstone, with little clay. High rainfall with free drainage is another factor. Acid-forming vegetation, such as heather or conifers, is the third, and this can tip a susceptible soil over the margin and start the process of podsolisation. Any combination of these three factors can cause a podsol to form. You can get one on a sandy heath in south-east England where the rainfall is low or on a heavier soil in north-west Britain where it's high. In general podsols are the characteristic soils of heaths and moors.

The next soil in the spectrum is the happy medium, neither excessively leached nor poorly drained. It's called a brown earth and is the ideal soil for agriculture. The topsoil is a rich dark brown, courtesy of the earthworms, which mix the humus with the mineral matter, while the subsoil may be a lighter brown or a reddish colour. It's interesting to compare a brown earth formed under deciduous woodland with one formed under grassland. In the woodland most of the organic matter comes from above in the form of leaf fall, whereas in the grassland it comes mainly from the constant turnover of roots within the soil itself. Thus the topsoil in a woodland is usually a shallow layer and very dark brown, while in the grassland it's not so dark but deeper and fades off gradually into the subsoil.

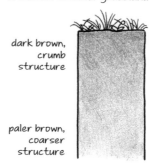

Brown earth under grassland

dark brown, crumb structure

paler brown, coarser structure

Next on the scale is a gley, a soil with a drainage problem. Gleys are invariably heavy soils, usually found in a low-lying place where the water table is high. The water table is the level within the soil below which it's waterlogged. Its position varies with the changing rainfall from summer to winter. The key characteristic of a gley is that the lower layers are a greyish colour, or in extreme cases blue. This is because iron, which is the red pigment in soil, turns blue or grey in the absence of oxygen. The middle layer of soil between the summer and winter water tables is usually grey with orange or brown flecks, showing where oxygen has penetrated during summer, often down the channel formed by a dead root.

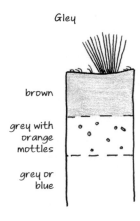

Gley

brown

grey with orange mottles

grey or blue

Earthworms, which can't stand waterlogging, are usually absent. Gleys are too wet to cultivate without major drainage work but they can stay moist through the summer, which is just what's needed for grass production. So traditionally they were used for meadow or summer pasture.

At the extreme wet end of the spectrum comes peat. It forms where waterlogging is so complete that air is excluded even at the surface. Without oxygen organic matter decomposes very slowly indeed, so a layer of partly decomposed organic matter builds up on the surface. This is peat. Peat formation is a self-reinforcing process, because peat itself retains an enormous amount of water, and peat bogs and fens can be many metres deep.

In reality these four types grade into each other. You can find a brown earth with tendencies towards a podsol or a gley with a peat layer at the surface. Every soil is an individual. But often the boundary between adjacent soils on the ground can be quite abrupt and signalled by a change in vegetation.

This is illustrated by a page from my notebook. I was teaching a permaculture course on a croft at Glenelg in the western Highlands. For the practical soil session we went up on the hillside with a spade and dug soil pits so we could see the profile in different places. I deliberately chose places where the vegetation was different, hoping to see a corresponding difference in the soil below. I was not disappointed.

Corrary, 10th August 1995

Four soil pits, high on the hill, steepish, each under a different kind of vegetation.

Vegetation	Soil Type	Notes
Bracken	deep brown earth, sandy loam, dark reddish	molehills
Grass	similar but sandy silt loam, slightly lighter colour	few remains of bracken rhizomes

51

| Rushes | similar to grass, but 15-25cm of black soil on top |
| Heather | very peaty and wet over an iron pan, sandy loam below |

This is a place with very high rainfall. Everything is dominated by water and even slight changes in the clay content or acidity can move the soil one way or the other along the spectrum of soil types.

The bracken, which can't stand poor drainage, was growing in the most fertile and best drained soil. The molehills were a sign that there were earthworms because moles mainly eat earthworms. Earthworms can't stand a very acid soil or a waterlogged soil, so they're a useful indicator. Where grass was growing the soil was just that bit heavier, which made it that bit wetter. The remains of bracken rhizomes (similar to roots) in the soil showed that it had grown there recently, perhaps in a drier-than-average year. Where the rushes grew, the layer of black soil on the top suggests a gley verging towards a peat. But this process is unlikely to go much further on a steep site like this. Peat tends to form on flat land and indeed there was an extensive peat bog close by where the hilltop flattened out into a plateau.

What was happening under the heather was perhaps the most interesting. The iron pan could only have been formed by podsolisation. Heather and podsols have a mutually reinforcing relationship: heather is one of the few plants that can live on a very acid soil and then it increases the acidity because its leaf litter is acid. This patch of soil must have been a little more acid than the others in order to start off the podsol process and then podsolisation will have made it yet more acid. In this case the pan has become such an effective barrier to water that a layer of peat is forming on the surface above it. This happens quite often and many a wet moorland owes its wetness to a hardpan.

Wherever people have cultivated soils from the extreme ends of the spectrum they've tried to turn them into something like the ideal brown earth. Gleys and peats have been drained. Podsols have been limed and had their pans broken by deep ploughing. Pan-busting has sometimes improved the drainage so much that old streams have run dry.

There's one more distinctive soil type which can be found in Britain and that's the thin limestone soil, or rendzina. This is one soil type which is defined not so much by its water relations as by its chemical makeup, which is of course very alkaline. Limestones soils are thin because they only consist of the impurities in the limestone, the lime itself having dissolved away. Chalk, being the purest limestone, has the thinnest soils. But it would be wrong to think of all limestone hills as covered from end to end by thin, alkaline soils. There can be drift deposits which are not thick enough to make an obvious difference in landform but which completely change the nature of the soil.

An example is the clay-with-flints, which often overlies the chalk. It can be very acid because the flints are made of pure silica, the mineral which makes granite so hard and so acid. If you're ever in the chalk country and come across a patch of bracken, honeysuckle and bluebell, either in a wood or on a roadside verge, you can be pretty sure you're on clay-with-flints. On pure chalk the herb layer of woods is typically dominated by dog's mercury.

Reading the Soil

One thing I recommend to any landscape reader is to learn the soil finger test. I always do it when I go to a new place for the first time, whether as part of my work or just to enjoy the landscape. The test tells you whether the soil is predominantly sand, silt or clay. With a little practice and with no tools other than your fingers you can quickly discover this important fact about any landscape. What you do is this.

Take about a teaspoonful of soil, knead it till it loses all structure and roll it into a ball. It needs to be at just the right moisture content to form the strongest ball it can. Too wet and it squidges, too dry and it crumbles. You may need to moisten it and if it's a bit too wet you can dry it by kneading it for longer. In fact if you knead a sample for too long you may need to re-moisten it. It's important to get the moisture content right because a sample that's too wet will feel more silty than it really is.

When you have the strongest ball you can make with that soil ask yourself this series of questions:

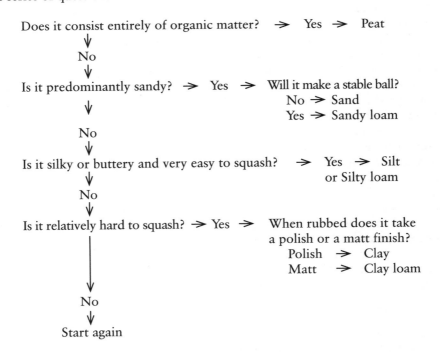

Does it consist entirely of organic matter?   ➔   Yes   ➔   Peat
    ↓
    No
    ↓
Is it predominantly sandy?   ➔   Yes   ➔   Will it make a stable ball?
    ↓   No ➔ Sand
                             Yes ➔ Sandy loam
    No
    ↓
Is it silky or buttery and very easy to squash?   ➔   Yes   ➔   Silt
    ↓   or Silty loam
    No
    ↓
Is it relatively hard to squash? ➔ Yes ➔   When rubbed does it take
                              a polish or a matt finish?
                                Polish   ➔   Clay
                                Matt   ➔   Clay loam
    No
    ↓
Start again

One mistake which is easy to make when you start using the finger method is to always answer Yes to the question 'Is it very easy to squash?' All soil is easy to squash and this question is relative. Once you've handled a few contrasting samples you'll get used to it. The feel of a really silty soil is unforgettable once you know it but hard to put into words.

This gives you the basic types, which are certainly enough to get started in landscape reading. But finer distinctions can be significant and you can take it a stage further. A soapy feeling tells you there's some silt in a soil which is predominantly sandy or clayey. This soapy feeling is quite different from the stickiness of clay. If you can feel it in your sample you can redefine it as follows:

Sandy loam	➤	Sandy silt loam
Clay loam	➤	Silty clay loam
Clay	➤	Silty clay

In the same way, if you can feel a bit of sand in a soil which is predominantly clay, you get:

Clay loam	➤	Sandy clay loam
Clay	➤	Sandy clay

In an arable field or garden it's fairly easy to come by a sample for the finger test because there's plenty of bare soil, but be sure to avoid untypical places like gateways or paths, where people may have added hardcore or gravel. Other things which can give a clue to the proportions of sand and clay in cultivated soils are cracks in the summer and muddiness in the winter. In a dry summer cracks can form in any soil but they're much bigger in clays. Winter muddiness is a clue to both clay content and structure. A sandy soil will hardly stick to your boots at all, a well-structured clay with plenty of humus in it will stick a bit, but a poorly-structured clay will turn your feet into lumps of mud the size of footballs after a couple of dozen paces.

Other things to watch out for in arable land and gardens are signs of erosion and waterlogging. The commonest form of erosion is sheet erosion, in which soil is removed evenly from the surface over a wide area. It can take some practice to recognise the signs of sheet erosion and the easiest way to tell it's been going on is often to look for signs of deposition at the bottom of the field. A deposit of eroded soil characteristically has a streamlined shape with a glossy surface when it's new, often with a scum of organic matter on it. Rill erosion is much more serious. It only starts when a lot of soil has already been lost by sheet erosion. Rills are little channels cut into the soil by the eroding water and they're quite unmistakeable. Another way to see the effects of erosion is to look out for a patchy growth pattern in a sensitive crop, and no crop is more sensitive than maize. A maize field which is suffering from erosion can often be divided into three: a flat area where the plants are normal, a sloping area where the plants are stunted and pale green or yellow, and the bottom where big, dark green plants indicate that soil is being deposited.

Maize is also sensitive to waterlogging, which deprives the plants of their vital nitrogen, and stunted yellow maize plants may be the result of nitrogen deficiency rather than soil erosion. If a maize field is patchy and the small yellowy plants are on the lower ground you can be pretty sure that the cause is waterlogging. Although other crops may not show it so clearly they will also suffer in areas of erosion or poor drainage. So maize can act as an indicator crop for conditions which need correcting in any case. The best time to see patchiness in a maize crop is before it's grown so much that the taller plants block your view of the shorter ones, perhaps in July.

Poor drainage is the commonest cause of patchy growth in any arable field or vegetable garden. A patch of bare soil in an autumn-sown cereal crop during spring or summer is often a sign that waterlogging or even standing water has killed the plants during the winter. Deep ruts where a tractor has hit a wet patch in the field tell a similar story. A pale appearance over all or part of the field in springtime tells a different story, one of winter rain leaching out the nitrogen rather than of waterlogging causing it to be converted to gaseous form. You're more likely to see this on a sandy soil than a heavy one and barley is particularly prone to it after a wetter-than-average winter. It's not unknown to see signs of water-logging and excessive leaching in different parts of the same field, especially in very wet weather. Occasions like this are a good opportunity to map out the different parts of the field so plans can be made to manage them differently.

Sometimes plants can tell you something about the soil by their stage of growth. Dry conditions encourage them to mature quickly and moist conditions encourage them to stay longer in the green, vegetative stage. If a corn crop ripens unevenly, green in one part of the field and golden in another, this reflects a different water-holding capacity in the soil of each part. It may be due to a difference in the sand and clay content or to a difference in the depth of the soil. Once again, it may suggest where the field should be divided in two with a new hedge so each part can be managed individually. In fact it may indicate where once there were two fields before an old hedge was grubbed out. (See photo 8.)

In grassland it's not so easy to find a bit of soil for the finger test. Moles and rabbits often turn up a bit of soil which you can sample but you have to be sure it's typical of the field. Rabbits may choose to burrow in one place simply because the soil there is lighter and thus easier to dig. I was once caught out by a sample from a rabbit burrow in a Cornish hedgebank. It was a sandy loam but another sample, from a molehill further out in the field, was a clay loam: the soil which had been used to fill in between the stones in the bank was not the soil from the field.

One thing to look out for in grassland is molehills because they indicate the presence of the moles' chief prey, earthworms. Apart from the good they do to the soil, earthworms indicate that it's neither poorly drained nor too acid, because they can't stand either of these conditions. If you look carefully you may also be able to see worm casts on the surface. On the other hand a thatch of undecomposed dead grass on the surface is a sign of grassland which is too wet for earthworms. If worms were there they would have taken it down into the soil and eaten it, turning it into humus in the process. Anthills are a sign of very dry ground because grassland ants are even more intolerant of poor drainage.

Poorly drained areas in a pasture are often shown up by poaching, the deep, muddy footprints which cattle leave when they walk on wet soil. The depth of poaching is directly related to the wetness of the soil and if the poaching is restricted to certain parts of the field it can be an accurate measure of the wet places and the dry. If it's spread all over the field it simply means the cattle were left out on the land too long before being taken in for the winter. During a dry summer wetter areas may stand out as dark green patches while areas of shallow soil will be indicated by browner grass.

In woods, opportunities to sample the soil are given by badger setts and the root plates of blown-over trees. Sometimes a root plate can hold a whole soil profile intact. The most impressive example I've seen was in a wood near my home where the roots of the tree held alternate thin layers of clay and limestone. The limestone looked for all the world like the regular cobblestones of a cobbled street, though much smaller. This was on land which had been farmed in earlier ages but was put down to plantation in the nineteenth century. With a soil like that it's easy to see why people stopped trying to cultivate it. Another time I visited the wood I spoke to a forestry worker who was planting trees there. He said he had to use a crowbar to plant every single tree. Having seen that root plate I wasn't surprised to hear it.

The colour of the soil can tell us various things. Those dark red soils, typical of some parts of Herefordshire and Devon, have a special quality to them. Years ago I remember a Devon farmer telling me "When our swedes come to market the wholesalers really go for them. With our red soil on them they know the quality will be good." Another Devon farmer told me his sheep always do well when he moves them onto a field with red soil. The red pigment is an oxide of iron. Iron is an important mineral nutrient but plants only need it in tiny quantities and almost all soils have enough, so it's not iron as a direct nutrient that makes these red soils so special. But it also has a key role in moderating the supply of all the other mineral nutrients which plants need from the soil and this may be what makes the difference.

Dark brown or black soils are usually high in humus. Old garden soils are often this colour because they've been manured and composted over the generations. But the black colour can also be due to coal. Ash was often put on gardens and while the ash itself has leached out of the soil the little bits of unburnt coal remain. Finding coal in the soil doesn't necessarily mean that it isn't also high in organic matter, as people who put their coal ash on the garden in the past probably put their organic waste there too.

Where you see black patches in a ploughed field alternating with ordinary brown or red soil you may be in an old coal-mining area and the black patches are the sites of former workings. But black patches of soil in a field can also show areas that were once waterlogged and now have been drained. In this case the black colour is the peaty layer which accumulates under the oxygen-deficient conditions of waterlogging. If the black areas of soil match the lower-lying parts of the field this is a clue that waterlogging is the cause.

These are some of the ways of observing the soil and they're all useful in the art and science of landscape reading. But the one which takes pride of place is looking for indicator plants and it deserves a section on its own.

Soil Indicator Plants

Everyone who spends time in the countryside gets to know some of the plants which tell us about the soil. You learn the most obvious ones simply by experience. Some years ago wardens in Snowdonia National Park made use of this fact to help heal erosion on over-used paths. They planted rushes in the most vulnerable places. Every rambler knows that rushes mean wet soil and are best avoided, so they made detours round them. Of course the rushes didn't persist for long where they'd been planted because the ground wasn't particularly wet. But by the time they died out the plants that grew up among them had grown tall enough to act as a deterrent to walkers.

Not many plants give such a clear message about the soil as rushes do. Rushes are what you call a strong indicator. Just the presence of this one plant is enough to tell you something definite about the soil. Very few plants are strong indicators. Most indicator plants have a preference for a certain kind of soil but can also be found on other soils. So, unless you happen to see one of the strong indicators, you have to look out for communities of plants, not a single species. Three or four indicator plants growing together will tell you something that you couldn't assume from seeing just one on its own. In fact the great majority of plants are not indicators at all. They grow in too wide a range of soils to tell us anything.

This page from my notebook records some of the plants I saw from the window on a train journey.

Waterloo to Templecombe, 23rd April 2003

Sandy country – oak, birch and bracken. Pine and heather in the very sandy bits.

Chalk – railway verges thick with hawthorn, some ash, occasional yew and the white pom-poms of wayfaring tree. Wide open fields. You can often see the chalk itself, both in scars made by rabbits in the railway cuttings and in arable fields.

Nadder Valley – vegetation very mixed; no clear message. Lovely intimate country, with alders and willows beside the little rivers.

Blackmore Vale – the spikes of dead elms sticking up through the hedges. Oak, ash, hawthorn and blackthorn.

The oak, birch and bracken community immediately told me I was in sandy country. None of these plants is confined to sandy soils, especially not oak.

But when you find yourself passing through semi-natural woods where all the trees are either oak or birch and the herbaceous layer is mostly bracken, you can be pretty sure the soil is sandy. The bracken, honeysuckle and bluebell community I mentioned earlier in this chapter as a sign of clay-with-flints is similar. They can all be found on other soils, but if you come across the three together on top of a chalk down the message is clear.

On the other hand pine and heather are strong indicators. They only grow spontaneously on soils which are very acid and very poor in plant nutrients. In the south-east of England this means pure sand. Planted pines don't count. The Scots pine is often planted on a wide range of soils and will grow well there. But in southern England as a spontaneous tree it's almost entirely confined to the very poor soils of former heathland.

There was no obvious change in the relief which told me that we'd passed from sand to chalk. First the plants told me of the change, then a few glimpses of the white rock itself confirmed it. The mix of trees and shrubs I noted on the chalk land includes both non-indicators and indicators. Hawthorn and ash grow almost everywhere, except in very infertile country, where ash is restricted to the richer soils. But yew and the wayfaring tree are plants which immediately make me say "Hello, we must be on the chalk now."

The Nadder Valley is perhaps more typical of the countryside at large than the other places I passed through that day, in that you couldn't tell anything about the soil from the mix of trees growing there. Perhaps if I'd got off the train and had a good look at the herbaceous plants I could have said something about the soil. But maybe not. A medium loam of moderate acidity with reasonable drainage doesn't have a highly distinctive flora.

The Blackmore Vale, at the point where the railway line crosses it, is a wide, flat sweep of land with few farms, no villages and massive thick hedges. As the train speeds across it you know you're in a distinctive landscape. The flat vale has the feel of heavy clay and this is confirmed by the dead remains of suckering elms in every hedge. The suckering elms, as opposed to the wych elm, are the traditional hedgerow elms of southern England. They have no defences at all against Dutch elm disease and always die in their youth. New saplings spring up from the roots but the bare poles of the dead trees are unmistakable – and a fairly sure sign of a clay soil below.

Of course from a train you don't see the half of it. Most of the herbaceous plants are too small and you can easily miss trees and shrubs if you happen to be looking the other way when you pass them. But it's fun. Not all journeys have such strong contrasts in soil as the one I took that day. But on most you can pick up the occasional clue about the soil you're passing over and add a little to your knowledge of the country at large.

There are a couple of rules of thumb about reading indicator plants. The first is only to take account of plants which are growing spontaneously, not planted by human hand. Even indicator plants will grow in a wide range of soils if we look after them and give them the conditions they like. This is what farmers, gardeners and foresters do for the plants they grow.

Of the three, foresters do least to change the conditions to suit their crop. They place more emphasis on choosing the species of tree to suit the site. But, as I mentioned before, many trees actually grow better in a more fertile soil than the one they grow in spontaneously. For example, oak as a timber crop is usually planted on heavy soils, because the quality of the timber can be poor on lighter soils. But in semi-natural woods, where the trees are self-selected, oak can only out-compete other species on infertile, acid soils, which in lowland areas means sandy ones. Thus, paradoxically, semi-natural oakwoods are mostly found on land which is bad for oak, at least from a forester's point of view. (A semi-natural oakwood is one where the coppiced trees, not just the standards, are oak. Almost all standards in coppice woodland are oak and they should never be used as indicators. See page 198.)

A page in my notebook illustrates how plantations can ignore changes in soil type while spontaneous trees and shrubs closely reflect them.

Teffont, Wiltshire, 12th November 2000

We went for a walk through the wood near Richard and Carol's. It seems to be a plantation on an ancient woodland site. The top end is beech and ash with bluebells, replacing an older hazel coppice which remains in one corner. There are a few rhododendrons in this part. The bottom end is mainly ash, sycamore and oak, with a very diverse edge of dogwood, spindle, wayfaring tree, native maple, hazel, hawthorn and even a bit of old man's beard. None of these appears to be growing inside the wood, which, as I said, has the appearance of an old plantation.

I took a soil sample on this edge (half a dozen sub-samples). As we walked back up I noticed that the edge gradually changed from the diverse mix to pure spindle, before the path entered the wood and I lost sight of the edge. I took another soil sample at the top end, near where the rhododendrons were growing. The two samples were both a fine sandy loam but the one from the bottom of the wood was pH7.25* and the one from the top was 5.5.*

* Alkaline and acid respectively.

Rhododendron is a strong indicator of an acid soil. It's also shade-tolerant and often self-seeds in woods where the soil is sufficiently acid, if there are seed parents nearby. The rich mix of shrubs on the edge of the lower part of the wood was probably the remains of the semi-natural wood that was there before the plantation. It includes some alkaline indicators and the diversity of plants itself indicates an alkaline soil. You could never have deduced the difference in acidity at the two ends of the wood on the basis of the planted trees.

Here we have examples of two different kinds of spontaneous vegetation. The rhododendron is an unwanted plant growing in a cultivated crop, in other words a weed. The shrubs on the woodland edge are the remains of the semi-natural vegetation, left over from the woodland that was there before it was converted to plantation. In general all the plants in semi-natural ecosystems, such as ancient woodland, moorland and unimproved grassland, can be regarded as spontaneous. On cultivated land, including gardens and plantations, only the weeds are spontaneous.

Sometimes plants growing in adjacent hedges or woods can be taken as indicators for the soil in a field, but you have to be careful about this. The very reason why one place has been made into a field and another left as woodland may be that there's a difference in the soil. Even the soil under a hedge may not be typical of the field as a whole. A hedge may follow the line of a ditch and a ditch is often deliberately sited in the wettest part of the land. In that case the hedge may well be on heavier soil than the rest of the field. The old West Country hedgebank can be deceptive, too. I once found several spindle trees, which are quite a strong indicator of alkalinity, growing on a hedgebank on a farm which I knew had a very acid soil, because I'd tested it. Presumably some alkaline material had been used in the construction of the bank, perhaps the remains of a former building with lime mortar in it.

The second rule of thumb with indicator plants is to ignore isolated individuals and plants which are growing poorly. Look for thriving populations and unless you can see a strong indicator, look for more than one species.

I was once cycling through an oak plantation on an acid podsol in the Forest of Dean when I came upon a single plant of old man's beard beside the track. It looked as though it was struggling. Old man's beard, or wild clematis, is a strong indicator of an alkaline soil, so what was it doing there? Well, the track I was on had been surfaced with limestone chippings and water draining through the chippings had made the soil just alkaline enough for an old man's beard plant to survive. So it was responding to the introduced limestone, not the natural soil. By contrast, a few days later I was driving through another part of the Forest and caught a glimpse of a vigorous group of old man's beard plants. Later I checked the geological map and there, sure enough, just where I'd seen the plants, the road crosses a narrow outcrop of limestone which cuts through the acid sandstone of the Forest.

That single plant of old man's beard, apart from being on its own and growing weakly, was growing in an untypical place, right beside the track. It's always important to check that the place where indicators are growing is not in any way special. For example, rushes can grow in compacted ruts where a tractor has passed while all around is well drained. Gateways, paths, old compost and manure heaps are all untypical places. Indicator plants growing in places like these will tell you about the place itself but not the surrounding field or garden.

Of course the absence of indicator plants doesn't prove a thing. There may be another factor which has a stronger influence than the soil. Where a piece of ground has recently been left to go wild there may be no suitable seed parents, either in intensively farmed land or in urban areas. The wild plants of towns are often determined more by what happened to be growing nearby when the soil became bare and available for colonisation than by the nature of the soil itself.

Where seed parents are not a limitation there may be a biotic influence which is stronger than the soil. Nettles, for example, can out-compete almost anything else where there's a rich accumulation of plant nutrients. But they can't stand repeated mowing, and regular grazing is usually enough to stop them getting established. So they may be absent from farmed grassland even though the soil is rich enough for them.

The specific things which indicator plants can tell us about the soil are: water content, acidity, level of plant nutrients and whether it's light or heavy.

The water content is perhaps the easiest of these to read. There are a number of fairly easily recognised plants which are faithful to a certain level of water content. For example: the two common species of buttercup grow on soil which is moist but not wet, rushes on soil which is wet for most of the year, and reeds where there is standing water for at least part of the year. Since soil compaction interferes with drainage many plants which indicate poor drainage can also indicate compaction. But there are also some plants which grow in compacted soils but not in particularly wet ones, such as the greater plantain.

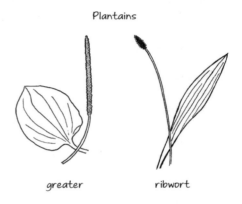

Plantains

greater   ribwort

Acidity is clear enough to read at the extremes: heather for acid, old man's beard for alkaline. But in the middle it's less clear so you need to look more carefully before you come to a conclusion. Quite often you can find plants which contradict each other. I once came across a magnificent stand of common spotted orchids on a steep hillside. They indicate an alkaline soil, so I was rather confused when I found equally abundant tormentil, which indicates an acid soil, growing among them. But there were other wild flowers there and they were all ones which tend to grow in more alkaline soils. So it was clearly the tormentil which was behaving out of character in this case. This kind of contradiction is always happening with indicator plants, and it just reinforces the importance of looking for a community of plants rather than one species.

As for plant nutrients, there's a definite group of plants, mostly garden and arable weeds, which are pretty reliable indicators of rich soils. For example, if you take over a new allotment and find that the weeds there include fat hen, chickweed, annual nettle, annual mercury and black nightshade, rejoice! Your soil is not only high in nutrients but probably fertile in other respects too. If, on the other hand, after cultivating and sowing, the soil between your newly emerging vegetables greens up with a carpet of little plants which look like grass but are not, you've probably got corn spurrey. This delicate little annual with thread-like leaves and small white flowers indicates a poor, acid soil. It's a sign that you have work to do, manuring and liming.

Soils which are poor in nutrients are often also sandy and acid, so there can be some confusion as to which of these three things a plant growing in such a soil indicates. Common sorrel is an example. It thrives on acid sands, but I've also seen it growing on chalk and it can be abundant on the alkaline peats of Somerset. I used to think it was an acid indicator but in fact it indicates soils poor in nutrients, however acid or alkaline they may be. The tiny sheep's sorrel, though, is an acid indicator.

It's often been said – and I've said it myself in the past – that all wild legumes, that is the clover family, indicate a soil low in nitrogen. They, like the alder trees mentioned earlier in this chapter, can take atmospheric oxygen and turn it into a mineral form in the soil. The reasoning is that because they can 'fix' nitrogen in this way they're specially adapted to live in a soil that's short of it. This turns out not to be the case in practice. Some do indicate low nitrogen and others don't, presumably because by the time you see them they've already enriched the soil with nitrogen.

A wide diversity of herbaceous plants indicates that the level of nutrients in the soil can't be very high. If it was, a small number of competitors would have taken over. This doesn't apply so much to trees. They don't respond to nutrients in the same way as herbaceous plants do. Even if they did it would take hundreds of years for the species composition to change, and most accumulations of nutrients have happened since chemical fertilisers became widespread during the past half-century. On the other hand low diversity doesn't necessarily mean high nutrients. There may be other reasons for it.

It's harder to tell whether a soil is sandy or clayey by means of indicator plants than it is to tell the other characteristics. This is because plants don't actually respond to sand or clay as such but to some characteristic such as the wetness of clay or the acidity of sand. So there are not many plants which are confined to sandy or clayey soils. But since it's so easy to test the soil with your fingers this is no great hardship.

There's one big problem with indicator plants: they can mean different things in different places. Plants which are fairly reliable indicators of one kind of soil over most of the country can grow in quite different soils in certain localities. For example, both suckering elms and creeping buttercup are both normally plants of heavy, wet soils. Yet they're common on the very sandy soils around the east Devon heathlands between the rivers Exe and Otter. I've also seen creeping buttercup on a sandy loam in nearby west Dorset and suckering elms on a gravely soil far away on the borders of Suffolk and Essex. These seem to be little pockets of country where these plants, for some unknown reason, behave quite differently from their norm.

This makes it hard to learn about indicator plants from books. Any list of indicator plants you read, including the one below, must be regarded with a degree of scepticism. These lists are often presented as a series of hard facts, like definitions in a dictionary, but the reality is less clear-cut than that. Some indicator plants, especially the strong indicators, are pretty consistent from place to place, but the majority are less reliable. This is another reason for looking for communities of indicators rather than reading too much into one species of plant.

So please regard the list below as a starting point, to be built on as your experience grows. In compiling it I've tried to keep to plants which are fairly consistent in all parts of Britain. I've also concentrated on plants which are either well known or easy to identify with the help of a guide. So it's nothing like a comprehensive list. On the other hand you don't need to learn all these plants in order to get started because not all of them will live in your area. If you don't yet know many wild plants the best way to use the list is to identify plants in the landscape first and then consult the list to see if any of them are indictors.

In some cases you need to be sure you have exactly the right plant. Closely related plants with similar names can indicate different soils. The buttercups are an example. The two common types, the creeping and meadow buttercups, both indicate moist soil but the less common bulbous buttercup indicates drier soil. The bulbous buttercup grows on dry limestone soils and is only common in the south and east. So in most places you can take a grassland with a yellow tinge to it in late spring as a sign of moist soil.

Buttercups are easier to tell apart by their leaves than their flowers

bulbous   meadow   creeping

(There are other buttercups but they're either very rare or grow in woodland.)

In the list overleaf I've used the words 'rich' and 'poor' to denote the level of plant nutrients in the soil. This doesn't necessarily correspond with the overall fertility. A fertile soil is one which supports abundant and healthy plant growth, not simply one which is high in nutrients. Broad-leaved dock, for example, is a good indicator of high nutrients but it can grow happily in a compacted soil.

An asterisk means that the plant is a strong indicator. This only applies to the soil condition marked by the asterisk. For example, bog myrtle is a strong indicator for wetness but may occasionally be found on soils which aren't particularly acidic.

SOIL INDICATOR PLANTS	
Trees and Shrubs	
Alder	*Wet, usually beside a stream or river Indicates land liable to flood
Beech	Well-drained; often suffers from die-back on poorly-drained soils
Bilberry or blaeberry	*Acid, *poor
Bramble	Well-drained but moist
Bog myrtle	*Wet, acid
Broom	Acid, sandy
Chestnut, sweet	Acid, sandy, but often planted on other soils
Elder	Rich, usually alkaline
Elm, suckering	Very typical of clay, but in some areas also on sandy and gravelly soils (see above)
Elm, wych	Alkaline, often on a clay soil over limestone
Gorse, common	Well-drained, usually sandy, poor; but increasingly found on a wider range of soils
Gorse, dwarf and western	Acid, not necessarily well-drained
Guelder rose	Moist to wet, usually alkaline
Heath, cross-leaved	*Acid, poor, wet
Heather, bell	*Acid, poor, dry
Heather, common or ling	*Acid, poor, moist
Old man's beard or wild clematis	*Alkaline
Rhododendron	*Acid
Rowan	Typical of light, acid soils, but very occasionally on limestone

Scots pine	Acid, but often planted on other soils
Spindle	Alkaline
Wayfaring tree	*Alkaline
Whitebeam	Well-drained, limestone or light sands
Willows, except goat willow	Wet
Yew	Especially common on chalk, but also on other well-drained soils

Herbaceous

Agrimony	Well drained
Birdsfoot trefoil	Low nitrogen
Bog asphodel	Wet, *acid and *poor
Bracken	*Well-drained, usually acid, usually sandy when abundant
Buttercup, bulbous	Well-drained
Buttercup, creeping	Moist, compacted or heavy
Buttercup, meadow	Moist
Chickweed	Rich
Coltsfoot	Heavy
Cow wheat, common and small	Acid
Orchid, pyramidal	Alkaline
Corn spurrey	Poor, acid
Cuckoo flower or lady's smock	Moist
Deadnettle, white & red	Rich
Dock, broad leaved	*Rich
Fat hen	Rich

Fleabane	Damp
Foxglove	Acid
Goosegrass or cleavers	Rich
Harebell	Well-drained
Heath bedstraw	Acid
Hemp agrimony	Damp to wet
Horseshoe vetch	*Alkaline, *dry; limestone or chalk
Horsetail	Wet subsoil
Kidney vetch	Well-drained, usually alkaline
Marjoram	Well drained, usually alkaline
Meadowsweet	Moist to wet
Nettle, annual or small	Rich
Nettle, stinging	*Rich, especially in phosphorous
Opium poppy	Rich
Orache	Rich
Pineapple weed	Compacted
Plantain, greater	Compacted
Pyramidal orchid	Alkaline
Ragged Robin	Wet, not very acid
Reed	Usually on soil flooded for at least part of the year
Rush, hard	*Wet, alkaline
Rush, soft	*Wet, acid
Salad burnet	Alkaline, dry
Sheep's sorrel	*Acid, dry
Silverweed	*Compacted or damp
Sorrel, common	Poor

Stinking iris	Alkaline
Thistle, creeping	Often compacted subsoil, fairly rich
Thyme	Dry, usually alkaline
Tormentil, common	Acid, poor
Willowherb, great hairy	Damp
Wood sorrel	Usually acid

Climate
and
Microclimate

I remember standing on the Isle of Erraid on a beautiful June day. The sky was that delicate china blue that's unique to the Hebrides. A soft wind blew in from the Atlantic bringing little puffy white clouds which shone in the sunshine as they sailed over the deep blue sea and the pale gold of the sands. Erraid is off the western tip of the Isle of Mull, just by Iona. From where I stood I looked inland along the low western peninsula towards Ben More, the highest point on Mull and the first big mountain of the Highlands. It was crowned with a thick black cloud and its flanks were hidden behind sheets of rain.

It wasn't chance that it was sunny by the coast and raining over the mountain. It always rains more on higher ground. As the wind blows towards the hills the air is pushed upwards and as it rises it cools. Cool air can't hold as much moisture as warm air, so clouds form and it rains. What I saw that day was a microcosm of the way the climate of Britain conspires with the geology to reinforce the difference between the highland and lowland zones of the country. When looking at the rocks of Britain I described how the whole island is tilted up in the north-west and down in the south-east. The Highlands, the Lake District and Snowdonia are all in the north-western parts of their respective countries, and in general the land gets lower and flatter the further south-east you go till it gradually merges into the sea in the Essex marshes.

By some cosmic coincidence the climate follows much the same shape. Most of our weather comes straight off the Atlantic Ocean. The prevailing westerly winds bring us frequent rain and almost constant clouds. Because they come from the west, this side of the country is wetter than the east. The weather systems which bring the rain are centred on the north of the island, which makes the north wetter than the south. So even if the whole country was flat there would be more rain in the north-west than the south-east. The land form intensifies the climatic gradient. Of course water is essential for life, but there's a happy medium in everything and here in Britain there's more often too much than too little. The rain cools down both soil and air, the clouds cut out the light which plants need in order to grow, and the humidity encourages fungus diseases. The rain leaches bases out of the soil, leaving them acid and short of plant nutrients. It frequently can't flow away as fast as it falls and soils become waterlogged easily.

The fact that the hills of the north-west are made of old, hard rocks only exaggerates the effect of the rainfall. Many of these old rocks are low in bases, so the soils they form would be poor and acid even without the rainfall. They're also impervious to water and this adds to the drainage problem. In some parts of the Highlands drainage is slow even on a permeable rock.

This was really brought home to me one day when I was doing some permaculture design work with a young couple on their croft in Wester Ross. Although not much above sea level and near the coast, the site is surrounded by mountains and very, very rainy. We were looking for a site for a new house. An area of flat ground, which lay in the bottom of the glen but well above the level of the river, looked attractive. But there were rushes there and you don't want to build a house where there's bad drainage. The soil was sandy and the rock below it was a loose river gravel. We were intrigued: how could such a soil be badly drained? But the rushes were plainly there.

We thought there must be some barrier to drainage lower down the profile, so we dug a pit. We dug and dug, down to a metre and a half, but found nothing but sand and gravel. It was just that the rainfall there is so high that even on such a well-drained soil rushes will grow wherever the land is flat. With that much rainfall, any flat areas overlying impermeable rocks become blanket bogs. (See page 55.)

So the climate serves to intensify the effect of the rocks, giving soils which can support little but moorland and the tougher species of trees. The resulting landscapes can't produce much in the way of human food and they're low in biodiversity. But tourists flock from all over Britain and the world to see the 'wild', uncultivated scenery of the Highlands, the Lakes and Snowdonia. The paradox is that their visit is so often spoiled by one of the two great forces which created the landscape in the first place – the rain.

Not all western landscapes are so rugged and infertile. Where the rocks are softer and richer in bases the hills are lower, the rainfall more moderate and the soils more fertile. This is grass country. A constant supply of moisture throughout the growing season keeps up the production of lush green leaves without a break. There may be more rain than you need in winter, but that doesn't matter too much because the cattle are snug in the barn eating the hay and silage that was made in the summer. This is why the west of Britain specialises in animal farming.

To the east of the hills and mountains there's less rain. The air is drier because it's left most of its moisture behind on the high land. It also warms as it falls, and warm air can hold more moisture than cold, so even if it is still moist it gives less rain. The relatively dry area on the lee side of high ground is known as a rain shadow.

In southern and eastern Britain the younger, softer rocks give more fertile soils and lower hills with gentler slopes which can be easily cultivated. All of it is in a rain shadow. You may not think so if you live there, but parts of eastern England have a rainfall which would class them as semi-desert in a hotter and sunnier part of the world. The dry climate has advantages for arable crops. Almost any work on the land, from ploughing to harvesting, needs dry weather, and the crops need a dry spell in late summer to ripen. Certainly grain crops need moisture to grow, but not so much nor so constantly as grass.

Understanding the regional climate is an important part of understanding a landscape, whether for the purposes of permaculture design or simply out of interest. In broad terms it tells you what the land is capable of. But understanding the microclimates within that landscape is one of the central skills of permaculture. Chosing different land uses to suit different microclimates is one of the main ways we can increase the productivity of the land without the use of external inputs. Reading the pattern of microclimates across a landscape is also a fascinating study for its own sake.

Microclimate

I live on top of a small hill surrounded by flat land. Today, as I write this, the hilltop is in bright sunshine but if I look down in any direction all I can see is fog. The hilltop and the valley have different microclimates. Beyond the flat land there are limestone hills and where there's a patch of unimproved grassland on those hills you'll find the wild thyme. It likes to grow on the anthills, because it appreciates the slightly drier soil it gets on these little grassy mounds. It tends to grow more often on the south or south-west sides of the anthills, the warm sides which get the noonday and afternoon sunshine. On the north sides of the anthills moss is more common. This contrast between the two sides of an anthill is also a difference in microclimate. There are microclimates on both these scales and every one in between.

One of the most striking examples I know of the effect of mircoclimate is Thatchers End, a tree plantation at Ragmans Lane Farm, planted on the very day in 1991 that Margaret Thatcher ceased to be our Prime Minister. It's sited on a steep bank overlooking the river Wye. In fact the site was chosen for tree planting because it's too steep for other uses and is best put down to woodland. The trees are a mix of oak, sweet chestnut, walnut and wild cherry or gean with a few self-seeded ash here and there. When the trees were planted they were not mulched to protect them from competition with grass, as newly-planted trees often are.

The trees have not all grown equally well. Far from it. The variation in growth between different parts of the field is so marked that I've long used Thatchers End as the basis for an exercise in landscape reading for my students. Their task is to identify the reasons for the patchy growth and try to decide which, if any, is the dominant one. It's not entirely a matter of microclimate. Variations in the soil and biotic factors have had their influence too. But then nothing we can see in the landscape is ever the result of only one factor.

I made the sketches you see overleaf in 2002, eleven years after the trees were planted. The difference between the trees in one part of the field and another is so great that some students find it difficult to believe that they were all planted on the same day. But they were. (See photo 11.)

(Incidentally, the unit of measure I used to estimate the height of the trees is a person-height. It's a handy one for people like me who find it difficult to judge the sizes of things in conventional measures like metres or feet. Even if you haven't got a person to stand beside the trees it's easy to imagine one. It doesn't really matter how tall the person, real or imagined, is because it's the comparison we're after here, not an exact measurement.)

Thatchers End

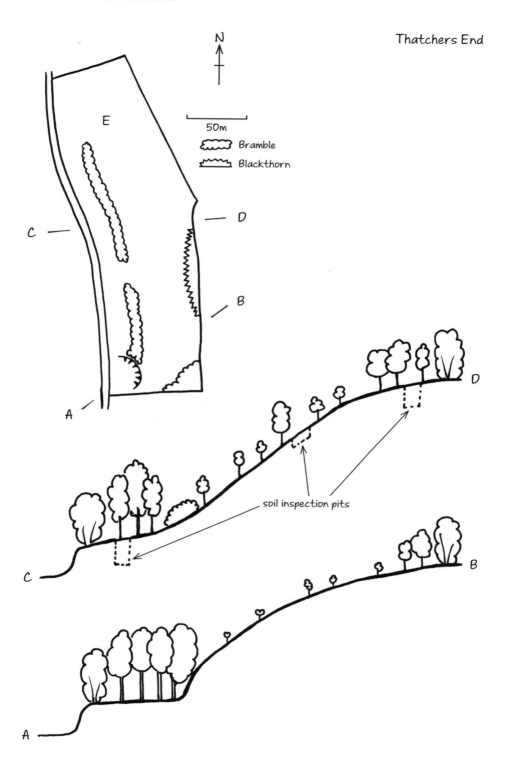

N

50m

~~~~~ Bramble

wwww Blackthorn

E

C

D

B

A

soil inspection pits

D

C

B

A

The soil inspection pits showed how soil has been eroded from the steep slope and deposited above the ancient hedge at the bottom, forming a deep lynchett of fertile soil. The tree sizes on both transects match the depth of the soil: medium-sized trees on the uneroded soil at the top, smaller ones on the eroded bank and taller ones on the extra deep soil of the lynchett. The variation in the growth of the trees is mainly a matter of soil moisture. Shallow soil holds less water than deep soil and sloping ground always drains more quickly than flat. But the soil doesn't account for the differences between the two transects. The trees on the steep part of the southern transect, A-B, are much smaller than those on the steep part of the northern one, C-D. Soil pits were dug on both transects and they showed the same depth of soil on both. There's clearly some extra factor at work here.

The answer lies in the way that Thatchers End fits into the landform of the Wye valley. The southern part of the plantation faces head-on to a stretch of the river which runs south-west. South-west is the direction of the prevailing wind and the river valley funnels that wind straight towards the southern bank and intensifies it. The middle part of the plantation, where I took the transect C-D, faces a low spur on the other side of the river, which gives it some shelter. The drying effect of the wind is at its maximum just in that area where the trees have hardly grown since the day they were planted. Wind is important because it increases the speed at which plants lose water from their leaves. The increased loss multiplies the effect of the limited supply of water from the soil.

The aspect of the slope also plays its part. The slope where I took the transect A-B faces south-west. Where I took the transect C-D it faces west. This means that the slope A-B faces straight into the prevailing wind. It also means it faces some of the hottest sunshine of the day, just after midday, while the slope C-D faces the cooler sun of late afternoon. The north end of the field, marked E on the map, slopes to the north-west. This is a moister aspect and in that part of the field there's no noticeable difference in tree size on the different parts of the slope.

The trees growing at the bottom of the slope on both transects have grown faster than any others in the field. They will have been helped by the shelter given them by the ancient hedge, a biotic factor, and by the exceptionally deep and humus-rich soil. But again there's a difference between the southern and northern transects. The trees in the hollow at the base of the slope A-B are bigger than the trees at the base of C-D. This is probably because the floor of the hollow is dead flat, so less water runs off and more sinks into the soil.

So, it's a complex situation and it's hard to say which is the dominant factor at work in Thatchers End. But now, three years later, things have changed, as they always do in the landscape. Where I took the northern transect, C-D, it's getting increasingly hard to see the difference between the trees in the top, middle and bottom parts. The difference is still there, but all the trees have now grown above head height, which makes it much more difficult to see what's going on. Meanwhile on the southern transect, A-B the short trees on the steep slope haven't grown at all, and look like they never will. They're still short enough to be seriously affected by deer browsing. This patch looks like staying as a clearing in the wood for some time to come, mainly because of its intense microclimate.

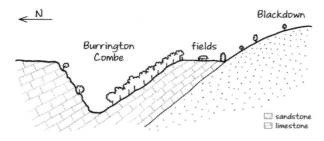

Eventually even that slope will become wooded, though almost certainly not with the planted trees. (See page 147.) To keep land permanently free of trees without grazing animals it takes a much more intense combina-tion of soil and microclimate. In our moist climate almost every patch of ground will eventually turn into woodland if it doesn't get grazed, cultivated or built on. However the photograph of Burrington Combe (see photo 12) shows just such a combination of soil and microclimate. It's a deep dry valley on the Mendips which cuts through the limestone rock from east to west.

I took the photo facing south-east. The north-facing side of the combe, in the middle distance, is an ash wood with some oak, sycamore, birch and hazel. Being on the dip slope of the limestone strata it's a little less steep than the south-facing side and so has a little more soil. But more important than that, it faces away from the sun. The south-facing side, in the foreground, catches the summer sun full in its face and, being on the scarp slope, it's steeper and has very little soil. All it grows is a little grass and the occasional whitebeam and yew, two trees which are well adapted to the dry conditions of limestone. There are in fact some parts of the south-facing slope which are wooded like the north-facing one. These are places where either there's an accumulation of eroded material at the foot, which allows a little soil to build up, or where the north-facing slope casts shade on it for most of the day. It's only the combination of both dry microclimate and thin soil which can keep the woodland at bay.

Behind the woodland you can just see a field. This narrow strip of flat land on limestone is the only part of this landscape which is farmed. Blackdown, the broad curving slope in the background, is on Old Red sandstone, a hard, acid rock, and it's a bracken moorland with occasional hawthorn and rowan trees. The boundary between fields and moorland is exactly on the boundary between limestone and sandstone, where the base-rich soil changes abruptly to the acid soil of the sandstone. The boundary between the fields and the woodland is exactly on the break of slope, where the flat ground suddenly becomes steep. Just a simple change in either the soil or the landform causes a complete change in the land use and vegetation. But when it comes to microclimate it takes a huge difference in aspect, backed up by a difference in soil, to have the same effect.

This is typical of the British landscape: aspect rarely has an obvious and visible effect on the vegetation. It can be important in landscape design, for example when siting a vineyard, but it only makes a clear visual impact in extreme circumstances. This is because we have a very equitable climate with no extremes of heat and cold or wet and dry. In more severe climates it's different.

In Norway, where cold rules, the aspect of a slope makes a profound difference to the vegetation. The area around the Oslofjord is on the boundary between the deciduous woods of the south and the coniferous woods of the north.

It's not a very hilly area, but even the slightest change in slope completely alters the woodland type. On south-facing aspects there's a mix of oak, ash, aspen and hazel. But just a tilt of five percent to the north is enough to see spruce and pine take over. In the Mediterranean lands water is the key to everything. Under the fierce skies of southern Spain I've seen similar slight changes in aspect make the difference between trees and no trees. A gentle undulation in Andalucía has the same effect as the gorge of Burrington Combe in cloudy England.

On the other hand, the microclimate changes more rapidly with altitude in our moist, cloudy climate than it does elsewhere. Imagine that you're in a sheltered valley, looking up at a high, treeless hill. It's a sunny day with a light breeze, and you're comfortable in a t-shirt. But you just know that if you climb to the top of that hill you'll need to take some warm clothes with you. It's not just that the breeze will be stronger up there but also because the temperature is actually lower at higher altitudes. For every hundred metres you go up the temperature falls by one degree Celsius. This doesn't sound like much, but multiplied by the wind chill factor it makes a very noticeable difference.

I'm talking here of a day when it's as sunny at the top of the hill as in the valley. But so often it isn't. As I saw clearly from the Isle of Erraid, it's usually more cloudy on the hills and it rains more. This means that the effects of altitude and wind are multiplied by lack of sunshine and the chilling effect of wet. On days when all four factors are combined it can be totally miserable on the hill while it's still quite pleasant in the valley. These differences accumulate over time and can have a marked effect on the vegetation, especially in mountainous areas, where the common pattern is moorland at high altitude and grassland lower down. (See photo 13.)

In merely hilly areas the difference may be less obvious, but if you look carefully you can see it. For example, orchards are notable by their absence from the hills. The unhealthy cocktail of wind, wet, cold and cloud presents fruit trees with more problems than they can handle. The damsons of the southern Lake District are an exception, but then damsons are exceptionally tough fruit trees.

While orchards make an obvious visual statement about the microclimate the effect of altitude on grassland is more subtle. In most western parts of Britain there's likely to be grassland on both hills and vales but there's a big difference in the length of the growing season. Much the same grasses may grow in the fields on the hills and in the valleys but they grow for fewer days in the year on the hills. This makes a big difference to the farming economy.

You can 'see' the length of the growing season by comparing the annual cycle of a single plant growing at different altitudes. Hawthorn is often the best choice. It grows almost everywhere and it's very conspicuous when it comes into bloom. Hawthorns which blossom earlier will have started growth earlier in the spring and will stop growing later in the autumn. In fact a difference of a couple of weeks in flowering time may represent several weeks' difference in total growing season.

Very often you can see a strong contrast between the hawthorns separated by a couple of hundred metres vertically, then travel a couple of hundred miles north or south and see very little difference between trees at the same altitude.

Such is the effect of altitude in the cloudy climate of Britain. But you must be sure to compare like with like. Firstly, a hawthorn in an exposed position will flower later than one in a sheltered sun-trap at the same altitude. Secondly, be sure that the hawthorns you're comparing are native ones. Most roadside planting is done with exotic plants. The seed, or even the plants themselves, may well be imported from eastern Europe, where labour costs are lower, and these exotic strains are usually earlier flowering. Thirdly, don't look at clipped hedges. Hawthorn flowers on twigs in their second year of growth, so a hedge which is clipped every year never flowers. But there are plenty of hawthorns in unclipped old hedges and scrubland, especially in hilly country.

Another microclimate factor which changes with altitude is frost. In general there are more frosts in the colder climate of higher land but at the microclimate scale this gets turned upside down: there are more frosts in the valleys than on the hills. This is because cold air is heavier than warm air and it sinks. On a morning after a light frost you can often see how the lower-lying parts of the landscape are frosty while the higher parts are frost-free. You may also see a contrast between flat land and sloping, where the cold air has rolled away down the slopes but hardly moved from flatter spots. Frost pockets are those places which always catch the frost on such a night of partial frost. They're always low lying places, valley bottoms and hollows where there's nowhere for the cold air to drain away to. They can be big or small. They can occupy valleys several miles across or measure just a few square metres.

They're places to avoid when planting fruit trees, as a late spring frost can kill the blossom and thus wipe out a whole year's crop. But in summer these very same places can often be warm sun-traps and being low lying they often have the best soil, so they can make good places for vegetable growing, despite the late spring. I know two gardens, close by each other and one just five metres higher than the other. The lower one is in a frost pocket and is consistently two weeks later in the spring than the upper one. But come the summer the lower garden catches up and then overtakes its neighbour, because it has a warm, sheltered microclimate in summer.

To see frost pockets you need to get up early in the morning after a night when there's been a light frost. If there's been a hard frost which has frozen everywhere you won't see any frost pockets. If you go out after an hour or two of sunshine you'll see some places which are frosty while nearby places are frost free, but these are not frost pockets. They're just in the shade. They're not more liable to frosts than other parts of the landscape on a night of light frost, nor are they colder than other places on a night of hard frost which freezes the whole landscape. They simply take longer to thaw out. But early in the morning after a light frost which isn't strong enough to freeze the whole landscape you'll see the low-lying frost pockets picked out in white rime.

There are other interesting frost effects which you can see on a morning after an overall frost. You still have to be up and out before the sun has had any effect, but in the depths of winter this isn't so early as it is in the spring, which is the main season for those light frosts. My notebook records such a morning.

Ragmans Lane Farm, January 2001

A morning of overall frost well illustrates how trees keep the ground warm. You can clearly see the contrast between white fields and frost-free woods. A frost-free strip extends into the fields where the trees overhang the woodland edge and it runs alongside the hedges, wide in proportion to how overgrown the hedge is.

The tree canopies keep the ground free of frost in just the same way as a blanket keeps you warm at night: they stop the stored heat in the soil from radiating away into space and reflect it back onto the soil surface. You can sometimes see the same effect on still water, where the only ice-free patch of water is under the canopy of a bankside willow.

On another morning of overall frost I made this sketch.

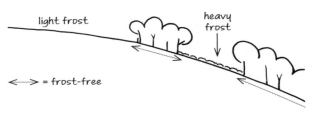

light frost

heavy frost

⟨——⟩ = frost-free

The field only received a light frosting because it's on a hilltop and the cold air can flow away from it. It's just the sort of place which stays frost-free on those nights when only the frost pockets freeze. But the clearing in the wood is a frost pocket. I could feel how much colder it was than the field as I walked into it, and the bracken fronds on its floor were coated with a thick layer of rime. It's a frost pocket despite being at the top of a hill because cold air can sink into it from above the woodland canopy but can't flow out of it. Cold air is viscous stuff and even leafless trees make a barrier to its flow. This is typical of a woodland clearing.

Since the patterns of frost pockets are topsy-turvey – colder in the valley bottoms and warmer in the shade of trees – you can't extrapolate them to other times of year. But snow is quite different. Both the patterns made by drifts and those made as the snow thaws can tell you much about the microclimate in other seasons. The great value of snow to the landscape reader is that it gives you a visual image of microclimates which otherwise can't be seen, only felt.

The exquisite sculpture of snowdrifts is all the more special because the perfection of their lines is so short-lived. But even when they've lost their pristine beauty they can tell you much about the patterns of wind in the landscape. So you don't have to hurry out early in the morning as you do to see the frost pockets.

Areas of thin snow or bare ground indicate the windiest places while drifts form in the sheltered places, where the wind slows down and drops its load of snow. The pattern you see on any one day will only tell you about the micro-climates produced by a wind from one particular direction. The south-west is the prevailing wind in Britain, so drifts formed by a wind from that direction are specially useful in predicting microclimates at other times of the year.

Snow blown by a north-easterly wind can be even more valuable. The wind doesn't blow from that direction so often but when it does it's colder and can do more harm to vulnerable plants. Both will show you where to place plants and structures which need the most protection and where extra shelter is needed. As well as giving you specific information about your site, studying drift patterns can help your general understanding of how winds interact with trees, hedges, buildings and slopes.

The other way of finding out about how the wind behaves in the landscape is to go to different parts of it and feel the wind. In fact your body is the ideal instrument for measuring the wind. Where you feel uncomfortable and cold an animal will feel much the same and a plant, even if it doesn't exactly feel, will grow more slowly and possibly be damaged. But you'd have to be everywhere at once to experience winds through the whole landscape simultaneously. Snow drifts give it to you literally frozen in time and easy to see.

When snow falls on a still day it covers the ground to an even depth. While the snow lies it tells you nothing but as it starts to thaw a kaleidoscope of microclimate effects is revealed in the pattern of thawed and unthawed patches. The places where it thaws first are the warmer spots and as with drifting you see the whole picture simultaneously. It's like having a map of microclimates drawn for you by nature.

Unless the land is totally flat you can see the effect of aspect, with more southerly slopes thawing faster than more northerly ones. The shade cast by hedges and woodland edges has a more localised effect but often a more intense one. It can either reinforce or counteract the effect of aspect. A north-facing slope with a wood at the top will be the last place of all to thaw out. During the rest of the year a slope like this may be one of the coolest and dampest places in the landscape.

On the southern side of a wood, hedge or wall the rate of thawing can be speeded up by sunlight reflected off the vertical surface. These sketches from my notebook

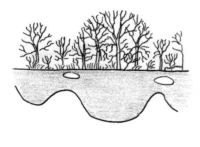

show a couple of examples. The thawed area matched the varying height of the trees fairly closely. I estimated its width was about two thirds the height of the trees. The two small patches of snow near the woodland edge were probably the remains of little drifts.

A day of thawing snow is like gold dust to a landscape reader. Frost, when it covers the whole landscape rather than just the frost pockets, can tell much the same tale as it thaws. But snow has depth, so it can distinguish grades of warmth by the time it takes to disappear, whereas frost thaws pretty well immediately once the sun strikes it. The only drawback to snowmelt as a guide to microclimates is that the thawing pattern may not be quite the same as the pattern of warm and cool places in summer. The sun is at a higher angle in summer, so shadows will be shorter and there will be less difference between north and south aspects, especially on gently undulating land. The general picture will be the same, but you'll have to check the detail by other means.

Once again your body is the ideal instrument. So often we ignore the micro-climate as we go about the landscape, our minds full of business. But pause for a moment in a spot, tune in, and you can experience the combined effect of heat, wind, sunshine and moisture in a way that no bank of weather recording instruments can.

The limitation of using your body is that it only tells you what's happening at one particular time, while the position of the sun changes from hour to hour and from day to day through the growing season. But you can often get an idea of the accumulated heat and light that each spot receives by looking at the flowering time of plants. If they're blooming in one place but not yet in another then you know that one place receives more sunshine than the other. Some flowers, such as dandelions are not just conspicuous when they flower but also when they go to seed. What a contrast there is between the glowing gold of the bloom and the ghostly grey of the dandelion clocks! Gone-to-seed ones have clearly received more sunlight than ones which are still flowering.

Often the shadier place is quite obvious. It may be on the north side of a hedge, for example. What's not so obvious is how far from the hedge the significantly shady area reaches, taking the average through the growing season and at different times of day. If there's a clear boundary between flowering plants and ones which aren't yet, or between flowering and gone-to-seed dandelions, it shows you how far the significant shade reaches.

Of course the same plant must be growing over a fairly wide area in order to make a comparison. Common wildflowers such as dandelions and buttercups often oblige. When they're in full flower over a grass field the areas where they're not flowering stand out clearly. It is possible that there are no blooms in one area because the plant itself doesn't grow there, and you have to check this. But this in itself may be useful information. If you're observing the microclimates in a garden you can deliberately plant the same flower in different parts of the garden and compare its development during the growing season. Nasturtiums are good for this. They also reveal the frostiest parts of the garden since they die off at the first hint of autumn frost.

Flowering times can reveal the effect of aspect as well as shade, as when crocuses come out on a south-facing bank but not on the flat. Sometimes you can also see them blooming underneath the bare twiggy canopy of a tree while beyond the reach of its branches they're not yet out. The soil is warmer under a tree due to the same blanket effect that I mentioned earlier. Of course once the tree's in leaf the position will be reversed and it will be cooler beneath its shade.

In late summer you can sometimes see the effect of shade on a field of corn as it ripens. There may be a green strip alongside a north-facing woodland edge when the rest of the field has already turned yellow. The width of this strip is typically two-thirds the height of the trees. If the summer has been a dry one, grass on southerly aspects is often visibly more parched than on flat or north-facing land. This dryness is partly a matter of the extra heat the ground gets from being tilted towards the sun and partly because a steep slope drains more freely. You have to be quite sure that any difference in the colour of the sward really is due to drying out. It may be that the grass on the flat has been grazed more intensively than that on the slope. This will make it greener, as grazing or cutting keeps grass in it's green vegetative stage, while gone-to-seed grass has a dry buff colour, regardless of the moisture content of the soil.

Another way in which plants can show the accumulated effects of a climatic factor is the wind flagging of trees. In these days of global warming this is perhaps a better bet than waiting for snowfall, although the picture you get isn't quite so detailed. In extreme cases a wind flagged tree looks as though it's been blown over by the wind. In fact it hasn't. It's just that the wind is so severe that the tree can only put on new growth on the lee side of its crown, so it only grows on that side. But most trees are somewhat affected by the wind, even if the flagging is so slight that you have to walk right round the tree to see it. It's worth doing this because you can only see mild wind flagging from a position at right angles to the direction of the wind. There's a whole range of intensity of flagging between the two extremes and they show you both the direction and severity of the prevailing wind.

Over Britain as a whole this is the south-west wind. But in hilly country the local landform can make the winds at ground level blow in quite different directions to those in the sky above. A narrow valley acts like a funnel and the wind will always blow up it or down it, regardless of the alignment of the valley. Wind flagging can indicate whether this happens or not in any particular valley. It's even possible for a tree or shrub to be flagged in a direction which isn't the most frequent wind but the strongest. I've seen this on Brean Down, a long narrow peninsula which sticks out due west into the Bristol Channel just south of Weston-super-Mare. (See photo 4.) The blackthorn bushes growing on the southern slope of the down are wind flagged from the south-west but those on the northern slope lean away from the north-west. The north-west wind is less frequent but when it does blow the shrubs on the north side are fully exposed to it, whereas they're sheltered from the south-westerlies by the bulk of the down.

It's in these unusual cases, where the prevailing wind in a particular spot is more influenced by the landform than the overall climate, that the direction of wind flagging is really valuable to the landscape reader. On open hilltops and plateaus the direction of flagging is true to the south-west and its main value there is to indicate just how windy the local climate is. In fact you can use the direction of wind flagging as a compass to guide you on a cloudy day. On Dartmoor, I'm told the locals find their way in the fog by the lean of the rushes and taller grasses. The moor is very exposed to the wind and also prone to prolonged mist and fog.

It's not just individual trees that get wind flagged. The canopy of a whole clump can be moulded into a single shape by a strong prevailing wind. These three

trees I sketched on the Mendip Hills are a sign of what the climate's like up there. You'd never see anything like it down here in the vale. Note how the most windward tree, which takes the full brunt of the wind, is the smallest.

The structure of a whole wood can be affected by the wind if it's on a steep slope, with short, wind-stunted trees at the top and tall ones at the bottom. A sketch from my notebook records the top end of just such a wood.

Great Stoke Wood, 22nd April 2005

As you walk up through the wood you can tell you're getting nearer the top because the trees get shorter. You can see more sky through their sketchy canopies. It makes me feel as though I'm climbing up into the sky. The shape of the trees is caused by them reaching for the light rather than by flagging because the branches reach towards the south-west, where the prevailing wind comes from. But their short stature is a matter of wind.

Evergreens such as holly and box can indicate the direction of the prevailing wind by dieback of the leaves or even whole dead branches. This is especially true within a few miles of the sea, where salty air can be blown inland by the strong winds of autumn and winter. No plants like salt but some can stand it better than others. On a headland you can often see a progression from the more salt-tolerant plants on the seaward side to the least tolerant inland. In practice this means a progression from grasses, through shrubs to trees. I describe an example of this in a later chapter. (See page 221.)

All these clues from plants are great aids to reading the microclimate. In fact you can often make very accurate guesses about where the different micro-climates will be just by looking at the landform: here a windy spot, there a suntrap, down there a frost pocket. But to be really sure there's no substitute for being on the spot when the wind blows, the sun shines or the frost is revealed in the cold light of dawn. Your mind can predict a microclimate but only your body can really know it.

The History
of the
Landscape

Having looked at how the landscape is formed by rocks, soil and climate, now we look at the fourth great factor, human beings. The amount of influence people have had is clear to see. Where once there was wildwood from coast to coast now there's a cultivated landscape of fields and hedges, woods and plantations. What's less clear is how long this has been going on. Is agriculture a new invention which has come in at the eleventh hour to disturb a wildwood which slumbered untouched by human hand for countless eons before that? Far from it. Our ancestors started turning wildwood into fields and hedges some seven thousand years ago, which was just about the same time that the wildwood recovered its full extent in Britain after the desolation of the last Ice Age.

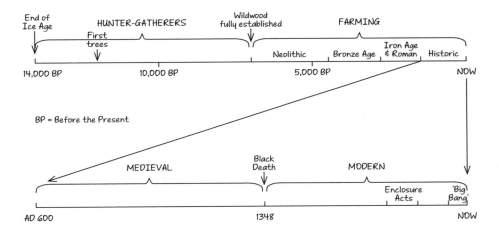

The Prehistoric Landscape

Approximately fourteen thousand years ago last Ice Age ended, leaving behind it a landscape of tundra. About twelve thousand years ago the temperature had risen enough for the first trees to appear in the south of the island. First came the cold-tolerant kinds like birch and pine, then the more temperate like oak and lime. Ash and beech arrived later and thousands of years passed before all these trees spread out and occupied the range they do now. Bit by bit the wildwood was formed. Around seven thousand years ago it had spread over the whole island except for a band of bog along the northern coast of Scotland, the outer Isles and the tops of the highest mountains.

Just how dense and continuous the wildwood was is not known for certain. There's no surviving part of the temperate world sufficiently untouched by humans to act as a model. The direct evidence comes mainly from pollen grains. In waterlogged deposits such as deep bogs and lake-bottom mud they can be preserved so well that the kind of plant they come from can be identified. They can be dated by the stratification of the deposit. But pollen study is not straightforward and it allows enormous scope for disagreement among those who study it.

The traditional view is that there were virtually no gaps in the wildwood at all. But then where have all the plants of open country come from? Could the grasses and grassland wildflowers all have immigrated with the first people who brought grazing animals and started opening up the wildwood? It is just possible. In North America almost all the pasture grasses and herbs are European immigrants. There are native grasslands there of course but the indigenous species aren't adapted to intensive grazing. The spear-like leaf of plantain, a common grassland herb, was called the white man's footprint by the Native Americans. As for the weeds, no agriculture means no weeds. A glance down a list of weeds in an American book reveals at least half of them to be familiar ones from this side of the water. Here too, many of the weeds may be immigrants, though from a much earlier time.

However that may be, most landscape historians now think in terms of the wildwood having more in the way of gaps and clearings than was previously thought. There's even a theory that the primeval landscape was one which periodically alternated between grassland and woodland, with large herds of wild herbivores playing a key role. This theory doesn't really stand up to close scrutiny and it's most likely that the pre-agricultural landscape of Britain was predominantly wooded. (See pages 132-133.)

There were people living in Britain at the time but they can't have had much effect on the landscape. There were very few of them and they lived by hunting and gathering. There is some archaeological evidence that they occasionally set fire to the wildwood, presumably to make hunting easier and increase the food supply for their prey. Burning can convert woodland to grassland or savanna, both of which provide more food for grazing animals and clear sight lines for hunters. This burning of the woodland seems puzzling because, except for the Highland pinewoods, British native woodland is completely non-inflammable. But the woods we know today are not the wildwood. They've been managed for centuries and trees have been harvested regularly. In a virgin woodland the trees die naturally and remain in situ, decomposing slowly. A large proportion of the total biomass at any one time can be dead wood. At the end of a long dry summer it might just have been possible to get a fire going hot enough to kill the live trees, or at least some of them.

Whatever the hunter-gatherers did, the effects must have been small and local. The real start of human impact came with the beginning of the Neolithic or

New Stone Age. The Neolithic people were the first farmers. They tilled the ground and kept animals. It was a pivotal moment, not just for the landscape but for humanity itself. The invention of agriculture can be seen as the most important event of human history. Before it we were just one species amongst many but with the power that agriculture has given us we're able to transform the biosphere, dominate it and, if we choose, destroy it.

Farming first started ten thousand years ago in what we now call the Middle East and spread slowly from there. Although it was also invented in other parts of the world, it came to this island direct from its first homeland and arrived here some seven thousand years ago. No-one knows whether it was the farmers themselves who migrated or simply the new ideas which spread from neighbour to neighbour and across the sea. But the crops which form the backbone of our agriculture to this day did indeed migrate here all the way from the Middle East.

That is a dry region with extensive deserts, and steppe in those areas with enough moisture for grasses but not enough for trees. Out of the steppe came wheat, barley and oats, grasses with seeds big enough to make them worth growing as food crops. The Neolithic farmers of Britain ate these along with the meat and milk from their herds. The hunter-gatherers of earlier times had eaten hazelnuts and the starchy tubers of bulrushes with their wild meat. But by the time agriculture arrived in north-west Europe the cereals had been improved by thousands of years of plant breeding and the methods of cereal farming were well developed. It was an efficient and well-known package. No-one went to the trouble to invent a northern form of agriculture using the indigenous edible plants. If they had the landscape might look very different now, perhaps more like the native wildwood and less like an imitation of the south-west Asian steppe.

Animal farming was part of the same package and here too exotics reigned. It wasn't the native roe and red deer but the imported cattle, sheep and goats which made up the herds. In fact it wasn't so much the kind of grazers which affected the landscape as the intensity of grazing. Domesticated herds of any kind have more impact on the landscape than wild herbivores because they can be kept at artificially high numbers. As we've already seen, trees can't survive intensive grazing but grasses can. (See page 2-3.) So the inevitable result was the gradual replacement of woodland with grassland. Over the thousands of years since the dawn of the Neolithic age grazing animals have destroyed many more square miles of British woodland than the plough ever has.

The key to keeping more animals is haymaking. Winter is the season when most wild animals die, weakened by hunger and cold. Making hay in the summer and feeding it in winter overcomes this natural control on the population of herbivores. It also means that two distinct types of grassland came into being: pasture, where the animals graze, and meadow, where hay is made.

As well as starting the process of converting wildwood to open country, the Neolithic people started managing it and turning it into the kinds of woodland we would recognise today. Right at the beginning of the Neolithic age they were already working some woodland on a regular coppice cycle. This means felling the trees every few years and allowing them to regrow from the stump. They regrow in a multi-stemmed form and the same trees can be cut

Coppice

again time after time. We know they did this because large quantities of unmistakable coppice poles have been found preserved in the peat of the Somerset Levels. They form part of the structure of wooden trackways. These trackways were not unlike the boardwalks that give access to some modern wetland nature reserves, but longer, linking up the islands with the surrounding dry land. The earliest of them dates back to some six thousand years ago, in the early Neolithic.

But why should the Neolithic people have bothered with coppicing when they had unlimited access to the wildwood which surrounded the modest clearings where they grew their crops and pastured their animals? Because coppice wood is much more accessible and easier to use. If you want firewood it's much easier to cut a crop of young poles at ground level than to fell giant virgin trees and chop up their huge trunks and thick branches, especially if the only tools you have are stone axes. The coppice poles will also be fairly straight and much the same size, perfect raw material for building a house, a fence or a trackway.

Farming communities spread all over these islands, including less favourable areas like the Scottish Highlands. In fact the Highlands had a higher population during much of the prehistoric period than they do now. Nevertheless the impact of the new way of life on the landscape may have been slight to begin with. The population was still very low and at first farming would only have supplemented hunting and gathering. There's very little you can see in the modern landscape that was made by the hands of the Neolithic people. Nevertheless their legacy is important because they gave birth to the semi-natural ecosystems which are with us to this day. This isn't to say that individual coppice woodlands and semi-natural grasslands have necessarily survived on the same sites from that day to this, although in some places this may have happened. It's more that these types of ecosystem have been in continuous existence since then. Whether they've stayed put or migrated around the landscape there's been continuity of habitat for the wild plants and animals which inhabit them.

Moorland and heath are two other important semi-natural ecosystems. Both of them are basically places where heather grows: moors in the uplands and heath in the lowlands. Most moorland was formed from woodland in much the same way as grassland, by grazing, though some of it does have a more natural origin. (See page 254-255.) Heath may also have been formed by grazing, but it's equally possible that it started out as abandoned arable land. The light sandy soils of what is now heath would have been attractive at first to people with the most primitive of farming tools. But such extreme sandy soils, with their scant supply of bases, would soon have become too acid and impoverished to continue cropping.

The age of prehistoric agriculture lasted something over five thousand years, from the beginning of the Neolithic to the end of the Roman period. We must include the Roman in the prehistoric because there are no written records from

Roman times which shed any light on the landscape. This is almost four times as long as the historical period, which started in the seventh century AD when the Anglo-Saxons learned to read and write. Throughout the prehistoric period people made fields, planted hedges, built walls and houses. Bit by bit the humanised area extended and the wildwood shrank. It wasn't a continuous process. Land came into cultivation, was abandoned and then re-occupied, sometimes after a gap of centuries. But by the time the historical period dawned there was very little wildwood left.

In some parts of the country there are almost certainly fields still in use today which were first laid out in prehistoric times. But more often prehistoric fields are overlain and obliterated by later reorganisations. The best places to see the remains of prehistoric fields are hill areas which were cultivated then, perhaps when the climate was warmer, but have remained uncultivated ever since. On older Ordnance Survey maps these were often labelled 'Celtic Fields' in gothic type. This was misleading as the Celts only arrived here at the end of prehistoric times, just before the Romans. On newer maps this has been replaced with 'Field System', which is more accurate but less specific as it's also used for abandoned medieval fields.

The most characteristic prehistoric fields are small and square. They can be clearly seen on some unploughed areas of the chalk downland of southern England. They make a distinctive chequered pattern with banks and lynchetts enclosing little fields of an acre or less. The square shape reflects the kind of plough used in those times. This was a scratch plough, or ard, which cuts into the soil like a chisel but doesn't turn it upside down like the later mouldboard plough. To use it effectively it was necessary to plough each field at least twice, in different directions. Hence the square-shaped fields. A long narrow field would have been easy to plough from end to end but tedious from side to side. You would have spent more time turning the plough than ploughing. 'Making short turns' is something farmers find tiresome to this day.

These little fields are often accompanied by the remains of lanes and farmsteads which suggest a landscape of small farms and hamlets. But a completely different landscape is suggested by the remains of a much more impressive kind of prehistoric field, the reaves. These are stony banks which march across the country in regular parallel lines, not quite straight, about a hundred metres apart, with occasional cross-reaves forming rectangular fields. The most extensive remains of a reave landscape is to be seen on the southern part of Dartmoor, covering tens of square miles in a pattern that can only have been planned by a strong central authority. They appear to have been used mainly for animal farming with a little cropping. Excavation of the associated buildings dates them to the late Bronze Age. There's a small set of reaves in Perthshire which are also Bronze Age while a set in County Mayo date to the Neolithic. They seem to have been far-flung in both time and space.

We can do no more than wonder about the social and political setup which created the reaves. Archaeology tells us nothing of such things. But we can say for sure that the country as a whole must have been quite densely populated if such effort was put into organised agriculture in such a marginal area as Dartmoor. Even given the milder climate of the Bronze Age, this infertile granite upland must have been a less favoured area. If people put that amount of effort into developing such marginal land there can't have been much space left in the country as a whole.

In some places reave fields may still be in use today. Part of the parish of Toller Porcorum in Dorset is laid out in a very reave-like pattern, which contrasts strongly with the irregular field pattern along side it. The straight field boundaries are certainly very old. Some of them consist of huge banks with ancient coppice stools on them. No-one built such big banks in modern times, and though they could be medieval the resemblance to the Dartmoor reaves is too close to dismiss lightly.

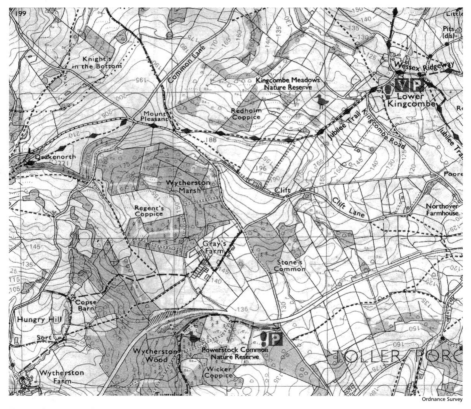

Toller Porcorum, Dorset

The Romans left their mark on the landscape with their straight main roads and the first towns to be built in Britain. Many of them are still in use, both roads and towns. The roads were like a modern motorway system, connecting the towns but hardly relating to the rest of the road network. This was made up of the winding lanes which had been there before. The towns were islands of urban life in a hardly-changing landscape of hamlets and little fields. Villas were built in the countryside, but they fitted into the existing pattern rather than replacing it. Villages did not yet exist. Nevertheless it was a densely settled landscape. Roman remains show up well in the archaeological record and they tell us that farmsteads and hamlets were spaced about half a mile apart over most of lowland Britain and more densely in some areas.

Here and there the Romans did impose order on the countryside. An example is the Gwent Levels, on the Welsh coast of the Bristol Channel either side of Newport. The Romans drained land in various parts of the country, including the Fens of eastern England. In most places their work has been overlaid by that of later generations but the present layout of the Gwent Levels has been shown by excavation to be Roman. It's a highly organised landscape, with long, narrow, rectangular fields bounded by straight, parallel ditches, evenly spaced and pointing to the sea. Time has softened this military regularity in places. Willows and other trees grow beside the ditches and give the landscape a feeling at times of intimacy, even mystery. Crossing the Levels in the train I've sometimes had the impression of looking down a series of long, leafy tunnels.

The fall of Rome marks the start of the Middle Ages. The economic collapse that took place when the Roman armies left Britain was total. The towns crumbled and export markets ceased to exist. People no longer had to grow a surplus of food to support the army and the civil service. Britain slipped back into a subsistence economy. People stopped using coins, and in the west of the country they even stopped making pottery. With so much less demanded of the land there must have been long fallows and plenty of opportunity for the fertility of the soil to build up. Trees spread back over abandoned farmland, at least in some places. Witness the fact that most large ancient woodlands in England have some prehistoric or Roman remains in them.

Nevertheless it was still a farmed landscape and in most parts a fully occupied one. When the Anglo-Saxons moved in to fill the power vacuum left by Rome they took over a landscape already organised into estates, not a wilderness from which they had to hack out new farms and settlements from scratch. There's evidence for this in some modern parish boundaries, which are often based on those of medieval manors, which in turn often date back to early Anglo-Saxon estate boundaries. Anyone carving out a brand new estate would surely use an obvious landmark, like a ditch and rampart which was recently made and ran more or less where the boundary was to be. On the Marlborough Downs in Wiltshire there's just such a rampart, called Wansdyke, which was built right at the start of the Anglo-Saxon period. Several parishes meet approximately along this line but the boundaries completely ignore the dyke. They're clearly older than it. The new Anglo-Saxon lords were taking over established estates with recognised boundaries. In other places parish boundaries ignore Roman Roads in much the same way.

There's more continuity in the landscape than we often think. Although there have been major reorganisations, most things are much older than they seem. Boundaries are often the oldest things of all. After all, boundaries are all about land ownership and this is a very touchy subject, one on which people have been reluctant to give an inch.

Woodland and Champion

In the Middle Ages we begin to see the different regions of Britain emerging, each with its own distinctive character. This is particularly so in England where historical records are more plentiful and landscape historians have been more active. The main division is into two landscape types, which used to be known as woodland and champion. In champion country there were big open fields, with each farmer's land holding split into a number of strips which were intermingled with those of his neighbours throughout the fields. All the farmers lived together in one village. Woodland country had small, hedged fields, each one privately owned. There were few villages and farmsteads stood alone or were grouped into hamlets of half a dozen houses or so. (See photo 14.)

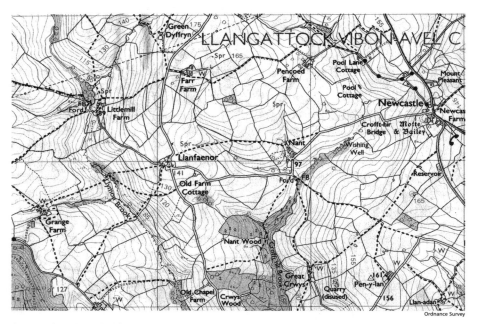

Ancient Countryside, Monmouthshire

The word champion may come from the French Champagne region, which is an extreme example of the type. Woodland comes from the abundance of trees in the hedgerows, compared to the wide hedgeless spaces of the open fields. Oliver Rackham has renamed them ancient and planned countryside. Indeed, the woodland countryside is ancient. This is how the landscape was in prehistoric times and any changes to it have mostly been piecemeal. The champion country was indeed planned. The open fields of each village were laid out in a highly ordered manner and the people who made them had to sweep away the previous layout. In their turn the open fields were swept away by the Enclosure Acts of the eighteenth and nineteenth centuries and replaced by large, rectangular fields.

No-one knows why the open fields were made, still less why they were made in some parts of the country and not in others. The planned countryside lies across England in a swathe from Yorkshire to Dorset, with ancient countryside on both sides. All of the planned countryside is in the fertile lowlands and it avoids upland areas like Wales and the Pennines, but the distinction is not all down to geology and climate. Why, for example, does the boundary make that dog-leg corner across the flat plain of East Anglia? Historians may have their theories, but the people who actually made the open fields are no longer around and they left no written record of their reasons.

The planned countryside of England

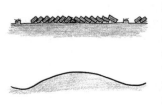

The open fields were very much an arable landscape, formed by the nature of the mouldboard plough. This is the plough which turns the soil upside down in a slice. Cross-ploughing is not necessary, so a long, narrow strip of land is an efficient shape to work as it keeps unproductive turning time to a minimum. The plough always moves the soil to the right, so as you plough up one side of your piece and down the other the soil is turned towards the centre from both sides. Thus the soil gets progressively piled up along the centre and removed from the edges and over the years this results in a series of permanent rounded ridges. These ridges were the basic unit from which the open fields were made up.

A strip was a parcel of one or more ridges belonging to a single person. A number of strips lying parallel together made a furlong and the furlongs were grouped together to make two, three or sometimes more fields. In the Midlands the field was the unit of rotation. A two-field village would have one field under crops and the other under fallow. A three-field village would have one under autumn-sown crops, one under spring crops and the third fallow. In East Anglia the furlong was the unit of rotation, so the number of fields in the parish was less significant.

A typical pattern of ridges and furlongs

The open fields would be available for common grazing both after harvest and in the fallow year and in addition there was common land for pasture. Meadow for haymaking was often the most valuable land of all. It was usually situated on the rich alluvial soils along rivers and streams. These soils might be too wet for arable farming but would have both the moisture and the plant nutrients needed for a good hay crop year after year.

There are four villages in England which still have working open fields: Laxton in Northamptonshire, Soham in Cambridgeshire, Portland in Dorset

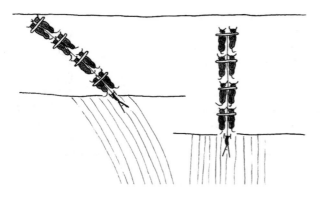

and Braunton in Devon. If you visit one of these you'll see that the strips are not always straight. Just as often they're curved, either like a C or like a reversed S. These shapes arose due to the practicalities of ploughing with a team of six or eight oxen. Once they'd stopped ploughing at the end of a furrow they could be turned, yoke by yoke, but until then they had to keep in line, pulling. This meant there was an unploughed, and thus unproductive, headland left at either end of each furlong. The width of the headland could be kept to a minimum if the team approached it at an angle, which is just what would happen if the ridge was curved. The ridges may have been deliberately laid out in a curved shape or they may have gradually grown curved through use. No-one knows for sure.

You may have noticed that one of the surviving open field systems is in Devon, well outside the planned countryside as shown on the map on the previous page. There were indeed open fields in some parts of the ancient countryside, though usually they only covered part of a parish, with hedged fields on the rest of the land. Mostly they were introduced relatively late and done away with relatively early compared to the open fields of the champion country. The origins of open fields can be traced right back to the dawn of the Middle Ages, and so can the distinction between champion and woodland country. The evidence for this comes from the Anglo-Saxon land charters. These were an early form of title deed and they described the boundary of the land in question by naming a series of landmarks. Oliver Rackham has done detailed research into these landmarks. His studies have shown that typical features of open field systems, such as ridges and furlongs, were used as landmarks significantly more often in what we now know as the planned countryside than in the ancient. The distinction between the two is very old.

Medieval grain yields were tiny by modern standards and very erratic from year to year. It was common for the farmers to harvest only two or three bags of grain for every bag of seed sown. In a bumper year it could be ten bags, in a bad year little or nothing. For comparison, present-day organic farmers get between twelve and forty to one. Production could be increased by changing from a two-field to a three-field system. This reduced the proportion of fallow land from half the arable area to a third. But it was the fallow which restored fertility to the soil so the change could be self-defeating in the long run. Grain was also grown in the ancient countryside but in many western and upland areas, where it's easier to grow grass than cereals, milk was a staple food.

The Middle Ages are generally held to run from about AD 500 to 1500. But from a landscape point of view 1348, the year of the Black Death, has more sense as the end point. These eight hundred and fifty years were a time of steady population increase. They also saw a steady increase in the proportion of land under open field. Perhaps in part this was a response to an increasing shortage of arable land. What later ages saw as inefficient the Medieval mind must have seen as efficient. It was very much a communal system, with everyone working together in two or three big fields. A plough and a team of oxen was far beyond the means of an individual farmer, so they clubbed together, those who could afford it providing an ox, those who couldn't only their labour. Matters such as the crop rotation and the date at which cattle could be let in to graze the fields after harvest had to be decided communally, and there was a manor court or village assembly to decide and enforce such things. In a dangerous world where death by starvation was a reality people saw safety in numbers and security in conformism.

Towards the end of the period land hunger became acute. Open field farming spread onto increasingly steep slopes. This led to the formation of large earthen terraces, which we call strip lynchetts. They usually occur in groups, like a huge staircase, and they're often marked on the OS map in gothic script. Unlike the lynchetts I described when discussing soil erosion (on page 93) they must have been deliberately constructed, at great cost in labour. They're a sign of desperation: there was no land left which was flat enough to plough so it had to be created. The country was full. (See photo 15.)

Around the year 1300 the population stopped growing. It even fell a bit, with runs of bad weather which the malnourished people were ill-fitted to cope with. Then in 1348 came the Black Death, the bubonic plague, which killed somewhere between a third and half the population. Suddenly the population pressure was gone. There was land to spare, not enough people to till it and little market

for surplus food. But an alternative land use was waiting in the wings, one which required little labour and had a ready market just across the English Channel in the growing cities of Flanders. It was wool production. Much of the land in both woodland and champion country went down to grass to feed the sheep.

During this period most of the open fields in the ancient countryside were enclosed by agreement between the occupiers. Often this was a matter of neighbours exchanging strips till each had enough in one

Strips and furlongs fossilised in the present landscape, Wedmore, Somerset place to be worth enclosing them with a hedge. If the adjacent strips which belonged to other farmers were still part of the open field system, the hedge had to follow the boundary of the ridges exactly. So the shapes of the old strips and furlongs are often preserved in the shapes of the fields we see today.

These patterns are often easier to recognise on a map than on the ground. In some places there are narrow parallel fields which are obviously based on former strips. They may be reversed-S-shaped, C-shaped or straight. In others just the occasional hedge or field wall with a reversed-S-shape gives the clue. (See photo 16 and 17.)

Another clue is field boundaries with seemingly irrational right-angled turns in them. These preserve the boundaries of the old furlongs, which were small and interlocked with each other in an irregular way. (See page 91.) As the furlongs were amalgamated to make larger fields these irregular shapes became dog-legs in the new field boundaries. Lanes often follow the outline of the old furlongs, giving them an irrational zigzag course. So can parish boundaries, which are shown on the Ordnance Survey 1:25,000 maps as dotted lines. These are relatively late boundaries which were only finalised when the open fields of adjacent villages spread out to meet each other. Lanes and parish boundaries are often better preserved than the zigzag field boundaries because it's easier to change the line of a hedge than a public highway or a muni-cipal boundary. Just the occasional dog-leg in a field boundary is not diagnostic because it could result from the amalgamation of ancient fields. As with almost every aspect of landscape reading, we're looking for a pattern. (See photo 16.)

As well as preserving the field boundaries, enclosure by agreement sometimes also preserved the ridges themselves. This happened where the enclosed land was put down to grass and never

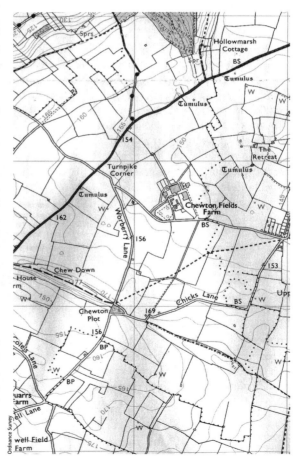

Dog-legs in field walls and parish boundaries, Chewton Plain, Somerset

ploughed again. Some of these ridges survive to this day, giving a corrugated surface to the land, composed of ridges, usually about ten metres wide and either curved or straight. It's known as ridge and furrow. Again, this is not diagnostic because the later, parliamentary enclosures could give rise to it too, but in that case the new field boundaries did not follow the boundaries of the old strips and furlongs.

In Scotland and some upland areas of northern England there was a different kind of open field system, the infield-outfield. The fields had the same kind of ridges and their remains may not look that different from the southern ridge and furrow I've already described. But the common arable of the village was divided up in a different way. The infield was cropped every year and fertility was maintained with manure from animals which were overwintered indoors. The outfield was occasionally cropped and grazed at other times. Both were enclosed by a large bank or wall, the head dyke, beyond which was common grazing land. In Scotland the term run-rig is often used for the system, though strictly this applies to the legal arrangements rather than the physical layout.

The history of woodland was also rather different north and south of the border. In the south, especially in the more populated parts, woodland soon became valuable. Wood was virtually the only fuel and coppice woodland was the main source of wood. In fact almost everything in medieval life, from houses to tools, was made of wood, though the great majority of woodland produce was used for fuel. Coppices were enclosed to keep out farm animals which would eat the new shoots which spring up after coppicing and thus prevent the trees regrowing. The barrier was usually a bank with a ditch on the outside and a hedge on the top. In the east of England these woodbanks can be massive earthworks. In the west they may be little bigger than a typical hedgebank, but they're usually distinctive enough to mark out the wood as ancient. Another sign of an ancient coppice wood is often a zigzag boundary, the result of successive generations of farmers carving roughly rectangular fields out of the woodland edge.

A completely different approach to woodland was to combine grazing animals and trees. This gave a savanna-type woodland with grass between the trees, now known as wood pasture. Coppice was clearly impossible in these woods but trees could be pollarded for a regular crop of poles. Pollarding is just the same as coppicing except the tree is cut two or three metres from the ground, out of reach of the animals. A new generation of trees might only get going at very rare intervals, perhaps when there was an epidemic of animal disease. Many wood pastures were commons or deer parks.

In Scotland the situation was different. Peat and coal were much more widely available as alternative fuels so woodland wasn't as highly valued and it declined more rapidly. Scotland may have had as little as five percent woodland in the Middle Ages, compared to perhaps fifteen percent in England. This lack of woodland can be seen in the way the Scots started to build in stone earlier than the English and Welsh, who used timber right into modern times. Some Scottish woods were managed intensively by coppicing and pollarding but most ancient woods north of the border which survived did so not because they were conserved but because they were remote from human settlement.

Parks were specialised wood pastures where private landowners kept deer. Like coppice woods, they had banks round them but in this case to keep the animals in rather than out, so the ditch was usually on the inside. The bank was topped with a wooden fence, known as a pale. This was expensive to maintain, so parks usually have the economic shape of a rectangle with rounded corners, which encloses the greatest area for the least perimeter. Sometimes the banks were never finished, or didn't keep to the oval shape because the emparker didn't own all the land he would have liked to. Occasionally a 'park pale' is marked in gothic script on an ordnance survey map, or a Park Farm or Park Wood will give a clue to the presence of a former park.

One of the best examples of a medieval deer park is Moccas Park in Herefordshire. It still has its herd of fallow deer and ancient trees of fantastic shapes, known as the Old Men of Moccas. In one corner of the park the trees are growing on ridge and furrow, which indicates that that part had been cultivated before it became a park.

Moccas Park, 4th June 2003

The Old Men of Moccas give me a tangible feeling of how old the countryside is. We gaily toss about words like Medieval and Neolithic, and hundreds of years run off our tongues like grams in a cake recipe. But these old, old trees, fuller of dead wood than live, sitting there, growing by tiny rings each year or perhaps dying a little bit more than growing, give a real visual experience to the word 'age'. And when I saw the ridge and furrow under their feet I had some tangible idea of just how long ago it was that those men and oxen made those ridges and furrows.

Another important part of the Medieval landscape was common land. This was the land left over when that needed for more intensive purposes had been fenced off. Commons could be grassland, moorland, heath, wood pasture or wetland. Their primary function was summer grazing, but they provided much else besides: wood, gorse or peat for fuel; bracken for animal bedding; wild fruits, herbs and fungi; wildfowl and fish. They belonged to specific communities and only members of the community had the right to use them. If overgrazing threatened to become a problem individual commoners were awarded stints, a stint being the right to graze a certain number of animals on the common.

Roadside verges were common land and where a road met a common it usually widened out gradually into a funnel or horn shape. This shape was formed partly as a result of new fields being enclosed from the common. The most economical way to enclose land is with a curved hedge or wall, and if new fields of this shape were taken out of the common on either side of a road it left a road which gradually widened as it entered the common. Another reason for funnels is that they're a useful shape for rounding up animals which have been grazing on the common. These horns can often be seen on surviving commons.

In the Middle Ages over a third of the land of England was common. The commons were a great medieval institution with a complex set of customs and traditions regulating the various uses that different people could make of them. The forests were another great medieval institution and they were often located at least in part on common land. A forest was not necessarily woodland. It was a tract of land where the king or some other great lord kept his deer. There were moorland forests, such as Bowland and Dartmoor, heathland ones, such as the famous Sherwood and even one or two in fenland. People sometimes look at moorland forests and say, "This was forest

A 'horn' on the edge of Dartmoor

in Medieval times, and just look at it now. It shows how much woodland has been lost in modern times." But the truth is they lost their trees thousands of years before they were declared forests.

Those forests which were based on woodland usually had a core of wood pasture and a much larger surrounding area of farmland which was also legally part of the forest. Here a special set of laws applied, ostensibly to protect the deer and their habitat but more realistically to give the king some extra income from fines. At the peak of forests in the high Middle Ages forest law held sway over vast swathes of the country but very little of that land was wooded. A forest was a place of deer rather than a place of trees.

Most forests were disaforested in early modern times. Plantations of oaks were made in some of those which remained in royal hands and the new meaning of 'forest' was born. However the real foundations of commercial plantation forestry were laid not by the state but by Scottish aristocrats of the seventeenth and eighteenth centuries. They were the first to experiment with exotic conifers. They've left us with notable forested landscapes, such as that between Blair Athol and Dunkeld on Tayside. But plantations didn't cover a significant area of Britain till the birth of the Forestry Commission in 1919.

Woodland, as opposed to plantation, remained an important economic resource through much of modern times. Charcoal was used more and more as the iron industry grew, till it was replaced by coke during the eighteenth century. Even into the nineteenth, when coal was taking over as the mass heating fuel, there was an increasing demand from industry for wooden goods such as bobbins, barrel staves and wheel spokes. Oak bark for tanning leather was a product which boomed in the early nineteenth century and the oak coppices of Argyll switched from charcoal to tan-bark production. Both of these markets led to reduced diversity in the woods as species other than oak were weeded out.

It's surprising how many people still come up with the old canard that woods were destroyed by industry. Far from destroying the woods, industrial demand saved them. Cutting trees down does not destroy a wood. Coppice woods have survived thousands of years of regular cutting, and pine can regenerate from seed. Woods are destroyed by grubbing them up for another purpose, usually agriculture, or by grazing them constantly at an intensity that prevents regeneration. The coppices of the charcoal age had an economic value so their owners had every incentive to keep them as woodland. The proof of this lies in the modern landscape: those areas which had the biggest charcoal-fired industries are now the areas with the most surviving woodland. Argyll, the Lake District, the Forest of Dean and the Weald are all examples. Iron masters didn't change to coke because there was a shortage of wood but because coke is cheaper. In a day's work a coal miner could produce more than twice the thermal equivalent that a woodcutter could.

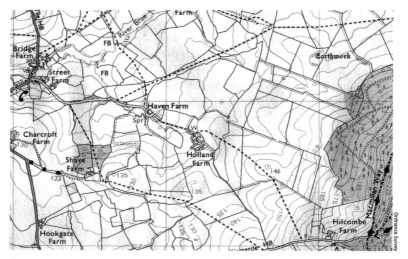

Ancient and modern intakes from the wood (which is now coniferised), Selwood Forest, Somerset

Some woods were lost. In upland areas there must have been losses due to intensive grazing by sheep, though woods which disappear in this way often leave few traces. In lowland areas they were grubbed out to make new fields. In ancient countryside the site of a lost wood often stands out on the map as an island of regular, straight-sided enclosures against a background of irregularly shaped fields. Sometimes there's a tell-tale name such as Southwood Farm in a place which is now far from any wood. But when the demand for wood products finally tailed off in the later nineteenth century woodland was not grubbed out wholesale because by then farming was in decline.

Enclosure

After the end of the Middle Ages the ancient countryside underwent modest changes. It lost its few remaining open fields through enclosure by agreement and bit by bit most of its common land was enclosed. By contrast the planned countryside was utterly transformed, from one kind of planned layout to another.

At the heart of the change were new crop rotations. Unproductive fallow could now be replaced with fodder crops, which not only gave fertility to the soil but also produced animal feed. Clover, which can enrich the soil with the nitrogen 'fixed' by the bacteria which live in its roots, was an important one. Turnips and swedes, which produce large quantities of sheep fodder and thus equally large quantities of manure, were another. In the Norfolk four-course rotation, crops of clover and roots were alternated with crops of wheat and barley. Other rotations alternated cereals and temporary grassland, sown with high-yielding grasses and clovers. Yields of both corn and animal produce could be much higher with the new rotations.

The snag was you couldn't introduce a new rotation on open fields. The very essence of the system was that everyone followed the same rotation and the same annual cycle, both of which had gone on from time immemorial. An integral part of this annual cycle was throwing open the fields to the whole community's cattle after harvest and during the fallow year. The new rotations didn't fit into this pattern. There was no fallow year and in most rotations one crop followed straight after another with no gap. There was no point in growing a crop of swedes for the winter if they would be polished off by the cattle of the whole village on the first day of communal grazing.

If the Medieval psyche had seen safety in numbers, the modern mind saw profit in individualism. The open field system had to go, at least in the minds of the landowners and larger farmers, who had the financial resources to capitalise on the new methods. Enclosure would mean exchanging their scattered strips for a private farm within a ring fence where they could do exactly what they pleased.

Not so the poorer land-holders who had just a few strips, the right to pasture one or two beasts on the common and the right to collect sticks for fuel. With these and the odd spell of work for their richer neighbours they could live a life of independence and dignity. Enclosure would mean they lost all this in return for a small field which wouldn't make a viable holding. Faced with the expense of hedging the new field and a pressing neighbour who wanted to buy it, it might not stay in their hands very long. To these small farmers enclosure meant becoming landless day-labourers, a class that was mercilessly exploited.

The enclosers resorted to Parliament to enforce their will. In the Scottish Lowlands a series of general acts of parliament in the seventeenth century allowed the run-rig to be enclosed. In the English champion counties a succession of individual enclosure acts picked off the open field parishes one by one over the eighteenth and nineteenth centuries.

The new layout of enclosed fields completely ignored the old pattern of fields and furlongs. The enclosure act landscape is one of big rectangular fields bounded by straight hedges of pure hawthorn. (See photo 18.) The winding lanes were replaced by straight roads with wide verges, which are sometimes mistaken for Roman roads. It's a mechanical landscape produced by an industrial age. Occasionally you will see a crooked hedge of mixed shrub species among the straight lines of enclosure act country. Often it's a parish boundary. These had always been hedged and, as we've already seen, boundaries are often the last

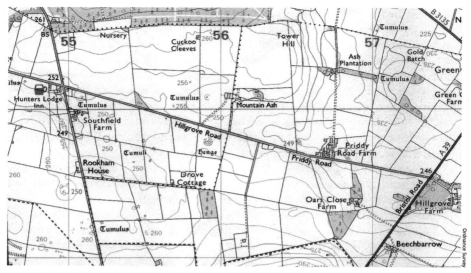

Enclosure act country, Mendip Hills, Somerset. Here the field boundaries are stone walls rather than hawthorn hedges.

thing to change in the landscape. Woodland was also left unaffected and often remains as irregular islands in the ruthless rationality of the modern chequerboard.

The large villages of the English planned countryside also remained intact after enclosure, though soon farmsteads began to appear on the new compact farms. Names like Trafalgar or Waterloo Farm often give a clue to the date of the new buildings, if not of the actual enclosure. But it was in Scotland that the Enlightenment ideal of an efficient modern landscape reached its most perfect form. On the north-eastern plain, between the firths of Tay and Moray, the former settlement pattern was swept away and replaced by model farms and cottages in a landscape that became among the most productive in Europe.

Along with the open fields went almost all the common grazing land in both planned and ancient countryside. In planned countryside it's hard to tell whether any particular piece of land was previously open field or common. The large, straight-sided rectangular fields are pretty well universal. But in ancient countryside enclosed commons often stand out on the map just like a former wood, as an island of rectangular fields in a matrix of irregular ones. If the common was wood pasture before enclosure some ancient pollard trees may remain, either scattered through the fields or incorporated into the new hedges. Commons on very sandy soils were often heathland. With the new farming methods heath could now be converted into productive farmland. Lime could be brought in to correct acidity and the new rotations could boost the humus level to make up for the lack of clay in the soil. Most of the heathland of England disappeared in the nineteenth century.

Moorland commons, of which enormous areas were enclosed in Wales and the North of England, were sometimes more intractable than heath.

Unlike lowland heath, high moorland has a severe climate to contend with as well as poor soil. There were attempts to turn moors into improved grassland but they were more successful in some regions than others. A modern version of ridge and furrow was sometimes used in vain attempts to turn moorland into arable. Its remains are quite easy to distinguish from medieval ridge and furrow. Apart from being found on moorland, where medieval ridge and furrow is rare, it's much narrower, typically five metres wide, and always straight.

The loss of the commons is one aspect of a general process of simplification which was going on in the landscape. Commons were often wild places with a mixture of trees, scrub, bracken, furze and marsh as well as heath and rough grassland. All this was replaced by fields of uniform grass and arable land. Simplification was a trend that was to gather pace in the twentieth century. (See photo 19.)

Enclosure wasn't the only consequence of the new farming technology. Medieval farmers could do nothing much to improve the plant nutrient content of the soil nor to increase its humus content, which would make it more drought-proof. So intensive arable farming was concentrated on clay soils, with their deep reserves of nutrients and their resistance to drought. Light soils and limestone uplands were mainly used for grazing or at most for extensive arable supported by large flocks of sheep. But the modern rotations changed all that. They could boost both the level of nutrients and the humus content. The league table of limiting factors to soil fertility changed. Poor drainage moved up to first place, so clay was looked on with less favour than before. A lot of land was drained with underground clay pipes but it also became attractive to move cultivation up onto the hills and downs and leave the heavy clay below under grass. In some parts of the country the landscape was turned upside down by this change. The ridge and furrow preserved in the permanent pasture of many clay vales is witness to the change.

In the Highlands of Scotland the notorious Clearances were the equivalent of enclosure. With callous cruelty the peasant population was thrown off the land and replaced with a monoculture of sheep, which brought in a greater return to the landowners. But within a generation the fragile fertility of the Highland soils had been exported in the wool clip and the land could no longer support the sheep. They were replaced by deer, which rich people would pay for the privilege of shooting. One gamekeeper can manage an area which would once have been home to hundreds of people and the Highlands have remained largely deserted ever since. The people emigrated to the colonies or to the industrial cities of the Lowlands. Terrible though the Clearances were, for every family evicted some ten families left of their own accord. For the first time in history there were alternatives to the hard way of life which their ancestors had followed for six thousand years.

It was the start of globalisation. On an unprecedented scale people could move out and food could be brought in. There had been international trade from Medieval times, when timber was imported from the Baltic and wool was exported to Flanders. Both had their effect on the landscape but Britain remained self-sufficient in food. Then in the 1870s railways opened up the interior of North

America and steamships shrank the Atlantic. The virgin soils of the prairies were cashed in for grain which could be sold on Liverpool dockside cheaper than local farmers could grow it.

Imported food became so cheap that farming went into a depression which lasted till the Second World War. There was a brief respite during the First World War but during the inter-war years almost every crumb on British plates was imported. Many farmers could do no more than keep their farms out of bankruptcy while the family survived on what their wives could get for the eggs, butter and vegetables they took to market each week. Much land reverted to scrub and rabbits roamed where the land lay idle.

The Big Bang

The enclosure acts took place over a period of centuries and, in England at least, only affected about a fifth of the land surface. But the changes in the British countryside since the Second World War have happened in half a century and affected the whole country. Compared to any previous period of change this one rates as an explosion.

It hasn't been unique either to Britain or to the countryside. Similar changes have gone on all over the world and in town as much as country. Everyone has been caught up in a race for ever-higher material standards of living. The twin engines which have driven this race are a technology which is developing faster than ever before in history and fossil fuels which are, in the long view, vastly under-priced. Farmers and foresters have done nothing different from what everyone else has done, yet they've been roundly criticised for it by people who happily eat the cheap food which the new methods have made possible.

The low price of fossil fuel hasn't only powered the new technology of tractors, fertilisers and sprays, it's also unleashed a level of globalisation which couldn't have been imagined a hundred years before. Technology has driven up world food production to the point where, until very recently, there was a constant surplus. Meanwhile the low cost of fuel means that food can be moved around the world at a fraction of the cost of producing it. The price for any commodity has become the lowest price at which anyone anywhere on the planet can produce it. The main response to falling prices has been to use more technology to bring costs down even further.

The combined effects of technology and globalisation have affected every part of the economy. But in the case of farming a third factor has been subsidies. Paid at first by the national government and later by the European Union, subsidies have attempted to cushion farmers from the effects of globalisation. They've never managed to make farming as profitable as other businesses but for a long time they guaranteed a steady return for anything which farmers produced. Farmers responded by going all out for maximum yield.

From a landscape point of view the most obvious effect of all this has been the loss of hedges. You need big fields to operate big machines and a bulldozer can remove in seconds what a man with hand tools could grub out in a day. Where hedges have gone soil erosion has increased and wildlife has suffered.

In the East of England, where the rate of loss has been ten times the national average, the whole character of the landscape has been changed.

Nevertheless it hasn't obliterated the difference between champion and woodland counties. Suffolk, for example, is still quite unmistakably ancient countryside. The hedges which do survive there contain a rich mixture of species and follow a curvy or crooked course, quite unlike the straight hawthorn hedges of parliamentary enclosure. There are old farmsteads standing alone amongst the fields, many of which date from medieval times.

As the fields have got bigger so have the farms themselves. Farmers have needed ever more land just to stay economically level with their urban counterparts. Greed is not always the motive for farm amalgamations, as this entry from my notebook relates.

Sprint Mill, Cumbria, 2nd Nov 1992

Hillary, a hill farmer from Tebay, said that most of the high fell farms are now abandoned and their land farmed by neighbours with farms at a lower level. It was electricity and the motor car that killed them off. They could support a family, but not with a car and a full range of electrical equipment. A young man couldn't go courting with any hope of success without these things.

Farms have not just become bigger but also more specialised. In the east, fertilisers and pesticides have done away with the need for rotations and the economies of scale mean that costs are lower on a pure arable than a mixed farm. In the west, the low cost of transport means that it's cheaper to buy straw from the other side of the country than to grow a little corn to provide winter bedding and grain for the animals. There's always been a predominance of grazing in the west and arable in the east but this was a matter of emphasis on what were essentially mixed farms. Now it's as rare to find a field of wheat in Wales as a cow in Cambridgeshire.

The moorland which the Victorians found beyond their power to tame could not withstand the might of twentieth century machinery and artificial fertilisers. Although plenty remains, much has been converted to improved grassland. On the North York Moors, with their dry eastern climate, it's even been converted to arable. Larger areas of moorland, especially in Scotland, have been converted to conifer plantation. Most of these Scottish moors are so wet that they need intensive draining before trees can grow there. Huge tractors are used to 'plough' a ditch between every single line of trees. These dark, monocultural plantations now far exceed the remaining semi-natural woods in area and the commonest tree in Britain is no longer the ash or the oak but the Sitka spruce. Most of the remaining lowland heaths have also been planted to conifers.

The landscape is getting simpler all the time. It's not just the loss of hedges, moors and ancient woodland or the increasing specialisation of farms. It's also a general tidying up and a form of triage in which land is either farmed intensively or not at all. You can see this in grassland country, where flat fields tend to be improved with fertilisers or by reseeding, while sloping fields, which are difficult to mechanise, are allowed to succeed to woodland. The great variety of grassland types is lost in this process. Coppice woods also become simpler when they're no longer managed. Where once there was a mix of patches in different stages of regrowth, each with its distinctive community of plants and animals, now there's a dense canopy overall. Little field quarries and redundant ponds have been filled in and small marshy patches have been drained. The rich mosaic of the countryside, so important to a diversity of wildlife, has increasingly become monochrome. (See photo F16.1.)

By the 1980s people began to wonder whether this single-minded pursuit of production at all costs was altogether what they wanted. Food surpluses had grown so great that disused aircraft hangers were hired to house the notorious 'grain mountain'. The cost of this surplus food in terms of lost wildlife was shown by a report from the Nature Conservancy Council which detailed the loss of semi-natural ecosystems during the previous forty years. The losses ranged from upland grassland, which had decreased by a third, to lowland flower-rich meadows, which had decreased by ninety-seven percent. Half of the ancient woodland which had been here in 1945 was gone. This was about the same amount of woodland as had been lost in the previous nine hundred years, since the great survey of Domesday Book.

One result of this was a great interest in nature conservation. Nature reserves had existed since around the turn of the century but since the mid-80s they've expanded by leaps and bounds. However nothing serious was done to stop the continued destruction of biodiversity on the rest of the land. There were some grants for planting hedges and woodland and a farmer might well be paid for replanting a hedge which his father had been paid for grubbing out. But in general the difference between nature reserves and the rest of the landscape grew ever wider.

The only kind of semi-natural ecosystem to receive any real protection was woodland. But it was too late. Although much of it had been converted to conifer plantation in the previous two decades, no-one was planting conifers any more. Globalisation brought the forests of the former Soviet Union onto the market and the price of timber plummeted. In many woods the value of the timber fell to less than the cost of harvesting it.

By around 1990 it became generally accepted that we were paying subsidies for the wrong things. They were either paid on the produce itself, which encouraged more production, or on the number of animals kept, which encouraged overgrazing. If the public wanted to pay farmers for anything it was for a beautiful countryside full of wildlife – the very thing which was being destroyed by over-production. It took fifteen years to make the change, and in 2005 subsidies were switched to a flat rate, with supplements for wildlife-friendly practices. Whether this will have the desired effect remains to be seen, but the results so far are not encouraging.

In fact subsidies on production were never able to keep up with globalisation. Over recent years prices have fallen so low that even the best of farmers find it difficult to make a living. A few have responded by taking their land out of food production altogether. On some farms cattle have been replaced by horses and on others the land is kept ticking over with a few head of sheep while the owners concentrate on putting on pheasant shoots. You can make a better living from providing recreation than from producing food.

OBSERVATION TIPS

Ancient and Planned Countryside

The main differences between the two are:

Ancient or Woodland	*Planned or Champion*
Hamlets or a mix of villages and hamlets, with old farmsteads standing alone.	Villages, with 18th-19th century farmsteads standing alone.
Fields of irregular shape, or showing the outlines of former strips and furlongs.	Fields of regular rectangular shape.
Hedges of mixed species.	Hedges mostly of hawthorn.
Lanes frequent, crooked and narrow.	Roads less frequent and straight, often with wide verges.
Frequent woods, including small ones.	Woods large or absent.

Two rules of thumb about fields and hedges are:

- Small, irregular-shaped fields are older than large, regular-shaped ones.

- The older a hedge is the more species of trees and shrubs there are in it. (See page 286.)

These are only general tendencies and there are exceptions.

Ridge and Furrow

There are two kinds of ridge and furrow, plus other kinds of earthworks which can be mistaken for it:

- Medieval ridge and furrow is usually about 10m wide with a rounded profile to the ridge. It may be straight or curved, either like a C or like a reversed S.

- Post-medieval ridge and furrow is narrower, usually about 5m wide, always straight, and usually on marginal land.

- Water meadows are always in a valley bottom, usually on land too wet for arable. Carrier ditches may still be visible along top of the ridges. (See pages 223-224.)

- Gutters or grips are straight, shallow ditches, spaced about 10m apart with flat ground between them, on land with obvious signs of intensive draining. (See page 276.)

- Where land has been 'ploughed' for a forestry plantation and later reverted to grassland or heath, the resulting ridges have an asymmetrical profile, with a narrow ridge on one side and flat ground on the other, typically 2m wide overall. (See page 190.)

Faint Traces

Ridge and furrow or other earthworks may be worn away to the point where they can only be seen when conditions are just right.

- In grassland the best time of year is winter or spring, when the grass is short and the sun is low. Sometimes there's a brief moment in early morning or late afternoon when the sun strikes a landscape of short grass at just the right angle and a whole medieval landscape jumps out at you in sharp relief.

- Melting snow can also highlight faint earthwork patterns, as can heavy rainfall, which can flood even the slightest of furrows while leaving the ridges dry.

- In ploughed land former earthworks can reveal themselves as soil marks. Darker soil indicates former furrows or ditches and lighter soil indicates ridges or banks. Where the ridge and furrow was broken down by steam ploughing, the dark and light soil may zigzag across the field where the soil was pulled alternate ways by the big plough.

- In growing crops, former ditches and banks can show up as crop marks, especially during dry weather. In the deeper topsoil of a former ditch the crops grow greener and taller than in the shallow topsoil over a former bank. A crop of barley, just as it begins to ripen, can reveal these contrasts best.

- Former hedges or field walls in grassland often show up as lynchetts if they lie across the slope. If they run up and down the slope or if the land is flat they may be revealed as low banks or not show at all.

Wild Animals and How to Recognise their Signs

In the drama of the landscape there's no distinction between actors and scenery. Everything, from rock to human, is both part of the scenery and an actor in it. Wild animals are more than mere inhabitants of the landscape, they also help to form it. Wild herbivores, by their grazing and browsing, help to form the vegetation which is their home. Predators in their turn affect the populations of herbivores and thus also the vegetation.

All wild animals run away from us, but the meetings we do have with them can be moments of delight. An unhurried fox glides over the ground, its feet hardly seeming to touch the earth, its fiery coat brilliant in the sunshine. Then at a certain distance it stops and looks back at you before going on its way. A roe deer in dense cover will sometimes have the same confidence, and a heart-shaped face will pop up among the leafy branches and peer at you curiously for a few moments before she bounds off with the rest of her family. To sit quietly with a wild mammal which is not running away is a rare experience. I've had it just once.

I was walking through a wood near my home when it started to rain. There are two huge old yew trees in the wood whose dome-shaped crowns keep out the rain like thatched roofs. I headed for the nearest one and sat in the dry, looking out through the trees to the fields beyond. A few moments later a hare came along with the same idea and sat in the dry just in front of me, unaware that I was there. For a while we sat there together, looking out at the fields. It was a rare moment of peace and intimacy. But being the fidgety, impatient person I am I couldn't sit still for long, and as soon as I moved the hare noticed me and lolloped off.

Such meetings can be peak experiences in our relationship with the land. But our indirect experience of wild animals, through their effect on the formation and functioning of the landscape, is stronger than we often think. Almost all of them are small invertebrates such as worms and insects. They work away largely unseen by us, removing debris, renewing the soil, pollinating plants and so on. As a general rule the smaller and less conspicuous a creature is the more important it is to the functioning of the ecosystem. They can manage quite well without us but we couldn't survive without them. To give even the briefest account of this diverse myriad of creatures would be beyond the scope of this book. Instead I'll concentrate on those wild animals which have the most direct and obvious impact on us, mainly by eating cultivated plants or killing domestic animals. Most of them are also key species which have a significant effect on the landscape as a whole. All of them are mammals. They are: deer, rabbits, voles, grey squirrels, foxes and badgers.

There have never been as many deer in Britain as there are today, either in the Scottish Highlands or in lowland Britain as a whole. The lowland landscape is ideal for them: wide fields of grass and cereals on which to feed and the occasional wood to lie up in. This is a much more productive landscape for deer than the all-pervading wildwood of prehistoric times. Most of the edible biomass in woods is in the tree leaves, which are mostly out of reach, while the shaded herbaceous layer is much less productive than grassland out in the open. In historical times the countryside was full of people, and deer are shy animals which shun human company. With the mechanisation of farming and the neglect of woodland the countryside is emptier of humans now than it's been for some four thousand years. These changes may not fully explain the deer explosion but they've certainly played a part.

The main deer species in lowland areas are roe, fallow and the tiny muntjac. One or more of them is found almost throughout the country, though there are some places which they haven't yet reached. These include parts of Wales and the English Midlands, the east of Kent and the west of Cornwall. They're lovely animals when you spot them on a summer walk. The misty white spots of the fallow and the stately antlers of their bucks suggest some medieval idyll, while the leaps and bounds of the chestnut-backed roe have the grace of an African gazelle.

They find most of their food in the fields but they do little damage there. They graze over a wide area, taking a bite here and a bite there. When they graze young cereal crops they may have no effect on the yield of grain because neighbouring plants compensate for the ones they eat by putting on extra growth. On grassland they do lower the yield because everything they eat could otherwise be eaten by farm animals but as the effect is spread over a wide area it's not very noticeable. It usually causes less concern than rabbit damage, which is more obvious because it's concentrated in definite areas.

Although the food which deer find in the woods is a small part of their overall diet it can have a major impact on the woodland itself. The kind of impact depends on the density of deer. Where the population is moderate they gradually change the species composition by selectively grazing the most palatable plants, both woody and herbaceous. In general the most palatable trees are ash, elm, hazel, pussy willow and birch. The least palatable ones include oak, chestnut and aspen. These preferences are shared by most browsers, including farm animals, though sheep will eat aspen. At ground level, selective grazing can turn a diverse cover of woodland wildflowers into a sward of tough, unpalatable plants such as pendulous sedge. But where the deer population is heavy they'll eat everything in the wood, less favoured species along with favourites. They can reduce the herb layer to bare earth and prevent the natural regeneration of trees altogether.

They also make coppicing extremely difficult. When trees are planted they can be protected from deer with individual plastic guards. Once they outgrow the guards they're safe from deer and can grow on to become mature trees for timber production. Self-sown seedlings can be protected this way too. Coppicing, on the other hand, involves cutting the trees down every few years and allowing them to regrow. They don't regrow as a neat single stem which can easily have a guard slipped over it but as a multi-stemmed spray of tasty young shoots.

These coppice stools are hard to protect individually and the most effective remedy is to put a deer fence right round the newly coppiced area. It needs to be two metres tall, so it's expensive and only used when the coppice is highly valued.

Ecologists sometimes use deer fences in unmanaged woods to exclude the deer from part of the wood. This not only protects some of the woodland but allows the effect of the deer to be seen and measured by comparing the vegetation on both sides of the fence. Sometimes the contrast is spectacular, though I do know one place where it's hard to tell the difference. Deer damage, like so many things in nature, is hard to predict.

Red deer are found in various parts of Britain but in the Scottish Highlands there are hordes of them. In fact they're encouraged, because deer stalking is a profitable business in an area where it's hard to make a living from the land by other means. In some places they've even been fed silage in winter in order to boost their numbers. Death by starvation over winter is the main control on the population of most wild animals, including deer, so a little feeding at the coldest time can make a big difference. Warmer winters due to the greenhouse effect must also be having an effect on the survival rate, as it is for all wild herbivores. Some people cite the extermination of the wolf, the only natural predator of the deer. But this happened long before the explosion in the deer population and I wonder whether it had much effect even then. The wolves would have taken mostly the easiest prey: the weakest young ones, sickly adults which wouldn't survive to breed and old ones past breeding age. When the prey animal is large and able to defend itself the effect of predators can be to keep the population healthy rather than to reduce it.

It's the high numbers of deer, along with sheep, which keep the Highlands in their moorland state. Newly planted trees are particularly vulnerable. One tug at the tasty tip of a little seedling can whip it right out of the ground. New plantations and natural regeneration alike must be deer-fenced if the trees are to survive and grow. Sometimes the effect can be spectacular. One of the most visible examples of this lies beside the road and railway which lead up from the head of Loch Lomond into the Western Highlands. Towards the end of the 1980s someone put a fence up around a small group of lonely old Scots pines which stood to the east of the railway, enclosing a hectare or so. Within a year or two you could see the dark triangles of young pines poking up through the heather and moorgrass. Among the little pines grew slender birch seedlings, their light seed blown in on the wind from further away. In a few years the land inside the fence began to look more like a young wood than a patch of moorland and another fence was put up enclosing a much larger area. I haven't been up that way for years now but no doubt there's now a flourishing pine wood where once there were a few forlorn old trees. (See photos 20.)

Rabbits are not native to Britain. The Normans brought them here from their home in the Mediterranean lands and farmed them for meat and fur. At first the rabbits had a hard time in their new northern home and had to be mollycoddled. They couldn't dig and had to have burrows constructed for them. Slowly, slowly they adapted to the climate, till some time around the beginning of the nineteenth century they were able to survive independently and they began to escape from their artificial warrens. As time went by they adapted more and more and bred 'like rabbits' till by the 1930s they dominated great tracts of countryside. They destroyed cereal crops and ate the grass so avidly that there was little left for farm animals in some places. Large areas around their burrows were reduced to bare earth.

Then in the 1950s the disease myxomatosis was introduced and in a few short months rabbits almost became extinct in Britain. Much land which had been kept as grassland or heath by rabbit grazing succeeded to woodland. But a few individuals had enough immunity to survive and gradually numbers built up again. Now they're an almost universal pest but very rarely a plague of pre-myxomatosis proportions. As readers of Beatrix Potter will know, they're partial to vegetables, and they harm young trees and shrubs by eating the bark, mainly in winter when herbaceous food is short. Vegetable gardens need to be rabbit-fenced if there's a warren nearby. Newly planted trees always need to be protected, either by a fence or with individual guards, except in those mountainous areas where there aren't any rabbits at all.

I once found myself in one of the rare places where rabbits have returned to their pre-myxomatosis numbers. It was a small farm on a gravely soil on the Essex-Suffolk border. The landscape was surreal. Going out quietly in the early morning I saw not hundreds but thousands of rabbits spread out evenly over the dewy grassland. The slightest sound from me and they were off down their burrows in a flash. The farm was all grass and there were no domestic animals grazing it but the turf was kept tightly cropped by the rabbits. The field nearest the main warren had been grazed so hard that there was little left but bare earth. The only plant that thrived there was the poisonous ragwort, that tall, shaggy-leaved plant with flat heads of golden-yellow daisy-like flowers, and it did very well. It was the one plant the rabbits didn't eat and being mainly biennial it appreciated the bare soil to seed into. On the rest of the farm the rabbits maintained a short, dense sward of long-lived perennial grasses by their constant nibbling. This left no space for a seed to germinate and there was no ragwort.

The hedges had been reduced to nothing but elder and nettles, which are both inedible to rabbits, the one because it's poisonous the other because it stings. All the other shrubs and herbaceous plants had been killed out, though the mature trees survived. There was also a small oak plantation with nothing growing beneath the oaks but elder and nettles. The landscape had something in common with an overstocked chicken run, which usually consists of bare earth and nettles plus any trees which were there before it became a chicken run.

Quite why this particular place had so many rabbits is a mystery to me. Certainly the loose gravely soil was ideal for digging burrows and that will have played a part in it. But not every place with such a light soil is so overpopulated with rabbits. As well as being an unusual landscape it was rather a tragic one. The rabbit population had almost certainly caused a crash in biodiversity. If they could have been harvested for human food it would have been some compensation and the damage could have been lessened by keeping the numbers down.

But since myxomatosis we've lost the habit of eating rabbit in this country and the wholesale price hardly pays the cost of the cartridges.

Like rabbits, voles mainly eat grass. They live in a network of above-ground tunnels which they make in tall, dense herbaceous vegetation. Wherever grass is left uncut for a year or more there will be a thriving population of voles, which in turn supports owls, kestrels and foxes. Their main food is the grass which they can access from inside their tunnels, but they're partial to the bark of young trees and shrubs too. They can kill newly planted trees and may play a part in preventing natural regeneration. Where voles are likely to be a problem special little plastic tree guards need to be used. Occasionally they have a population explosion and then they can wreak havoc in any garden which lies next to their grassy habitat.

While deer are the biggest problem for coppice production, grey squirrels can wreck a timber crop. Timber trees are tall, single stemmed ones grown for planking, in contrast to the multi-stemmed coppice, which yields a crop of poles. Squirrels do their damage by stripping patches of bark off the trees. Being mainly nut eaters, their hungry time of year is not the winter but the summer. In autumn they eat their fill of nuts and bury some to tide them over the rest of the year. But their buried stores can get a bit thin by summertime and one alternative food supply is tree cambium, a thin layer of tissue which lies just underneath the bark. To get at it they strip off the bark and this kills the wood underneath, ruining it for timber. They go for youngish trees and most often strip bark off the trunk about two-thirds of the way up the tree. If most of the circumference of the trunk is stripped it dies at this point. The tree goes on growing from side branches, and one of these may take over and become the new leader. But that gives a crooked tree which is useless for sawing up into long, straight planks. Even if the trunk doesn't die, bark stripping leaves a patch of dead wood in the trunk which will spoil its value as timber.

Grey squirrels are probably the biggest limitation on growing broad-leaved trees for timber in those parts of Britain where they live. They're an introduced species, originally from north America. They've steadily spread from their points of introduction and are now found virtually all over England and Wales south of Cumbria and in a few places in Scotland. The trees they damage most are beech, sycamore and sweet chestnut, though they will go for others if their favourites aren't there. Wild cherry is the one they trouble least.

Though bark stripping has economic importance, the biggest effect squirrels have on the landscape is on the reproduction of hazel trees. Some people think they've made it into a relict tree, one which can no longer reproduce and will die out when the present generation of hazels comes to the end of its life. They completely strip the trees of nuts and they often do it when the nuts are unripe, so any which they bury and then forget about are too immature to germinate and grow. Only where nuts are unusually plentiful will any escape the squirrels and have a chance of becoming new trees. This can happen in a big wood with a high proportion of hazel in it. The squirrel population doesn't go up directly in proportion to the number of hazel trees in the locality because nuts aren't the only thing squirrels need in their habitat. Where hazels are superabundant some other factor will put a limit on their population. A big hazel wood can have enough nuts left over by the squirrels to support a population of dormice. In the same way, a good-sized hazel orchard, or platt as they're known, will only lose a proportion of the crop to squirrels. But if you have two or three hazel trees in your back garden you'll be lucky to harvest a single nut in most years.

However, in years when there's an unusually heavy nut crop the squirrels can't keep up with it. Then there can be nuts left over both on garden trees and in woods and hedgerows. But whether these are enough to keep hazel going as a wild tree is not yet known. There are many more pitfalls between nut and mature tree, from rodents to browsing deer. Only a really abundant supply of seed is enough to ensure the reproduction of any tree.

Foxes can make a living in any terrain, rural or urban, including the most inhospitable. No wonder they have such a special place in our folklore. Reynard is admired for his cunning but also condemned for his cruelty. When a fox gets into a hen coop he usually kills every one of the hens, not just the one or two he can carry home with him. This, some people say, proves that he takes delight in killing. But this is to judge foxes as though they were human beings. A fox is not a sadistic human but a wild animal. Hunting is a difficult way to win your food and by no means every attempt ends with success. The instinct to make every possible kill is a valuable survival trait and foxes which don't have it don't survive. In the chicken coop the fox meets a totally unnatural situation: a bunch of birds bred for production rather than survival, trapped in an artificial box with no way out. The fox can hardly be expected to suddenly turn off its instincts.

Their main food is mice and voles. They do take lambs, though probably most of these are the weak or sickly ones which wouldn't survive anyway, and of course chickens. Some people reckon that they only take these larger prey items when one of a pair has been killed. When they're raising a litter both the vixen and the dog fox hunt to feed the cubs. The two don't meet. The dog fox leaves food at a hand-over point and the vixen comes and retrieves it later. If the vixen is killed the dog fox finds his gifts rejected. He reacts to this by bringing larger and larger prey items in the hope that they'll be accepted. Conversely, a vixen deprived of the help of her mate may be forced to take larger prey in order to get enough food for her cubs in the time available to her. If this is true, killing foxes is counter-productive. In fact foxes may on balance be the farmer's friend, as every vole they kill means more grass left to be eaten by the cattle and sheep. It takes a lot of voles to feed a fox and the constant toll they take must have a significant effect on the vole population.

Badgers seem big and formidable compared to foxes, but they get most of their food from an even smaller prey, the humble earthworm. At night, when the badgers are out and about, the worms are above ground too. In the humid night-time air the worms can pop the top half of their body out of their burrows to find grass and other leaves to drag below for their food. The badgers snuffle the worms up wholesale. If the soil is dry and firm the worms can put up quite a resistance, and then the badgers have to dig a little hole to get them out of the earth. They also dig for other soil-living creatures and plant foods. Pignut, a small plant of the cow parsley family with carrot-like leaves, has tasty tubers and where it's abundant badger feeding holes can pock-mark the ground in spring. But earthworms are their staple.

If the soil is so dry that the badgers can't dig the worms out they may have to go in search of other foods. One of their favourites is maize and this is often just at the tasty sweet corn stage when a summer drought makes worm hunting difficult. In some places they raid people's gardens for the super-sweet varieties, which they can reach from the ground. The kind of maize grown for silage is much taller and they have to knock the plants over to get at the cobs. The present fashion for growing maize for silage is at least partly responsible for a rise in the badger population. They can take it right up till it's harvested in October and it can give a big boost to their diet just before winter when they need supplies of fat to see them through.

They're not much interested in other cultivated vegetables, at least not in my experience. I used to have a small patch of lawn in front of my house which they often visited to catch worms. In summer their little holes appeared regularly. When I turned the lawn into a vegetable patch I expected trouble. But no, they never took the vegetables and the only time they ever disturbed them was to get at some woodlice which were living under a layer of mulch in the leek patch. Of course if I'd grown sweetcorn there it would have been another story.

Occasionally badgers take poultry. This seems to be a very local habit. Badgers live in families and it may be that once they acquire a taste for poultry they teach it to their offspring and it becomes a family tradition. Where they do get the habit it can be a strong one. I know a farmer in Devon who is simply unable to keep free range chickens because of badger predation. Shutting them up at night is no help. The badgers will come out during the day if needs be. The farmer solved the problem by keeping the birds out on the pasture in a series of movable arks – an ark being a combined chicken house and enclosed run. This way he can give them fresh ground each day though their freedom of movement is somewhat restricted.

One more animal which deserves mention is the wild pig. They were exterminated in Britain in late medieval times but in recent years people have begun to keep them as a speciality meat animal. Inevitably some of them have escaped and established themselves in the wild. There are now populations on the Kent-Sussex border, in west Dorset and the Forest of Dean. Like badgers they have a great love for maize. I've heard that in France, where they've never been eradicated, they can be a problem for farmers who grow it. One farmer in central France whose land is surrounded by extensive woodland reported losing a third of his maize crop to them. Whether they could be such a problem here in Britain, with our much less wooded landscape, is hard to tell. They also root up the ground for food, especially for pignut where they can get it. In a well-wooded part of west Dorset I was told that they'd fully rooted up a third of a pasture field one springtime. It had greened up again by June, from regrowth of the existing plants, but a lot of grass production was lost. Where wild pigs have established themselves farmers have made strenuous efforts to exterminate them, but they're shy and secretive animals and some always survive.

Animal Signs

Since wild animals go out of their way to avoid us you can't rely on actually seeing them to find out about the local populations. You have to look for their signs. Footprints, droppings and the homes they make can all be distinctive, but the commonest and most visible signs are usually those of feeding.

Deer leave little sign of grazing in the fields but their impact on woodland is easily recognised, though the first sign of deer in woodland may not be browsing but rubbing. Stags and bucks rub their antlers against trees and bushes for two reasons. Firstly, they need to rub the velvet off the new set of antlers which they grow each year. Velvet is a thin layer of skin and fur and any small tree or shrub makes a good rubbing post. Red stags often use a small shrub, which they can more or less demolish in the process. Secondly, roe bucks mark their territory by rubbing a scent gland at the base of their antlers against young trees or coppice shoots. This can wear away the bark in a vertical slash with shaggy edges. You may see these rubbing posts in a wood where they've recently arrived before any browsing becomes noticeable. In fact it's not inevitable that they will take up residence. It may just be the sign of a tentative exploration.

Light browsing by deer shows up in damage to individual tree seedlings or coppice shoots. Even if deer are only passing through the wood occasionally they may repeatedly browse the same shoots. This is especially so in a mixed-species wood with a range of tree sizes, as they will go for the most accessible shoots of their favourite species. Browsing the same shoot again and again can give the characteristic shape shown in the drawing. Whenever the terminal bud is bitten off regrowth comes from a side bud, giving the repeated curves. The shoot has every chance of eventually growing out of reach of the deer but it will make a crooked stick. If it's a coppice pole it will be useless for craft purposes. If it's being grown for timber the crookedness can be lost in the centre of a much larger trunk but such a setback to its early growth will probably mean the tree never makes it to timber size.

Tree shoots can also be nibbled by rabbits and hares, and if there have been winter snow drifts they can reach the same height as deer. But it's easy to tell whether they or deer are responsible for a bitten twig as rabbits and hares have two rows of teeth and make a clean bite, whereas deer only have a bottom row and make a ragged bite. Sheep also have one row of teeth so it can be hard to tell the difference between sheep and deer browsing.

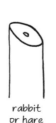

rabbit
or hare

sheep
or deer

In general the presence of farm animals in a wood masks the signs of wild animals. Farm animals may be deliberately let into a wood for shelter or they may get in because of bad fencing. You can tell if cattle or horses have been there because their droppings are much bigger than those of deer and quite different in shape. If you get a good look at their footprints these are also a giveaway. Horses and ponies have a big, rounded, single-toed footprint. Cattle have the same cloven hoof as deer but it's much bigger.

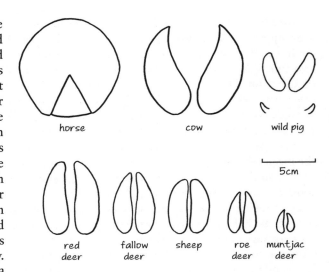

Note: Size of tracks varies with the age of the animal and horse tracks vary with the breed of horse or pony

Sheep tracks and droppings are very similar to those of deer. But when sheep are in a wood they always leave their calling card: little wisps of wool stuck on thorns or on the wire of the boundary fence. If farm animals have been in the wood it doesn't mean that deer haven't, just that it's hard to tell.

The rubs and nibbles of the occasional visitor are one thing, the signs of a resident deer population are quite another. Heavy deer browsing can keep seedling trees permanently stunted as gnarled little specimens some twenty to thirty centimetres high. These can be a major component of the ground-layer vegetation in some woods. But the most obvious sign of a resident deer population is a browse line, which they make by eating the leaves and smaller twigs off trees and bushes up to the height they can reach with their mouths. Above the line there may be dense foliage, below it nothing but bare poles and trunks. The more deer there are in the wood the clearer the browse line is. Roe and fallow deer make a line about one and a quarter metres above ground. This is below the line of sight of an adult person and you may not see it as you walk through the wood. But go down on your haunches and it comes sharply into focus. It's quite unmistakable. The browse line on ivy often shows up more clearly than a general browse line, especially in a wood with little undergrowth. Deer love ivy and often leave a crisp browse line on the ivy-covered trunk of a tree at the height where their necks are fully extended.

Other animals can make browse lines too, but the height will vary. Rabbits and hares can make one at just over half a metre high, the little muntjac deer at one metre, cattle at almost two metres, and ponies, with their long necks and agile heads, even higher. Sometimes you can see two distinct browse lines in the same piece of woodland.

Another kind of browsing which is quite distinctive is bark-stripping by grey squirrels. While most herbivores will take a bit of bark from time to time, squirrels are the only common ones that climb. Any patch of stripped bark two-thirds of the way up a tree of twenty to forty years' growth must be the work of a squirrel. They do also strip bark from the bases of trees, and sometimes from other parts. But even then it's fairly easy to recognise squirrel damage. Most other animals go for younger trees and squirrels tend to take off a large patch of bark with a neat, unframed edge. The only other animal which takes a similar kind of patch is the occasional pony which develops a taste for bark. But ponies usually leave the marks of their large incisor teeth while squirrels leave a smoother surface.

If you don't see any signs of bark stripping it doesn't mean there are no squirrels around but they may leave a clue in the remains of nuts and cones which they've eaten. Various species of mammals and birds eat hazel nuts and they all have their own way of opening the shell. Some rodents make a neat little hole, others gnaw the shell half away. Birds wedge them into crevices in tree bark and punch holes in them with their bills. Only squirrels split them roughly from end to end. Their treatment of conifer cones is also distinctive. Birds get at the seeds inside by prising open the scales of the cone but not removing them. Mice gnaw the scales off neatly. But squirrels rip them off, leaving a rough and ragged central core, quite unlike the more methodical work of the mice.

Moving out of the woods into the fields there's just one wild animal with a really distinctive pattern of feeding and that's the rabbit. The kind of rabbit grazing I described earlier in this chapter is extreme. There they'd eaten the grass down short over a large area and completely destroyed it in a smaller area around the warren. There are rarely enough rabbits around to do that. Where there's a more normal population of them they 'farm' the grass: they graze it down tight over a limited area near their burrows and leave it more or less untouched elsewhere. There are two reasons why they do this. Firstly, they like eating short grass. Keeping it short keeps it in the young, vegetative phase of growth in which it's much more digestible and more nutritious. Tall, yellowing gone-to-seed grass is pretty well inedible to a rabbit. The second reason is defence. Every predator in the countryside is partial to a bit of rabbit and rabbits have developed a whole lifestyle based around running away. Short grass gives them good visibility and hopefully a head start when one of them sounds the alarm.

Sometimes a rabbit lawn can show up at quite a distance, as this sketch from my notebook shows. The hill is a famous local landmark. It's usually grazed by sheep but for some reason it wasn't that year and this revealed the effect of the rabbits.

Glastonbury Tor, 28th August 1999

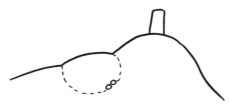

Rabbit lawns have a distinctive kind of turf, usually very short and springy. Where there's high grazing pressure from the rabbits there's often a lot of moss in it because moss is more resistant to close grazing than grasses and herbs. In places it can even get close to a hundred percent moss. But where the population of rabbits is more moderate a rich sward of grass and sun-loving wildflowers can develop. Even if there are no other grazing animals rabbits alone can prevent succession to scrub and woodland by eating any tree or shrub seedling which germinates in their patch. Where the rabbit-grazed area extends under trees or a hedge there's usually bare soil. The combined effects of rabbit grazing and shade have more impact on plant growth than either one on its own.

Rabbits also have a distinctive way of grazing a field of cereals. As in grassland they graze a limited area and leave the rest untouched but here the grazed area is usually a narrow strip along the boundary, often crescent-shaped. This shape is all about insecurity. Safety lies in the hedge, so they don't venture far into the corn field. The ultimate safety lies in the burrow, which is usually in the middle of the grazed strip, and here they venture a little further out into the field, giving the strip its crescent shape. They eat the plants when they're young and green but the effects can still be seen in the mature crop as a strip of sparse, stunted plants.

When badgers take maize cobs they usually concentrate on the edges of fields too but this is surely a matter of convenience rather than fear. Knocked-down maize plants are a sure sign of badgers, except in those limited areas where there are wild pigs. The number of plants knocked down is not just a measure of the number of badgers living in the area but also the dryness of the summer. The harder it is for them to take earthworms the more they go for maize. But far and away the commonest signs of badger feeding are the little holes they make to take earthworms and other ground-living prey. Rabbit scrapes, which are a territorial mark made by the males, can be similar but they're wider and shallower. Wherever there are rabbits you'll see their droppings but finding their droppings beside a little hole doesn't necessarily mean it was made by a rabbit. If both animals are living in the same area there's no reason why rabbits shouldn't leave their droppings where badgers have been feeding.

Wild pigs also dig the ground to find their food but in a completely different way. Instead of making individual holes they turn the soil over, making a bare patch thirty to sixty centimetres wide to a depth of two or three centimetres, though it can appear to be twice that depth because of the depth of the spoil. This is distinctly shallower than the rooting of a domestic pig and it has a more even and methodical appearance. The rooted patch is soon colonised by annual plants and young perennials. It can present one of the few opportunities for annual wildflowers to grow in areas of permanent pasture. In spring these patches may stand out as a splash of bright green against the duller hues of the old pasture. In woods wild pigs searching for acorns may turn over the leaf litter rather than digging the soil.

While rabbits scatter their droppings all over the area where they live, badgers have special latrine areas. They deposit each turd in a little hole not unlike their feeding holes and the latrine area becomes honeycombed with them.

A badger latrine can't be mistaken for the work of any other animal. Foxes have quite the opposite habits. They leave their droppings singly and far from making a hole they leave them on a prominent high spot such as an anthill or a tussock of grass. Both are territorial signals and perhaps the different dunging pattern of each species reflects the different lifestyles of the family-oriented badger and the solitary fox.

Equally unmistakable are badger setts. Usually dug into sloping ground in woods or very wide hedges, they're major earthworks. Piles of fresh spoil jut out from the mouths of the holes if they're in active use. Elder trees and nettles often grow near the entrance to a sett. These two plants are usually the sign of a high level of plant nutrients in the soil. Maybe badgers are in the habit of urinating near the sett. It's not that they're the only plants to survive, as in the case of the severe rabbit infestation I described earlier, because the badgers don't destroy plants as the rabbits do.

Rabbit holes are much smaller and often taller than they're wide, in contrast to badger holes which tend to be wide and low. In light soils they can be found anywhere, but in heavy soils they're usually in the lynchett of an old hedge, especially one which has accumulated a lot of new soil in the past few decades. Here there's a good depth of friable topsoil, which is much easier to dig than solid clay subsoil, and the drainage is good. Foxes are not enthusiastic diggers and prefer to take over an unused rabbit burrow or badger sett, so there's never much sign of earthworking at the mouth.

Paths are a clear sign that some animal is about but it's not always easy to tell which. Badger paths tend to be direct and very well used. Badgers are creatures of habit and will go on using the same path determinedly, year after year. If you're lucky enough to see a footprint it's easy to tell a badger's from that of a fox or a domestic dog. A badger has five toes in a row along the front of the foot while the fox and dog have four. The badger also has long claw marks. Unless there's snow, or you find a footprint in the clear smooth mud where a puddle has recently dried, you'll be lucky to find a whole one. But you can often recognise a badger print just by the claw marks left where it's scrambled over the bare soil under a hedgerow.

Follow the path and you may come upon another clue. I know a place on the Mendips where a badger path

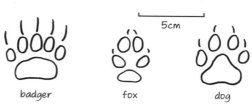

badger   fox   dog

Tracks of some claw-footed animals.
Note: Dog tracks vary greatly in size according to breed.

crosses a dry stone wall. The broad brown stripe made by their muddy tummies where they climb over the wall could not have been made by any other animal. It may be that the badgers established a path here when the wall had tumbled down and refused to give it up when it was repaired. But it is just possible that this is an ancestral path which dates back to before the 1790s when the Mendips were enclosed from open common. They're very persistent animals.

1. The influence of landform on land use: woods on the steep slopes, mixed farming on the gentle slopes and meadow on the flood-prone riverside. Wintours Leap, lower Wye Valley.

2. The upper Dart Valley is a variant on the combe and plateau pattern of landscape. (See page 6.)

3. Cattle, the sculptors of the British Landscape. Herefordshire.

4. These shrubs are wind-flagged by sea winds. (See page 80.) They also show a browse line, indicating that grazing animals are present. (See page 19.) Brean Down, Somerset.

5. The U-shaped profile of this valley indicates that it was carved out by a glacier during the ice ages. (See page 37.) The upper Towy, Mid Wales.

6. The secret of river erosion is revealed in a spell of dry weather when the water level has dropped. Glen Brittle, Isle of Skye. See page 37 for explanation.

7. This lime kiln is a sign that the underlying rock is limestone. Mendip Hills, Somerset.

8. The varying colour of this ripening cereal crop shows how the depth of the soil varies from one part of the field to another. Chilcomb Down, Hampshire. (See page 55.)

9. The whitebeam tree indicates a dry soil, as here on the limes tone of the Mendip Hills, Somerset.

10. Anthills and rushes respectively pick out the dry and wet parts of this field.

John Adams

11. Thatchers End Plantation. The little trees in the foreground are exactly the same age as the taller ones behind. See pages 71-73 for explanation.

12. The bare, south-facing slope of this narrow valley contrasts with the dense woods on the north-facing slope. Burrington Combe, Somerset. (See page 74.)

13. The change from hedged fields in the valley to unenclosed moorland on the hilltop is mainly a matter of the difference in microclimate. Near Ffarmers, Mid Wales. (See page 75.)

14. Typical ancient countryside: an old farmstead standing alone, small, irregular-shaped fields and frequent woods. Ragmans Lane Farm, Gloucestershire.

15. Terraces on two different scales: the big ones are medieval strip lynchetts (see page 93), and the small ones are terracettes, formed by the feet of cattle (see page 235). Glastonbury, Somerset.

16. A typical enclosure-by-agreement land-scape, with vestiges of ridge and furrow in the foreground and curved or dog-leg hedges beyond the farmyard. The rectangular fields further off are clearly later enclosures. Glastonbury, Somerset. (See pages 93-94.)

17. An S-shaped hedge, a relic of medieval strip fields enclosed by agreement. (See pages 92 and 94.)

18. Planned countryside. Although well outside the main area of planned countryside, this parish was enclosed by act of Parliament. Stringston, West Somerset. (See page 99.)

19. Simplification of the landscape: hedges are lost to excessive browsing by sheep, while heathland succeeds to wood through lack of grazing. Simmonds Yat, Herefordshire. (See page 101.)

20. Fencing out the sheep and deer has allowed this Highland pinewood to spring up around a few old seed parents. Near Crainlarich, Strathclyde. (See page 109.)

21. The 'urban common' at the Sustainability Centre, Hampshire, in the first year I knew it. (See page 139.)

22. The flat land is farmed intensively, while the steep land succeeds to woodland. (See page 142.) The white fields in the distance are suffering from soil erosion. (See page 47.) South Downs.

John Adams

23. Lady Park Wood in the lower Wye valley. This stand has been untouched by human hand for a hundred and forty years. (See page 153.)

24. Native ponies play a part in bringing former plantation back to semi-natural grassland. Note the drawn-up shape of the surviving tree. Dundon Beacon, Somerset. (See page 155.)

Where a path goes under a wire fence there may be a few hairs caught in the barbs of the wire. Coarse grey hairs caught on the underside of the fence indicate a badger and soft red ones a fox. Red or brown hairs caught on the top strand may indicate a young deer that couldn't clear it easily. Once, walking through some woods on the edge of Dartmoor, I saw a little tuft of red hairs caught on the upper side of a top wire that only came up to my knee. It was the remains of a fence that divided the old woodland from the recently sprung up woods around it. I followed the path on quietly till I came upon a doe and fawn, grazing peacefully in a small patch of grass, surrounded by the green willows of the new wood.

Niches, or How Plants and Animals Make their Living

Every plant and animal in the landscape lies at the centre of a web of relationships with other plants and animals. Take a grass plant in a field. The most obvious relationships it has with other plants are competitive. It competes with every neighbouring plant, whether of the same species or another, for space, water, light and mineral nutrients. But some of its relationships may be more co-operative than competitive. Clovers, for example, fix atmospheric nitrogen and the grass plant may be able to make use of some of this nitrogen if its roots intermingle with those of a clover. Bracken, on the other hand, may actually harm a grass plant with the poisonous chemicals it uses to suppress the growth of competitors. A whole range of herbivores, from caterpillars to cattle, will eat it and it will play host to parasites and diseases.

The plant also has relationships with its non-living environment. As a grass, it will be pollinated by the wind rather than by insects. It will tend to grow in one kind of soil or another, according to its species. It will have a tolerance for a certain level of shade and a certain range of temperatures. Unless the field is a monoculture, sown with just one kind of grass, you'll see a different mix of grasses and herbs in each part of it, reflecting the changes in physical conditions. Some plants will be found all over the field. Others will be concentrated in specific parts of it, perhaps under the shady branches of a hedgerow tree or in a wet patch of soil. Some will be more common at the top of a slope and others at the bottom. You can see this clearly at buttercup time in many sloping fields, where the green sward on the top of the slope gradually fades to bright yellow at the bottom, reflecting the increase in soil moisture as you go down.

Tall, medium, short, early, mid-season, late, climber, nitrogen-fixer and semi-parasite are just some of the roles that plants can take in the grassland sward

If you get on your hands and knees and look at the sward in profile you'll see that there are differences in the vertical distribution of plants just as there are in the horizontal plane. Some are tall with upright stems while others are short and bushy. The tall grasses and herbs have to devote some of their energy to a strong internal structure which will hold them up, while the short ones have to suffer the lower light levels in the understorey. Tallness and shortness both have their advantages and disadvantages. There are differences in annual cycles too. Some plants do most of their growth early in the season, others later on. The early growers enjoy the moister soil of springtime while the later ones take advantage of the warmer days of summer. By specialising in this way, as tall or short plants, early or late growers, the different species of plants in the field avoid head-on competition and so are able to co-exist.

The web of relationships doesn't stop at ground level. Some plants have deep roots, others have shallow ones, some a single tap-root, others a fibrous mass. This means they reduce competition below the ground as above. They also link up with other organisms in the mutually beneficial relationships we know as symbiosis. For instance, although we say that the clovers and other members of the legume family fix atmospheric nitrogen, that's not strictly true. What they do is trade some of their own organic food for nitrogen fixed by bacteria which live in nodules on their roots.

The relationship between clovers and the bacteria in their root nodules is fairly well known. What's less well known is that almost all plants have a similar beneficial relationship with fungi. The roots of the plant join up with the slender threads which make up the vegetative body of the fungus. These fungal threads are only one cell thick, thousands of times thinner than a root. So for the same expenditure of energy a fungus can produce a network which can penetrate a vastly greater volume of soil than a plant and thus garner much more water and nutrients from it. The plant exchanges some of the organic food it produces by photosynthesis for some of the water and nutrients gathered by the fungus. One very visible example is the relationship between the fly agaric mushroom and birch and pine trees. The showy red mushroom with the white spots is just the fruiting organ of the fungus. By far the larger part of its body is below ground, a network of microscopic threads like cotton wool which can occupy great volumes of soil.

It's a good deal for both partners but the trading doesn't necessarily stop there. Different plants, indeed plants of different species, can be linked up with the same fungus and so foodstuffs can be passed from one plant to another. It's not impossible for nitrogen fixed by one plant to end up benefiting an unrelated plant living nearby. The fungal network is like a bustling market where mutually beneficial deals are being done every day. In the soil of a mature ecosystem there's a wide and intricate web of fungal connections that we're only just beginning to understand. It's hard to see what's going on under the ground but bit by bit a whole new picture of how ecosystems work is emerging from below.

The sum total of the relationships which each species has with its neighbours and its non-living environment is known as its niche. The niche of a species is the way it fits into the ecosystem, the way it makes its living. To make an analogy with human beings, if the habitat of a plant or animal is its address, its niche is its occupation.

That analogy may at first seem to apply more to animals than to plants. An animal eats a certain range of foods and going out to find these foods each day looks very much like ourselves going off to work. Badgers and foxes, for example, although they both eat much the same range of foods overall, are specialists within that range. Badgers mostly eat earthworms and foxes mostly rodents. Although there is some overlap, the difference in their occupations is clear. Plants, on the other hand, all make their food in the same way, through photosynthesis. But food is only one resource among the many which are needed for life. Just as animals specialise by eating different parts of the total food resource, so plants specialise in their use of other resources. For example, as we've just seen in the case of the grassland, they share out the resources of space and time by virtue of their differences in shape and annual cycle.

Eating food is one side of the coin; being eaten is the other. The relationships a species has with predators, grazers, parasites and diseases are very much part of its niche. This isn't just a matter of which species are trying to eat it but also how it avoids being eaten. Plants may discourage grazers or browsers by: having thorns, stinging, being poisonous, tasting unpleasant or bearing their leaves up out of the reach of browsers (as trees do). Or they may be able to rebound from being eaten by growing especially fast or by bearing their buds safely below ground so their potential for regrowth is not diminished (as grasses do).

In all these ways and others each species carves out a unique niche for itself. Having a unique niche isn't just an advantage. It's essential. Ecologists have observed that no two species can live together in the same ecosystem at the same time if they occupy the same niche. Sooner or later one or the other will lose the competition and die out.

This is just what's happening now with the two species of squirrel in Britain, the native red and the introduced grey. The grey is slowly spreading through the country and wherever it's taken up residence the red has died out. There's no direct confrontation between the two. It's just that the grey is better suited to the squirrel niche in Britain than the red is. Exactly why this is so isn't fully understood but two aspects of it are probably food and disease. The red squirrel is in fact a conifer specialist. This may seem strange for the native squirrel in a country with so few native conifers, but Britain is only part of its range and some other parts of Europe have more conifers than broadleaves. The branches of conifers are thin and flexible, so reds have evolved a light weight, adapted for moving around in a world of whippy twigs. Broadleaf branches are stiffer so the greys have been able to evolve a heavier weight. This extra weight probably gives them the edge when it comes to surviving the winter. The reds are also quite incapable of digesting acorns, one of the most abundant foods available to a squirrel in this country. But perhaps the decisive factor is a disease called parapox virus. It seems that the greys carry it without suffering from it but it's fatal to the reds.

In contrast to the squirrels, the three British woodpeckers live alongside each other quite happily. All three climb trees, picking insects out of the wood as they go, and all three excavate holes in the trees for their nests. But there are significant differences in their niches. The green woodpecker, much the biggest of the three, doesn't confine its insect-hunting to trees but also feeds on the ground, both in the woods and out in open country. Although it's not the commonest woodpecker it's the one you're most likely to see, perhaps strutting across a pasture, demolishing one old cowpat after another in search of grubs and bugs, or flying back to the woods with its characteristic undulating flight. The other two woodpeckers only feed in trees, but the lesser spotted tends to forage higher up the trees and on smaller branches than the great spotted. There's also some difference in the choice of nesting sites between the three. The green often nests lower down than the others, the great spotted at the middle level, and the lesser spotted more towards the tops of the trees. Overall there are enough differences to reduce competition to the point where they can live side by side.

The natural world is full of groups of animals and plants which, like the woodpeckers, are similar but different enough to co-exist. It's this differentiation of niches which makes possible the diversity of species we see around us. Without it there would be only one kind of tree in the woods with one kind of shrub, one herbaceous plant, one bird and so on. The difference between niches is the very key to the diversity of species.

It's interesting to compare a very diverse ecosystem with a very simple one. Take, for example, a semi-natural meadow and a bed of nettles. The semi-natural meadow may have as many as a hundred species of plants living in it, each one avoiding competition by specialising at something slightly different from its neighbours. But nettles don't avoid competition. They meet it head on and carve their niche in life by being more competitive than other plants, given an abundant supply of nutrients. This is why there are so few of these competitive species compared to the uncompetitive ones, because they all occupy much the same niche, based on consuming a lot of resources and growing faster than their neighbours. It's a game of winner-takes-all that in extreme cases ends up with a pure stand of one species. That's why soils which are low in nutrients support much more biodiversity than fertilised ones.

Niche is a theme that runs right through landscape reading. In fact much of this book is about niches. The description of Blawith Common at the beginning of the book is all about the differences in niche between bracken, heather, grass and trees. The soil indicator plants described in Chapter 3, The Soil, are useful to landscape readers because they have a 'narrow' niche, at least in respect of soil. That's to say they only grow in a narrow range of soil types. Most plants tell you much less about the soil because they grow in a broader range of soils. Hawthorn and ash trees are examples of plants with a broad soil niche.

In Chapter 6, Wild Animals, I described the niches of several mammals, concentrating on those aspects of their niches where they interact with ourselves. Mammals tend to have relatively broad niches in most respects, but not so all animals. For example, many butterflies can only feed on a very narrow range of plants at the caterpillar stage of their life cycle. An example is the brimstone, that pure yellow butterfly that's the first on the wing in early spring and possibly gave us the word 'butter'fly. Its caterpillars are entirely restricted to buckthorn bushes. Considering how uncommon the buckthorn is it's surprising how often you see brimstones.

In general the narrower the niche a species occupies the more it can tell us about the landscape. Butterflies are choosy about both microclimate and food plants and they're sensitive to chemical pollution. In other words their niche is fairly narrow all round. Because they're so exacting they're sometimes taken as indicators of the ecological health of a landscape.

Pioneers and Stayers

Broad or narrow is one way to characterise niches. Another is according to whether the species in question is a pioneer or a stayer. The distinction is more obvious in plants than in animals. Pioneer plants are good at occupying new space whenever it becomes available. A typical pioneer produces large numbers of seeds, each one very small, and disperses them far and wide in search of unoccupied soil. It grows fast and has a short life. A typical stayer puts much less effort into reproduction. It produces fewer, larger seeds and has no special method for dispersing them. It grows slowly, has a long life and is good at holding on to whatever space it occupies. In fact most plants fall somewhere between the two and have some characteristics of both pioneers and stayers. Nonetheless there are some extreme types and a good illustration is the contrast between the birch tree, an extreme pioneer, and the beech, an extreme stayer.

The home of the birch is wild highland landscapes with poor, acid soils and a cool, harsh climate. The trees produce masses of tiny, light seeds, winged like little butterflies to help them fly on the wind. They're so abundant and travel so far that birch seedlings can pop up almost anywhere where there's a bit of bare soil for them to germinate in. Being so small they must have that bit of bare soil, however tiny, because the little seedlings can't compete with established plants. Whether these seedlings turn into trees depends on two things, the grazing pressure and the amount of shade. They're palatable to grazing animals so only their abundance and their quick growth gives them a chance of survival where there are grazing animals. Birch is quite intolerant of shade and can't grow where there's already a canopy of other trees. This means that while birch readily forms new woods it can't maintain itself on the same site in the long run. A pure birch wood can only exist for as long as the lifespan of a birch tree, typically some seventy years. Quick to spring up and quick to disappear, they're temporary affairs in the moorland landscape. But they can leave a more permanent legacy behind them if other, more shade-tolerant trees become established within the birch wood during its lifetime.

Birch is as much at home on the light, acid soils of lowland heaths as it is on upland moors. But in recent years its niche has started to broaden. Now it can be found in many lowland woods wherever there's a gap in the canopy to let in enough light, even sometimes on heavy, alkaline soils. It seems at least in part to be moving into the niche once occupied by oak which, as I will soon relate, has fallen on hard times.

Beech is often thought of as characteristic of the chalk but in fact it grows on a wide range of soils so long as they're well drained. As a native tree it's confined to the south-east of Britain. But this may be because it arrived here late in prehistory and its natural expansion was halted by the growth of agriculture. Cultivated land is a barrier which beech is ill-adapted to cross. Where it has been planted in the Scottish Highlands it self-seeds freely on well-drained sandy soils, so it probably could have colonised the whole country under natural conditions.

In many ways it behaves in completely the opposite way to birch. Firstly, it has rather heavy seeds which fall to the ground and germinate near the mother plant, so it's slow to colonise new territory. Secondly, it never grows spontaneously outside woods. It seems to need the preparation of other trees before it can get established. Thirdly, it's the most shade-tolerant broadleaved tree in Europe, both as a seedling and as a mature tree, and casts a very deep shade itself. This combination is quite common: shade-tolerant trees usually cast the heaviest shade. It means that beech can both get going in the shade of an existing woodland and once established can prevent other trees growing beneath it. In fact the understorey of a closed-canopy beech wood is virtually bare of vegetation. When an old beech dies and leaves a gap it's usually filled by another beech. Other trees such as ash may spring up at first, but they can't compete with beech for long in the narrow, shady gap left by a single tree. This means that once beech has occupied a site it can be hard to shift it.

Birch and beech represent the two extremes of the spectrum. An interesting example of a tree with a combination of pioneer and stayer characteristics is the oak. We usually think of it as a stayer. It certainly is a very long-lived tree. There's no image that better evokes stability, longevity and continuity with the past than an old oak, like those I saw in Moccas Park that day. (See page 96.) But they're not tolerant of shade and find it hard to compete with shade-tolerant trees like beech, lime and sycamore. They can't hold their ground in competition with these trees in the long term, unless a poor soil or a tough climate gives them the edge. In other words they can take the role of stayer where conditions are hard but not where they're easy. They do indeed dominate most semi-natural woods in the soft, fertile lowlands of Britain. But that's because they've been favoured by the woodsmen of the past, not because of their own competitive ability.

The key to the oak's ability to act as a pioneer lies in its relationship with the jay, the smallest and most colourful member of the crow family. Jays eat acorns, and what they can't eat immediately they bury for later use. They have a preference for burying them well away from the parent tree, outside the woodland, and of every four they bury they only recover one. All large tree seeds, such as hazel and beech nuts, are taken by birds. But it's the special habits of jays and their preference for acorns which give the oak its ability to spread, if not as far as a birch, at least as surely.

Once sown by the jays, oaklings are well adapted to survive in grassland. An acorn is enormous compared to most seeds and the seedling can live on the food it provides for a long time before it has to rely on its own photosynthesis. This means it can compete with the resident grasses. It also means that it has the opportunity to put down a deep tap-root before it sends up its shoot. This can be crucial in helping it to survive occasional browsing, as the root is able to grow a new shoot from its own reserves. Its ability to survive and grow in grassland is just as important to the oak's pioneering ability as its relationship with the jay.

Seedlings of beech, left, and oak, right. Both have a large store of food from the seed, but note how the oak's is less vulnerable to grazing – another advantage as a pioneer.

Another pioneer characteristic of the oak is that it's not tolerant of shade. This means that acorns which fall straight to the ground from a woodland oak and germinate there have little chance of growing into a tree. In times past the occasional one would make it and this was enough to maintain the oak as a component of most kinds of woodland. But in the year 1908 oak mildew disease arrived in Europe from North America. Oak seedlings growing in the open are able to survive it and grow into mature trees. But in the woods the combined stresses of shade and mildew are too much for them and these days oak almost never regenerates inside woods.

Herbaceous plants can be either pioneers or stayers just as trees can. But there isn't the same simple continuum from one end of the scale to the other because herbaceous plants aren't such a homogenous group as trees. While trees and shrubs are all perennials, herbaceous plants can be annual, biennial or perennial. Herbaceous perennials live for several years and if they fail to reproduce in any particular year that's not a problem because they'll have more chances. But annual and biennial plants only flower once in their lives. An annual completes its whole life cycle in a single year while a biennial lives for two years but only flowers and seeds in the second year.

While the niche of a perennial whether herbaceous or woody can lie anywhere along the spectrum from pioneer to stayer, all annuals and biennials must be pioneers. Their survival depends on reproducing every year, so their seeds must find a place where they can germinate and grow each year. Most plants need at least a small patch of bare soil in which to germinate and bare soil is both scarce and short-lived in nature. So annuals and biennials have evolved great powers of colonisation just in order to keep going from one year to the next.

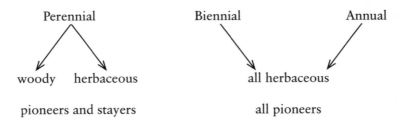

Perennial		Biennial	Annual
woody	herbaceous	all herbaceous	
pioneers and stayers		all pioneers	

In a totally natural landscape they must have been rare but in a cultivated landscape they're common. Most of our important food crops are annuals or biennials, including cereals and root crops. Those broad landscapes of rippling corn that cover much of the east of Britain represent a powerful symbiosis between ourselves and our crop plants. A symbiotic relationship is like a barter deal, with each party offering something the other needs. In this case the crop plant offers abnormally large edible parts: seeds in the case of cereals and roots in the case of carrots and sugar beet. In return we agree to provide the bare soil they need and to look after them throughout their life cycle. It's very much like the deal between the oak tree and the jay, except that we've gone further than the jay: we've changed our crop plants so much by selective breeding that they could no longer survive without us. But as long as we humans are around, being a crop plant is one of the most successful niches on the planet.

Weeds are just as dependent on us as crops are. In fact most of the annual plants in the British landscape which are not crops are weeds of arable fields and gardens. Some are dependent on arable fields for their existence and these have become very rare as weed control on farms has been revolutionised by seed-cleaning machines and herbicides. Other weeds have alternative habitats.

Herb Robert, for example, is a wild flower of the cranesbill family which grows in ones or twos in whatever bare soil it can find on woodland edges and hedge banks. Sometimes one of its seeds will germinate in the bare soil of a vegetable garden. If that plant is allowed to flower and set seed its progeny will come up in masses the following year. Like the fox in the henhouse, it's genetically programmed to struggle for every piece of bare ground it can find. When it finds bare soil in unnaturally large quantities it grabs as much as it can.

That's why the key to controlling annual weeds it to kill them before they set seed. 'One year's weeds means seven year's seeds' is a true saying, though of course it only applies if you allow the weeds to set seed. The seven years doesn't only reflect the amount of seeds produced but also their longevity. In fact many species' seeds can stay dormant for decades. In some cases they only germinate when they're exposed to the light by digging or hoeing. This is another adaptation to the rarity of bare soil in nature: it stops them from germinating when there isn't any. This means they can lie inert in the soil for years without wasting their effort on trying to grow in undisturbed vegetation. Then, if they're lucky, a wild pig may come along one year and create a seedbed. They'll know it's time to germinate by the sudden light they're exposed to as the pig roots up the soil.

Goosegrass, or cleavers, is an annual with a successful niche. Anyone who spent their childhood in the country will remember the endless games you can play with it. The whole plant, including seeds, leaves and stems, is covered with minute velcro-like hooks which stick to people's clothes. Seeing someone walk along unaware of the green tail stuck to their back is an endless source of amusement for children. What sticks to clothing sticks equally well to the fur of wild animals and this stickiness is the mechanism goosegrass uses to spread its seeds far and wide.

Another important aspect of its niche is its tolerance of cold. It can germinate at lower temperatures than most other plants, which means it can start growing in the middle of winter. There's more bare soil around in winter than in summertime when everything's in full leaf. One place where goosegrass often finds bare soil is directly underneath a hedge. Perennial plants don't grow there because of the shade cast by the hedge in summer. But in winter the hedge is leafless and you can often see the two-lobed seedlings of goosegrass springing up there in January. As summer comes on it can scramble away from the hedgerow and into the field, climbing up the crop plants and dropping its seeds into the fertile soil below. One of the best ways to control goosegrass is to leave a strip of perennial grasses at the edge of the field, a barrier which the goosegrass finds hard to cross.

Goosegrass makes its living by doing something which most other plants can't do, growing at low temperatures. The others don't do it because it comes at a cost. All the tissues of a goosegrass plant need to contain anti-freeze chemicals and it takes energy to produce these, energy which could otherwise be used for extra growth, more seeds or some other useful function. Each species invests its energy in a different enterprise, an enterprise that enables it to avoid competition with its neighbours.

Biennial plants have the advantage of two years in which to complete their life cycle and this enables them to grow bigger than annuals. In the first summer they concentrate on producing leaves. The food they produce is passed down to the roots where it's stored over winter, often in a single large tap-root. This means they start their second year with a capital sum of energy to spend on reproduction. Many of our root crops, such as carrots and beetroot, are biennials. We harvest them at the end of their first growing season, before they can use the stored energy for their own purposes.

Many biennials use part of their stored energy to send up a tall stem, as much as two metres high in some species. This gives the seeds a head start in long-distance dispersal, whether by clinging to the coat of passing animals or by flying on the wind. Burdock and spear thistle are two biennials which do this. They also invest part of their energy in specialised dispersal mechanisms. In burdock the dispersal mechanism is the burr, a mass of seeds equipped with big, barbed hooks. They take longer to remove from either fur or clothes than the sticky little balls of goosegrass, so they probably travel further before they're released. The large purple flowers of spear thistle mature into thousands of small seeds, each one equipped with its own parachute. They can float for miles on the warm air currents of summer, seeking out bare soil wherever it may be found.

The other common thistle, the creeping thistle, is perennial. It also sends out thistledown to colonise bare and disturbed ground but it doesn't have the same sense of urgency about reproduction. It has shorter stems and smaller flowers and produces much less seed. It can afford to relax, not just because it's perennial but also because seeds are not its only means of reproduction. It also reproduces vegetatively. Vegetative reproduction means that part of the plant's vegetative body can become detached from it and form a new plant. Creeping thistle does this by means of vigorous horizontal roots which spread out from the parent plant and give rise to new vertical shoots, known as suckers. Each of these new plants can become a viable individual which can live independently and in its turn produce new offspring. Eventually a single plant can expand vegetatively into a wide-spreading clump of creeping thistle. As sexual reproduction has not taken place the whole clump is genetically identical. In this sense the clump can be considered to be one plant and is often referred to as a clone. After a couple of years the horizontal roots die away, leaving independent daughter colonies. But although they may be separate from their parent, from a genetic point of view they're the same plant.

No annuals or biennials can reproduce vegetatively and only a minority of perennials can, but those which do tend to be quite successful. Couch grass and creeping buttercup are two garden weeds which do it, couch by means of root-like rhizomes under the ground and creeping buttercup by above-ground runners. Unlike the annual weeds, which are successful because they produce masses of seed, these two succeed as weeds because they're difficult to eradicate. Just a tiny piece of couch rhizome or buttercup stem left in the soil can regrow into a new plant and from that into a whole new population without a single seed being shed.

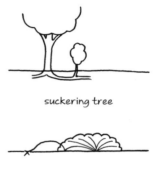

suckering tree

bramble

Trees and shrubs which reproduce vegetatively mostly do so by means of root suckers. Wild cherry, most elm species and blackthorn are examples of suckering species. Above-ground runners are less common in woody plants but one example is the bramble, which forms a new root system wherever one of its arching canes touches the ground.

One of the great advantages of vegetative reproduction is that the new plant is much better able to compete with the existing vegetation. New suckers of creeping thistle and blackthorn can come up through a dense grass sward with no problem. Unlike a seedling, supplied with only what it could bring in the tiny knapsack of its seed, a sucker can call on the energy resources of the parent plant. This is an efficient way of colonising new ground but it's not real pioneer stuff. Compared to the widespread dissemination of thousands of tiny seeds, the slow but steady spread of a blackthorn clump is modest. In fact clonal plants can be either pioneers or stayers. An example of a suckering pioneer is rosebay willowherb while dogs mercury is a typical suckering stayer.

Those clumps of tall purple-red flowers which brighten the roadsides in high summer are rosebay willowherb, a plant as flamboyant as its name suggests. In North America it's known as fireweed, because it's one of the first plants to come up after a forest fire. Here it seeks out urban demolition sites, railway banks, clear-felled plantations and even newly coppiced woodland. But wherever it grows it must have plenty of sunlight. A single plant can produce up to a hundred thousand seeds, each one borne on a feathery sail which carries it far and wide in search of a suitable home. The large number of seeds means that they can come up thickly where they find such a site.

Dogs mercury, by contrast, is a plant you could easily miss. Although it's common and abundant in some woods it's a drab plant which acts as a background to the more colourful wildflowers rather than as a visual focus itself. It's about shin-high with dark, 'leaf-shaped' leaves and inconspicuous greenish flowers. It's very shade-tolerant and is rarely found outside of woods and thick hedgerows. It produces little seed, most of which is sterile, and has no special mechanism for dispersing it. But it does quietly reproduce by means of underground rhizomes. Bit by bit it covers the ground and once established it can hold its own with tenacity. Like a miniature beech tree, dogs mercury casts a heavy shade. The plants emerge early and die down late in the year so they cast this shade throughout the growing season. Mercury grows best in a soil which is neither too wet nor too dry and on the alkaline side. On these soils it forms a pure stand, while on less ideal sites it's usually mixed with other herbs. Under a lightly-shading tree canopy it can grow especially vigorously. The ash tree casts a light shade and in pure ash woods mercury can grow so densely that it actually suppresses ash seedlings and prevents the tree reproducing. Other trees, such as wych elm, yew or beech, are more shade-tolerant as seedlings and are not completely suppressed by it. A pure ash wood with an understorey of mercury can change into a mixed wood if seed parents of these species are nearby. Thus a humble herb can decide the fates of mighty trees just by quietly holding its own.

Animals can be pioneers or stayers in the same way that plants can. On the whole, herbivores tend to lie towards the pioneer end of the scale and carnivores towards the stayer end. Rabbits and rodents grow faster and breed more prolifically than foxes and badgers. Song birds, most of which are omnivorous, raise more clutches per year and more chicks per clutch than birds of prey. The same is broadly true in the world of invertebrates. In the garden the herbivorous pests recover their population much more quickly after being sprayed with an insecticide than their insect predators do – a good reason for not using broad-spectrum insecticides. But plant-eating invertebrates can be stayers too. In fact some of the most extreme species at both ends of the spectrum are invertebrates. Two extreme examples are aphids and stag beetles.

Aphids have incredible powers of reproduction and dispersal. In the summer time when food is abundant they speed up their reproductive rate by doing without sex and cutting the egg stage out of their life cycle. Females reproduce without mating and give birth to ready-hatched young. In fact the whole cycle can speed up so much that the young already have the embryo of the next generation in them when they're born. There can be several generations during the growing season, and in a few short months a single female may give rise to thousands of descendants. When this massive rate of reproduction begins to overwhelm their food supply they divert some energy from reproduction to dispersal and produce a generation with wings. Like birch seeds or thistledown, they can ride for miles on the wind in search of a new host plant. When autumn comes some males are born, mating takes place and the resulting eggs wait quietly for the following spring.

Stag beetles are curious little creatures with antlers like male deer, and they're real stayers. While aphids support their extravagant lifestyle by tapping into the rich flow of food below the soft surface of sappy young plants, stag beetles eat dead wood. There's not much nutrition in wood and it's hard to digest, so they're slow growers. The females lay their eggs in rotting logs and the larvae can take as much as five years to mature into adults. The adults live for a brief summer of mating, enlivened by the occasional jousting between males using those magnificent antlers. They can fly but they're heavy and clumsy fliers and rarely travel far.

The stag beetle is now a rare and endangered species. How can this be if it's supposed to be a 'stayer', so capable of holding its ground? The answer is that the stag beetle evolved in a very different world, as did all the other animals and plants in the landscape. It was the world of the wildwood, that primeval woodland which was once as extensive as open country is now. One thing which was always available in the wildwood was dead wood. There was no-one to harvest the trees, so when they died they slowly rotted where they fell. A creature which needed nothing more than a reliable supply of dead wood was secure. If its reproduction rate was slow and its powers of dispersal negligible it didn't matter at all. There was always another log nearby.

Not so for aphids. Each species of aphid is restricted to a narrow range of host plants, some to a single species. In the diverse ecosystem of the wildwood there was never a guarantee that there would be another host plant at hand. In the constant hunt for new hosts an abundance of highly mobile young was the key to these aphids' survival. If most of them failed to land on the right plant it didn't matter much. Just one aphid which did so could rapidly multiply into a new colony which could provide thousands of new young to seek out the next host plant.

Their wildwood origins is revealed in the behaviour of plants as well as animals. It explains why plants which need light are pioneers and those which tolerate shade are stayers. Rosebay willowherb and birch are both light-lovers. In a world where closed woodland was the norm they had to migrate or die out. Dogs mercury and beech, on the other hand, were surrounded by their favoured habitat just as fish are surrounded by the sea. Energy spent on producing thousands of seeds and sending them far and wide would have been energy wasted. It's more worthwhile for a stayer to invest in fewer, larger seeds. This equips them to compete with the established vegetation, which is always a feature of a stayer's stable habitat. A beech nut is many times the size of a birch seed and has a much greater store of food to see the young plant through its infancy. Vegetative reproduction represents even more investment of energy in each offspring and the chances of each one growing to maturity are correspondingly greater.

When discussing the history of the landscape I mentioned that there is some disagreement over whether the wildwood was predominantly woodland or a landscape which alternated between trees and grassland. The niche differences between pioneers and stayers lends a lot of weight to the view that it was predominantly woodland. The fact that pioneers are mainly species of open country and new woodland while stayers are creatures of mature woodland suggests that mature woodland was the norm. Pioneers needed their powers of reproduction and dispersal to find a habitat that was infrequent, and stayers didn't because their habitat was always close at hand. It's not a conclusive argument because it's possible that the conditions which reigned during the past few thousand years are not typical of the much longer timespan over which the plants and animals have evolved. But it certainly tips the balance in favour of the prehistoric landscape being largely woodland.

Today the situation is reversed. Open country is the norm and undisturbed woodland is rare. Aphids, which once needed extraordinary powers of reproduction just to survive, are now presented with monocultures of host plants. This makes them formidable pests. Annual plants, which once struggled to produce enough seeds to have a chance of finding that elusive patch of bare soil, are now surrounded by acres of it every year. In the wildwood they must have been rare but now we know them as weeds.

Creatures like the stag beetle, which never needed to bother much about finding new habitat, have had a rude awakening. Not only has woodland been greatly reduced but dead wood has become even scarcer. The human hunger for firewood in past ages and the modern passion for tidiness have reduced their habitat to a few remote islands. The stag beetle is just one of a whole suite of dead-wood-eaters which are either rare or already extinct. In the humanised landscape they're the ones which need to be able to breed and fly like an aphid. But evolution works on a much slower timescale than human history and they're stuck with the traits and abilities which suited them in the wildwood.

As though that wasn't enough, the stayers now have to cope with a rapidly changing climate. As the world warms it's most unlikely that complete, mature ecosystems will steadily migrate north. Pioneer species move fast and they'll have no problem colonising new territory as it becomes warm enough, but not so the stayers. In most cases the speed at which they can migrate is slower than the rate at which the climatic belts are moving northward. They'll be left behind. They're also ill-adapted to cross seas, cities and areas of intensive farming and forestry. Stuck in a climate which is becoming too hot for them, it's likely that most of them will become extinct.

As climate change intensifies, many of the more diverse ecosystems, like semi-natural woodland and unimproved grassland, may hang on for a long time. Though they will lose some of their species, their very diversity makes them resilient, maybe resilient enough to keep going in temperatures which are really too high for them. But eventually they will succumb. Landscapes with less resilience will probably break down much sooner. What replaces them will be pioneer communities made up of the most mobile species from the south. Like all pioneer communities they will be low on diversity. They will wait for the stayers to catch up and turn them into mature ecosystems, but they will wait in vain.

How fast these changes will happen and how far they'll go depends on whether we start to take climate change seriously. At the time of writing nothing like enough is being done about it. We're too late to preserve the landscape in its present form. But there's every reason to suppose that if we make an all-out effort now we can save something which is not too far removed from the countryside we know and love.

Succession, How Landscapes Change through Time

Landscapes are constantly changing. Sometimes we humans are the cause of the change but landscapes also change in spite of us. Mostly we try to keep things more or less as they are, whether as arable fields, grasslands, gardens and so on. But all the while natural succession is waiting in the wings, ready to take the vegetation forward to a more mature stage as soon as we relax our grip. The whole gamut of succession runs from bare soil, or even bare rock, at one end of the scale to mature woodland at the other.

Imagine a field that is ploughed and then abandoned. The bare soil will be seized upon by pioneer plants, including many annuals and biennials, eager as ever for any unoccupied space to seed into. These form the initial vegetation which for simplicity's sake we call the annual stage. Once all the available growing space is taken up, the annuals and biennials find it increasingly difficult to re-seed and then perennials have the advantage. This brings on the herbaceous perennial stage. There may be trees and shrubs present from the start, small and insignificant among the herbaceous vegetation, or they may become established later. Either way, when they begin to emerge above the canopy of herbaceous plants succession moves on to the scrub stage, a mixture of herbaceous and woody plants. Once the tree canopy closes the land has reached the woodland stage. At first the woodland is composed of pioneer trees but eventually these are replaced by stayers, or climax species as they're sometimes known. Shade-tolerant woodland herbs are slow to colonise and their presence in a wood can usually be taken as a sign that succession is well advanced.

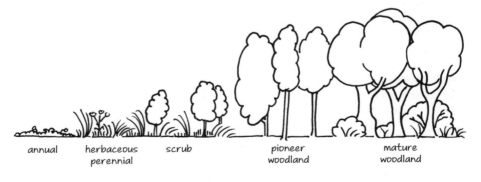

| annual | herbaceous perennial | scrub | pioneer woodland | mature woodland |

The stages of succession

That's a simplified picture of succession. In practice there are lots of variations on it. For a start, not every succession begins with bare earth and not every one makes it as far as mature woodland. But the general trends are always the same: from bare soil to plant cover, from annual to perennial, from herbaceous to woody, from smaller plants to larger, from short-lived to long-lived, from light-demanding to shade-tolerant and from pioneers to stayers. Just as the plants change so do the animals, with a general trend from pioneer species in the earlier stages to stayers in the later ones. Many animals are quite dependent on one stage of succession. The stag beetle is an example. Others, like the fox, make use of a variety of different ecosystems within their range, from open fields to ancient woods, each one representing a different stage of succession.

Since the plants of each stage are longer-lived than those of the one before, succession slows down as it goes along. The change from annual to perennial usually takes no more than a couple of years. The changes from grassland to scrub and from scrub to woodland are much slower, though quite fast enough to be witnessed in a single lifetime. But the change from pioneer woodland to mature woodland is measured in the lifetimes of trees rather than people, centuries rather than decades. This makes it difficult to be sure what's really going on in woodland succession. We can get glimpses of the process but never see the whole thing. Nor can we piece together the story from what previous observers have recorded because reliable ecological records don't go back very far in time. Another problem with observing woodland succession is that all the mature woods in Britain, indeed almost all in Europe, have been heavily influenced by people. Just how a particular ancient woodland would have developed if it had never been coppiced or used as wood pasture is something we can speculate about but never know for certain.

Succession can go into reverse. This happens when a wood is grazed intensively enough to prevent the regeneration of trees. If the grazing goes on for long enough without a break the existing trees will die off one by one and the woodland becomes grassland. This reverse succession isn't a natural process because the grazing animals are put there by people. The resulting grassland can be regarded as semi-natural in that it's been created by human action but not actually sown by us. In fact no process of succession in a heavily humanised landscape such as Britain's is wholly natural because the starting point is never wholly natural. A woodland may spring up on an abandoned field without any direct human input. But if the field has been ploughed and fertilised in the past and the sources of tree seed are woods and hedges which have been managed for centuries, it's not a natural wood. It will never be the same as a natural wood which might have occupied that site. In matters of succession, as in all else, the landscape is a blend of natural and human influences.

From Bare Soil

Successions which start with bare soil are not that common. We usually regard exposed soil as an affront to good management and plant it up as soon as possible. From a productive point of view this is a good idea as uncovered soil is always vulnerable to soil erosion. But for wildlife the odd patch of it is no bad thing. Some invertebrates appreciate it, particularly those which need heat. Bare soil is warm because vegetation cools the soil down both by shading and by the cooling effect as water vapour evaporates from plants' leaves. Seed-eating birds also benefit from a hands-off approach to bare soil. The first wave of wild plants which colonise it are mostly copious seed producers. That's how they found the patch of bare soil in the first place. They're the kind of plants which used to be common in any field corner but have become much less so in these days of herbicides.

Just which plants do spring up first depends in part on the history of the site. If it was a vegetable garden or an arable field, the same mix of weeds which were there when the land was cultivated will come up when it's abandoned. Where the soil seed bank is more limited there may be very little diversity. I once saw a building site which had been left unattended for a few weeks during the summer covered with a pure stand of chickweed. Chance plays a great part in succession. In this case there must have been chickweed seeding nearby just at the time when the builders all went off to catch up on another job. Sometimes vegetative reproduction can provide some of the first plants on a bare site, springing up from the remains of plants which were growing there before. On urban demolition sites a patch of mint can sometimes mark the spot where there was once a back garden. Creeping thistle may do the same in other locations, as may Japanese knotweed.

Light-seeded pioneer plants will colonise from far off and on many sites they will be in the majority. Annuals and biennials like spear thistle may be prominent but there are plenty of perennials too, such as the lowly coltsfoot. Its small, pale dandelion-like flowers come straight out of the ground in early spring, way before the leaves, which are shaped like the sole of a horse's foot and give the plant its name. In a more mature ecosystem coltsfoot is an indicator of moist or heavy soils but in the open conditions of a denuded site it can grow in any kind of soil. Later, as the space is taken up and the plants start competing with each other, it will gradually die out on drier, lighter soils where other plants have the edge. Early colonisers can't be used as soil indicators because the main selection pressure on them is not their soil preferences but their ability to colonise. A common pattern at this stage is a mosaic of single-species patches. Each patch is centred on a single plant which got in early and then spread quickly outwards from that initial point, either vegetatively or by seed, while there was still very little competition.

The more time passes the more a plant's suitability to the soil and micro-climate tells. Even the slightest advantage over a neighbour with a similar niche will tell with time. A tough, persistent plant like coltsfoot may hold on for quite a few years in a soil where in the long run it will die out. But eventually the melee of competition will be fought out to its end and the mosaic of patches will give way to a mix of plants which varies from place to place according to soil conditions and microclimate.

Woody plants may be amongst the first colonists. Light-seeded trees and shrubs such as birch, pussy willows and buddleia have just the same power of colonis-ation as herbaceous plants like coltsfoot and spear thistle. Oak, ash and sycamore may also be there right from the start if there are seed parents nearby. But for the time being any trees are a minor part of the community, relatively slow-growing and often not noticed among the herbaceous growth.

In the countryside successions which start from bare ground are fairly unusual, especially large scale ones. There may be the odd patch of bare soil, such as the spoil heap of a new pond, but it's very rare that a whole field is ploughed then abandoned. In urban areas, on the other hand, there's often a gap of a few years between the demolition of old buildings and the redevelopment of the site. The late Oliver Gilbert made a study of the ecosystems which develop on these sites and gave them the name urban commons. Where other people saw waste land he saw a unique type of ecosystem. Urban commons are unique because they're made up of a mixture of native and exotic plants which can be found nowhere else on Earth. They also follow their own distinctive path of succession.

He called the first stage of this succession the Oxford ragwort stage, after the exotic annual which originally escaped from the botanic garden of Oxford University. This stage consists of a mix of annuals, perennials and light-seeded trees and usually lasts some three to six years. Gradually the taller perennials shade out the lower-growing ones, bringing on the tall herb stage, which may last some five years or so. This is succeeded by the grassland stage, which gradually gives way to scrub, as brambles, bushes and trees gain the upper hand. Urban commons can develop into woodland but they rarely get that far because sooner or later the site is redeveloped. How soon this happens depends on the state of the economy. When it's buoyant not many urban commons get past the tall herb stage but in the Thatcher years, when much of British industry went to the wall, there was a big increase in urban woodland.

Urban commons lift my heart. When I'm travelling through a city by train I always get a little flip of joy when I see one of those bits of land which have been left to go their own way. It may be a patch of rosebay or a bank of brambles, sometimes with a family of foxes sitting there watching the train go by. In some places urban commons make a play space for children which is far more interesting than a park, one with an edge of freedom and adventure. They can also make an interesting study for anyone interested in wildlife. Like any kind of semi-natural vegetation, no two of them are the same. One of the strongest influences on their plant composition is the raw material they grow on. Bricks and mortar is the most common material but sometimes it's industrial wastes. In extreme cases these can be so toxic in their raw state that they only become habitable for plants after the leaching action of rain.

Pulverised fuel ash is an example. It's a waste product of coal-fired power stations. Initially it's highly alkaline, toxic and sterile. In the 1960s it was routinely dumped in heaps near power stations and to begin with these heaps were as bare as a moonscape. The first plants to colonise were salt-marsh species from the coast, even on sites far inland. After more leaching alkaline-loving legumes such as clovers and vetches were able to move in. These nitrogen-fixing herbs are well adapted to raw, young soils. Nitrogen is only stored in the organic fraction of the soil, not in the mineral matter, and as young soils contain little organic matter they're deficient in nitrogen. The legumes were followed by orchids.

These also do well on infertile soils as they have a specially strong symbiotic relationship with fungi which provide them with nutrients in exchange for organic food. By the 1990s a whole range of herbaceous plants had colonised and some sites were developing into orchid-rich woodland. Although they're botanically important, sites like these rarely get any official protection. It's hard for people to recognise that something so urban, so industrial in its origin, is a jewel of biodiversity.

The two key features of an urban common are, firstly, that the natural soil has been completely replaced by something human-made, usually demolished buildings, and, secondly, that the plants are a mixture of natives and exotics. Just occasionally this combination of factors occurs on a rural site and in recent years I've got to know such a site at one of the places where I regularly teach. It's a former naval communications base on the South Downs in Hampshire, now converted into the Sustainability Centre. The Navy left behind a block of all-weather sports pitches made of asphalt and this has since become something very similar to an urban common.

Shortly after the centre opened, the then manager was offered £40 a load to take a large quantity of 'topsoil' which a builder needed to get rid of. This seemed like a heaven-sent opportunity to do something about the sports pitches and be paid for it. It could simply be spread over them and left at that. Asphalt isn't a total barrier to drainage, or even to the strongest plant roots, and beneath it on this site is free-draining chalk, so plants could grow on it without any further remedial work. Well it was a bit too good to be true. When the soil arrived it turned out to be of very mixed quality and contained a liberal sprinkling of rubble and broken glass. It also contained the seeds and vegetative remains of numerous garden plants. The resulting plant community was a spectacular mix of these exotic ornamental plants and local natives. Cultivated daffodils, Spanish bluebells, evening primrose, monbretia and even some unusual opium poppies with enormous floppy blooms held court among the more lowly ribwort plantain, eyebright and birdsfoot trefoil. One or two plants in the mix could have come from either source, such as tutsan, a pretty native undershrub which is often grown in gardens.

By the time I started my regular visits to the centre some four years later this young ecosystem had already started to change. The exotic plants were losing out to the natives. Overall diversity was high, with a quick count yielding fifty-four species of plants, but of these only six were definite exotics. Perhaps in an urban location the exotics would have held on better. There wouldn't have been the same pressure of native plants in the surroundings waiting to move in, and the microclimate would have been much warmer. The site lies high on the ridge of the South Downs and, although partly surrounded by plantations, it's exposed to the prevailing south-westerly winds. By contrast, urban environments are usually warmer than the countryside and this tips the scales more in favour of the exotics. Even so, on a calm sunny day the site is a suntrap and there are often several kinds of butterfly there, enjoying the bounty of both native and exotic flowers. On one occasion my students and I came across a pair of adders basking in the summer sun. (See photo 21.)

22nd July 2003

buddleia
nettle
creeping thistle

buddleia

bramble
nettle
sycamore seedlings
dock, mullein
great willowherb

silverweed (lots)
mayweed (lots)
rush, dock
fleabane
purple loosestrife (one)

tall oat-grass
monbretia
sycamore seedlings

The adders were taking advantage of the warm south-west side of a large mound of earth. There were several of these mounds. Once the 'topsoil' had been spread evenly over the asphalt the loads kept on coming and the late-comers were just left where they were dumped. This gives some variety to the structure, which in turn has influenced the vegetation. The sketch from my notebook shows a particularly strong contrast, where the two main mounds stand on either side of the one poorly-drained part of the site. Ground level here is distinctly below that of the surroundings and surface water has nowhere to flow away to. It looks like this is where all the lorries tipped their loads and their weight both caused the dip and compacted the substrate. There's very little sign of poor drainage on the other flat parts of the site.

The suite of plants on the low spot were all plants of wet or compacted soils while the mounds were home to most of the woody plants on the site. The most prominent of these was buddleia, that famous exotic shrub which is such a favourite with the butterflies. All the biggest buddleias were growing on the mounds. On the flat areas it was only growing where there were little pockets of thicker soil and the plants were much smaller. I suspect it needs that extra bit of drainage that a thicker layer of soil gives it. It could hardly need the soil for any other reason as it often grows in the cracks of brickwork, where there's no soil at all. Scattered around the site there were sycamore seedlings, a common tree in the surrounding plantations and perhaps a sign of the direction the succession will take later on.

The following year I visited in spring. I saw that part of the flat area was developing into a flower-rich sward, grazed by rabbits and deer. This area is next to the bushy edge of a plantation which gave the animals a nearby refuge. There was a distinct deer browse-line on the shrubs which form the edge of the plantation. Grasses were in a minority and ribwort plantain was prominent. Plantains are tolerant of soil compaction so they're well suited to a thin layer of soil over asphalt. I visited the site again that summer and the sward looked completely different. Masses of tiny eyebright had come up through the other plants and was growing in dense clumps, covering over half the surface area, while the tall ragwort made a conspicuous upper storey. Much of the ragwort was being defoliated by the black and orange caterpillars of the cinnabar moth. They're specific to ragwort (not being mammals they're not affected by its poison) and they eat it so voraciously that they probably have a controlling effect on its population. Both eyebright and ragwort are plants which come into their own later in the year and this was a good example of how appearances can change with the seasons.

Two years later there was a drastic change. Although some of us who work at the centre insisted that ecologically this was the most interesting and diverse part of the site, the manager couldn't quite shake off the idea that it was a mess that needed clearing up. She decided it should become extra camping space. This spring the mounds were bulldozed into crescent-shaped bays and all the flat parts of the site were covered with a layer of municipal compost to protect the campers from the broken glass in the soil.

Despite our misgivings it wasn't such a disaster. After all, the site developed its original diversity partly because there was bare soil available for colonisation, so the bare soil of the remodelled mounds is a new opportunity. If the compost spread on the flat areas had increased the level of plant nutrients in the soil this would have reduced plant diversity. But municipal compost is low in nutrients, as it's largely made from hedge and shrub trimmings which people take in to the local recycling centre. The compost was spread in May and by the time I visited in August a lot of plants had pushed their way up through it, especially the plantains. Tall plants had been left standing and the compost spread around them. These included both shrubs like buddleia and tutsan and herbaceous plants like teasel, that tall biennial whose seeds are so much appreciated by goldfinches. Other plants had seeded into the surface of the compost. The most abundant of these was the scarlet pimpernel, that bright little annual known as the poor man's weatherglass because its flowers open when it's going to be fine and close when it's going to rain.

As for the garden plants, only the most tenacious species were still in evidence. I saw two clumps of mint, spearmint and pineapple mint, and there was a bright orange splash in one corner where monbretia was making its mark. Already naturalised in the West Country where winters are mild, monbretia seems to be spreading eastwards as the climate changes. Maybe it will become a long-term resident here. Meanwhile on the other edge of the site two new plants had made their appearance. These were yellow wort and the golden-brown carline thistle, both typical chalk downland plants. Far from being exotic, these are representatives of the historical vegetation of the area. There are some small remnants of chalk downland nearby to act as seed sources, and this may be the direction which succession will take on this site if it's prevented from succeeding to woodland by regular cutting. In the fullness of time the former sports pitches could possibly go some way towards replacing the wide acres of downland which have been lost.

From Grass to Scrub

Most often the starting point for succession in the countryside is not bare soil but grassland. When farming takes a downturn marginal grassland is usually the first land to be let go. Even when there's no recession in farming, steep fields may be allowed to scrub up simply because they're too steep to drive a tractor on. This means that they can't be fertilised, manured or harrowed, let alone cut for hay or silage, so it's impossible to farm them intensively and get a high yield. Even if the farmer is content to use steep land just for extensive grazing, he can't go round with a mower once a year to cut whatever herbage the animals haven't eaten. This is called topping. In the days of cheap labour it was done by hand on steep slopes but these days the value of the grazing would be less than the cost of the labour. Topping doesn't only cut uneaten grass and thistles, it also catches any tree or shrub seedlings that have come up in the pasture. (See photo 22.)

Clearly a field which is regularly mown or topped won't scrub up and one that's completely abandoned will. Whether a field which is only grazed will scrub up is less certain. It all depends on the intensity of grazing. Most young woody plants can survive the odd nibble now and then but none of them can stand being repeatedly bitten down to the ground. How often they get bitten doesn't only depend on the total number of animal-days during the year but also on the pattern of grazing. If a large number of animals are put on the land for a short time then taken away and put back again later in the year, they'll eat everything that's growing there each time. But if a smaller number of animals are left there continuously they can pick and choose. Unless they happen to be goats, they'll eat the grass rather than the shrubs and the field is more likely to scrub up.

Scrub formation isn't necessarily a continuous process. A few shrubs may get going one year, perhaps when there are fewer animals on the farm than usual or grass growth is particularly good. Then if grazing returns to normal no new shrubs may appear for years. The ones which are already there will probably stay because it's much more difficult for animals to kill established shrubs than to prevent new ones from getting started. The growth of the existing shrubs may be restricted by browsing but they won't go away unless they're deliberately cut down. Scrub formation can go on like this in fits and starts and a field may stay at the scrub stage for many years. But it's pretty well impossible to reverse the process by grazing alone. Once it's started, succession to scrub usually carries on to its inevitable end, however slow the process.

Apart from steep land, another place you may see the beginnings of succession from grassland is on the edges of towns where fields are bought up for development. A few years often elapse before construction begins and the land isn't usually grazed during that time. With the hedges no longer trimmed and the grass no longer cut the landscape takes on a new look. The billowing, curvy hedges and tall, waving grass can look like an idealised vision of the countryside for a year or two, but it doesn't stay like that for long. Soon any large, vigorous herbaceous plants in the sward begin to take over from the grasses.

The great advantage which grasses have is their ability to regrow vigorously after being defoliated. As soon as the field's no longer grazed or mown they lose that advantage. The plants which have the edge now are the ones which can cast the most shade on their neighbours. These are mainly plants which are tall, have broad leaves, or both, such as stinging nettles and docks. Usually these plants will have been there in the grassland but kept in check by regular defoliation. Now they can expand at the expense of the grasses.

Nettles and creeping thistle can spread vegetatively and if they're present they'll start to form clumps which grow ever larger and progressively exclude other plants. Docks and hogweed, although they only spread by seed, may take over other parts of the field. Wet fields can become dominated by rushes, which are more sensitive to cutting than grasses but have the competitive edge in wet soil. Hedge bindweed may weave its way out from the hedgerows. In high summer it will cover the drab, gone-to-seed thistles and grasses with a layer of bright green leaves, dotted with its white trumpet-shaped flowers. The big blue blooms of meadow cranesbill and the frothy white flowerheads of meadowsweet may appear in parts of the field. Despite their names (a meadow being a grassland which is cut for hay) both of them seem to do better when mowing and grazing stop. If big brother bracken is waiting in the wings it may rapidly take over the whole field. But otherwise an abandoned field often develops into a mosaic of patches, each dominated by a different herbaceous plant.

It's interesting to note that this change from grass to broadleaves is just the opposite to what Oliver Gilbert observed in urban commons, where the tall herb stage comes before the grassland stage rather than after it. Just why urban successions should work the opposite way round to rural ones at this stage, I haven't the faintest idea.

The time it takes for the herbaceous stage to give way to the woody stage on an ungrazed grassland can vary enormously. I first realised just how long it can take one day when I was walking on Priddy Mineries, a local beauty spot on the Mendip Hills. It's fifty hectares of rough grass and heath on former lead workings. I'd known it since childhood but it had never occurred to me before to wonder why it hadn't scrubbed up. Now, with my landscape-reader's hat on, I thought, "Surely grassland always succeeds to woodland. This place hasn't been grazed for decades, but just look at it." Although willows have sprung up in the boggy places and on the edges of the two ponds, on the grassland there's hardly a tree, just the very occasional pine or hawthorn. One or two of them are young but others are quite old and show no signs of reproducing.

The Mineries are not grazed because the lead content of the herbage would poison the animals. No-one I've spoken to can remember a time when they were grazed and there's no reason to suppose they ever have been since lead working stopped over a hundred years ago. It's not the lead that's halting succession, because most trees are quite indifferent to it. It's not the lack of seed parents either. There's a large conifer plantation just over the road and a species-rich self-seeded woodland on the ruins of the lead works nearby. At least part of the answer is that the grass itself is preventing the trees from taking root.

In fact 'taking root' is literally true in this case. The grass is predominantly purple moorgrass, a tough, tussocky grass which produces a thick mulch of dead leaves and stems at its feet. It effectively prevents tree seeds which fall from above getting to the soil where they could germinate. Any seed which did manage to reach the soil would be unable to grow through the thick mulch above. It also seems to be capable of preventing the succession to broadleaved herbaceous plants I described above. This is an infertile upland soil and none of those plants would grow here, at least not well enough to challenge the tough purple moorgrass on its own ground.

No grassland can keep the trees at bay like this forever, at least not in our climate. Eventually something will happen to expose a little soil, perhaps the digging of some animal. Once a single shrub gets going it starts to shade the grasses around it, creating the conditions for other shrubs or trees to germinate. But the rate at which this happens varies greatly. Priddy Mineries, with its dense sward of tough, tussocky grasses, is definitely at the slow end of the scale.

An example of the fast end of the scale is what happened when the rabbits were virtually wiped out overnight by myxomatosis. On many of the chalk downs of southern England there had been such overpopulation of rabbits that bare soil showed through everywhere. This gave plenty of opportunity for trees and shrubs to germinate and once germinated there was little to stop them growing. Large areas of the downs scrubbed up rapidly.

Wherever grazing stops suddenly the key to how quickly scrub forms is what's there in the year when grazing stops. If there's some bare soil or if there are woody plants which got going before grazing stopped, scrub can start to form immediately. If not, a mulch of ungrazed grass can develop as it has at Priddy Mineries. Even though it may not be as thick or as tough as the mulch of purple moorgrass it will usually hold up the process for several years. Much depends on the species of trees and shrubs involved as some are more able to get established in thick grass than others. This entry from my notebook illustrates some of the dynamics which can go on. The scene is a small field adjacent to our town cemetery. One half of it is occupied by a strip lynchett, both 'step' and 'riser', while the other half is more generally sloping.

Glastonbury, 27th August 2003

The lynchett field just below our house has been bought by the cemetery and not grazed for four or five years. It soon grew over with brambles and a few trees. Now they've strimmed the brambles but left the trees, mostly oak with some hawthorn, one holm oak and one ash. (There's no sycamore, although it's the second commonest self-seeder in the area after ash.) In the generally sloping part of the field the trees are scattered, but on the lynchett part they're entirely confined to the riser. This doesn't seem to be because they cut the trees on the step, as there's no sign of stumps. It seems to be because the riser is the only place they've managed to grow.

This could be because: a) the sward was more open on the riser, as it usually is on a steep slope; or b) the trees got started before the end of grazing and were successful on the riser because animals tend to graze less on steep slopes. Of the two (a) seems less likely because trees have regenerated on gentle slopes in the other half of the field, slopes which wouldn't have a noticeably open sward. (b) fits the case well because the trees are thick on the lynchett riser, scattered on the gentle slope and absent on the flattest land, the lynchett step. In other words, the steeper the slope the more young trees.

This tendency for animals, especially cattle, to graze more intensively on the flat than on a slope has a big influence on the pattern of succession. It's much more comfortable for them to stand or walk on flat land, where all four feet are at the same level. So they spend most of their time on flat ground and graze it most intensively. The steeper the slope the less it gets nibbled and the more chance trees and shrubs have to get going. This reinforces the tendency of farmers to farm the flat land intensively and let the steeper land scrub up. A lightly-grazed field that includes both flat and steep land will usually have at least some scrub on the steep but none on the flat. The steep part may also have one or two mature trees on it, the relic of a previous episode of regeneration.

This note illustrates the same effect on a gradually increasing slope.

Downland near Heytesbury, 20th August 1999

The flat ground at the top is mostly grass. The shrubs, which are all hawthorn, get bigger and more frequent as you go down the hill and the slope gets steeper, till they form a solid band of scrub which merges into the hedge. I can't be sure that microclimate doesn't also play a part here.

In some places microclimate seems to be the dominant factor in the pattern of scrub formation, particularly on exposed, windy hillsides. This is the case on Blawith Common, which I described at the beginning of the book. A common pattern of scrub development can be seen on undulating hillsides where little ridges and valleys run up and down the slope. Very often the little valleys scrub up while the ridges don't. I've seen this pattern in places as different as the chalk downs of Sussex and the brackeny mountainsides of Mid-Wales. Shelter seems to be the key factor here. The soil may play a part, as it will be deeper in the valleys than on the ridges. But this may either help or hinder scrubbing up according to what the herbaceous vegetation is. Deeper soil certainly helps trees and shrubs grow faster, as we saw at Thatchers End. (See page 73.) Faster growth helps them to survive their first vulnerable years in bracken, but in grassland deeper soil may actually work against them. Grasses respond to extra soil fertility much more vigorously than young woody plants. This increases competition from the grass and actually reduces the growth of shrubs.

Once they're established the shrubs will certainly grow better in the deeper soil, but in terms of actually getting established a deep, fertile soil can be a disadvantage.

Going back to the lynchett field, there's another clue which points towards the grazing pattern rather than the openness of the sward being the key factor in the distribution of scrub. This is the predominance of oak. Of all the woody plants, apart from those which reproduce vegetatively, oak is the one which is least bothered by a mulch of grass. A mulch doesn't hinder the jays from burying acorns and the huge food reserves in the acorn give the young plant the power to push up through it. But not even oak can stand the repeated browsing it will get in the animals' favourite hang-out area. Ash and sycamore are avid self-seeders and there are seed parents of both nearby. In fact ash is one of the most troublesome weeds in my garden which is only a hundred metres from the field. But there's no sycamore and only one ash in the field. They have less ability than oak to grow in a closed grass sward, or to resist browsing.

Alongside the native oaks was a single holm oak. This is an exotic evergreen species from the Mediterranean. The occasional exotic is typical of a site on the edge of a town, where most of the planted trees are exotic. There was some hawthorn in the field, too. Hawthorn is perhaps the most successful woody pioneer of all. It's certainly the most widespread. You can see it all over the country, coming up in grassland and bracken on almost any soil and in any climate but the very harshest. It's spread by birds which eat its bright red fruits and excrete the seeds. This gives it long range without the necessity of a very small seed. But even more importantly it has a special ability to compete with grasses.

As anyone who has experience of planting trees will know, the worst companion for young trees is grass. Trees planted into grass struggle for the first few years and a proportion of them die, while neighbouring trees planted through mulch or into a ground cover of some other plant grow faster and few of them die. Much the same is true of self-seeded trees: a vigorous sward of grass is the worst environment for them to start life in. The main factor in the conflict between tree seedlings and grass is probably competition for water. Grasses have a finely-branched fibrous root system which is very efficient at extracting every last drop of available water from the soil. Tree seedlings have far fewer roots but hawthorn is something of an exception, with a more fibrous root system than most others.

Two other shrubs which often colonise grassland by seed are gorse and brambles. They're almost as successful at seeding into grass as hawthorn, though their niches are somewhat narrower. Gorse is commoner on dry, light soils and bramble on more fertile clays and loams. Brambles have the added advantage of vegetative reproduction. Wherever a bramble seedling comes up it can quickly spread like a growing cumulus cloud, its arching canes striking root wherever they touch the ground. Meanwhile other brambles may spread out from the hedge, bounding further into the field each year and joining up with the bushes which have grown from seed. (See photo 35.) In a few years the field can be turned into nothing but a mass of brambles two metres high or more. But almost always there will be the tip of a tree here and there, poking above the dark green pillows of bramble, harbingers of the next stage of succession. When these trees grow up and their branches spread, their shade will shrink the thorny pioneers back down till they become no more than a minor component of the woodland floor.

Blackthorn is another shrub which invades grassland vegetatively. It rarely spreads by seed but its powerful suckers make seeding almost superfluous. It will spread out inexorably from a hedge into pasture, even in the face of quite heavy grazing. Only regular topping or mowing for hay or silage will keep it in check. If left to grow it will form a dense, impenetrable thicket, armed with the most dangerous thorns of any British plant. They're long, strong and sharp and the wound they make often turns septic.

Vegetative reproduction is a great advantage to any woody plant which wants to colonise grassland. The daughter plant has the constant support of its parent, which can keep it supplied with food and water along the pipeline of the blackthorn's root or the bramble's cane. Competition with the established sward is much less daunting with this sort of backup. In fact it's a contest which the woody plant invariably wins. Vegetative reproduction enables trees and shrubs to move into difficult areas where seedlings wouldn't stand a chance. Returning again to Thatchers End (on page 71-73), I have no doubt that one day the area where the planted trees have failed to grow will become wooded. Blackthorn is steadily advancing from the hedge above and brambles from below. Once these shrubs have replaced the grass the way is open for trees to follow, even on this inhospitable site.

There are plenty of other shrubs which can colonise grassland and turn it into scrub but these four – hawthorn, gorse, bramble and blackthorn – are the most common and most successful. It's no coincidence that all four of them are thorny. This is obviously an important part of the niche of any plant which wants to establish in the face of browsing but it doesn't give them total protection. Blackthorn only produces thorns in its second year so its first-year shoots are unprotected. Hawthorn, bramble and gorse do form thorns on first-year twigs but the thorns take some time to harden and become sharp, so the shrubs have a vulnerable period each spring when their shoots are quite tasty. This is why hard grazing can prevent scrub increasing but can rarely get rid of it.

On the whole farmers hate scrub. Even though present-day economics may mean that sometimes they can't help letting go of the steepest land, it goes against all their instincts. Certainly it's less productive than either neat fields of grass or an orderly tree plantation but it is good for biodiversity. Insects and other invertebrates benefit from the combination of shelter and sunlight which the semi-wooded structure provides. The shrubs give nesting sites for song birds while the rough grassland provides food for seed eating birds and habitat for small mammals. Most shrubs, including the ubiquitous hawthorn, only produce flowers and fruit on second-year twigs. In a hedge which is regularly trimmed, as most hedges are, they never flower or set fruit. But in scrub they're left alone and provide a rich store of pollen and nectar for insects and berries for birds.

Most of these species are the kind of opportunist, mobile species which aren't particularly rare. Nonetheless most of them are in decline in the modern countryside and there are some rarities too. These include several species of orchid which tend to crop up just at that moment in succession when grass succeeds to scrub. An outstanding example is the very rare military orchid. Scrub is essentially a changing ecosystem and these plants seem to thrive on change. So do some animals. Some invertebrates need a variety of different early-succession habitats for different stages of their life cycle. All of these need to be present at the same time but all of them are transient. This presents a problem for nature conservation. The nature reserve approach to conservation works against change. It's based on identifying the best wildlife sites and preserving them as they are. But you can't put a fence up around scrub and preserve it. It just turns into woodland.

From New Woods to Old

An idea that used to be more popular than it is now is that each stage of succession prepares the way for the one that follows, the plants of one stage creating the conditions which the plants of the next stage need. Perhaps the notion of plants co-operating with each other is out of fashion in these days when the right-wing ideal of competition takes centre stage. Competition in nature is very obvious. You can see it where plants compete for space to grow and hear it when territorial birds fill the spring air with song or the roar of stags echoes round the glens in the autumn. But underneath this showy exterior there's a quiet world of co-operation, less visible but in a way more vital. Without the co-operative relationships which go on in nature life could hardly exist.

A prime example is the relationship between nitrogen-fixing microbes and their host plants, such as clovers and alder trees. The nitrogen eventually becomes available to other plants and provides one of the main sources of nitrogen for the ecosystem as a whole. Another is the below-ground link-up between plant roots and fungi, which performs a similar function with regard to the nutrients which come from the mineral fraction of the soil.

The web of mutual aid extends above ground too. Gorse, for example, doesn't only have a symbiotic relationship with nitrogen-fixing bacteria. It's also pollinated by insects, which in return receive nectar and pollen for their food, and its seeds are dispersed through another symbiotic relationship, this time with ants. The seeds are borne in a pod, like a pea pod, which snaps open when it's ripe, catapulting the seeds away from the plant, but the ants take them that bit further. Each seed has a little package of highly nutritious food attached to it. The ants bring the seeds back to their nests in order to share this food with their fellows and then leave the seed in the nest. So they not only disperse the seeds but sow them too.

All these relationships are mutually beneficial but gorse's relationship with trees could be described as self-sacrificing. It actually helps trees to get established and grow but, as a sun-loving plant, this means death for the gorse once the trees grow up and their canopy closes. It does it, firstly, by suppressing grass.

Young trees survive better and grow faster in gorse than they do in grass, as long as they keep their heads above the gorse canopy. Secondly, it protects them from browsing with its thorns. Thirdly, it improves the soil for them. The sandy soils where gorse usually grows tend to be acid and poor

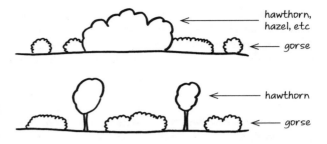

Patterns of succession, Hoggs Cliff, Dorset

in nutrients. Gorse helps to counter both these deficiencies, both by fixing nitrogen and by bringing bases up from the subsoil, which can enrich the topsoil when its leaves fall and decompose. Among the bases it accumulates is calcium, the element in lime which is so important in countering acidity.

Of course terms like 'self-sacrificing' are anthropomorphic and have no place in a discussion of the habits of plants. Gorse is only doing what's good for itself. If trees take advantage of that it's presumably because the shrub hasn't evolved a means of preventing them. Gorse isn't the only shrub which performs this role in succession. In fact shrubs preparing the way for trees is the rule rather than the exception. Trees can follow directly on from the herbaceous stage, especially where there are patches of bare soil. But the most common path from grass to trees is via a shrub stage. Brambles take the role of gorse on loamy and heavy soils, although they neither fix nitrogen nor, as far as I know, accumulate bases from the subsoil. But they do an excellent job of suppressing grass and keeping off browsers with their thorns. In my part of the country a very common sight on lightly-grazed pasture is clumps of bramble with ash saplings growing up through them. Ash is perhaps the foremost pioneer tree round here but you rarely see it growing directly in grass.

There may be more to what brambles do than meets the eye, if my own experience is anything to go by. I once planted five hazel trees, cultivated varieties for nut production. I protected them from deer and rabbits with guards of wire netting, I manured them and eliminated grass competition with an effective mulch. On the whole they grew very slowly, perhaps because I'd planted them too close to existing trees, but one of the five took off like a rocket. It was near a bramble patch and in the couple of years after planting it I allowed the brambles to grow round it. In every other respect it was in just the same situation as the other four. In theory my mulch and wire netting had done everything that brambles can do. But in practice there seems to be another factor which we don't know about because now that hazel tree is over twice the size of the others.

Bracken is another plant which can act as a nurse to young trees, either alone or mixed with bramble. This isn't necessarily as 'self-sacrificing' as it is for gorse because both bracken and bramble can survive in the shade. Like gorse, bracken's a great soil improver. With its strong network of deep roots it can bring minerals up from the subsoil and even break up a hardpan. Meanwhile it adds a generous helping of sweet humus to the soil surface each year. By these means it can gradually change a podsol into a brown earth and a much wider range of trees can grow on the more fertile soil. A walk I took on the hillsides near the Highland village of Kinlochleven illustrates the role of bracken in succession.

Kinlochleven, late October 1992

The ravines on the hillside are full of a variety of trees: birch, oak, ash, beech, holly, rowan and wild cherry. The slopes around them are either treeless or pure birch. The birch is growing mostly on the bracken-covered parts of the hillside rather than the grassy ones. There are tiny birches in the grass but few saplings or larger trees. As they grow the trees tend to shade out the bracken, which slightly obscures the pattern. The grassy areas seem to be wetter than the bracken. Where it's very wet bog myrtle is mixed with the grass but it doesn't mix with bracken.

I don't think the wetter soil of the grassy areas directly affected the establishment of birch trees. It wasn't that wet. It effected it indirectly because bracken can't stand poor drainage and on this site the young birch trees needed the protection of bracken. They could germinate in the grassy areas but they couldn't escape the attention of the sheep and deer. The tall, inedible bracken saved the trees simply by hiding them from the eyes of the animals. As so often, grazing was the dominant factor. This is reflected in the contrast between the open hillsides and the ravines. The ravines are difficult to access. I know because I tried to cross one of them, climbing from crag to crag, and failed. Over the years this has been enough of a deterrent to grazing animals to enable woodland to survive in the gullies and develop to maturity.

The diversity of trees in the ravines makes a stark contrast to the pure birch of the newly-forming woods on the open hillsides. It's the richest mix of trees I ever saw in that district, where the harsh climate and poor, leached soils restrict the range of tree species. The ravines have a sheltered microclimate and a richer supply of mineral nutrients than the surrounding slopes and this is the main reason why a wider range of trees grow there than on the open hillside. (See page 254.) But another reason is the length of time that trees have grown undisturbed in the ravines. In general, a new ecosystem of any kind is simple and it becomes more and more diverse as time goes by. It's quite normal for a new wood to consist of just one species of tree. As the years go by more and more are added, and diversity on its own can be taken as an indicator that the wood has been a wood for a long time.

The trees you can most often see forming new woods in pure stand are birch, ash, sycamore, alder and pussy willow. To some extent these pioneers prepare the way for the stayers which will come later. Birch, like bracken, improves the soil by bringing bases up from the subsoil and adding them to the topsoil through its leaf-fall, and by breaking up pans. The light soils which it naturally favours can easily become podsols and it can start the process of turning them into brown earths. Meanwhile alder and willow can pump water out of an excessively wet soil. Above ground, the pioneers create the humid woodland microclimate which trees like beech seem to need in order to get established.

Young woods aren't always single-species. If they're next to an existing ancient wood the species mix may be very similar to that in the existing wood or it may not. Much depends on what happened in the year when tree colonisation got started. It could be that one kind of tree had a really good seed year or weather conditions might have favoured one kind over the others. A more diverse mix of species may indicate that the trees got established over a number of years rather than all at once.

Animals can have an influence too. I know a place where sycamore and ash self-seeded directly into the bare soil of a former pig paddock. There are far more mature ashes nearby than sycamores and at first most of the seedlings were ash. But there are also rabbits there and they much prefer the taste of the ash. Every ash seedling has had its bark eaten. Not many of them have been killed outright but they've all been weakened and the sycamore have suppressed most of them. Now the saplings are past the age where they're vulnerable to rabbits but only one ash has survived and the old pig pen is on its way to becoming a pure sycamore grove.

It's not just the trees in a wood which become more diverse as time goes by but the whole woodland community, including herbaceous plants, animals, microbes and fungi. At first there's little in a new wood but the trees, as very few herbaceous species can survive the change from light to shade. It can seem an almost sterile place. New colonists must come from other woods and, since most of the species which fill woodland niches are stayers rather than pioneers, colonisation can be a very slow process indeed.

During the lifetime of the first generation of trees many of the woodland niches still don't exist. The thicket of young saplings may cast such a dense shade that even shade-tolerant woodland plants find it hard to grow under them. Only when the trees start to age will there be knot holes which can provide nest sites for tits and nuthatches or dead wood for the insects on which woodpeckers feed. Only when they die and fall to the ground will there be habitat for stag beetles and their like. But one generation of trees is a short time in the life of a wood. It takes much longer for the full suite of species to move in and some real stayers may never arrive at all. A rich mix of woodland wildflowers is a sure sign of an ancient wood.

The first herbaceous plant to colonise the little wood I planted some twenty years ago is lords and ladies, the wild arum. It's more a hedgerow plant than a woodland one and has migrated into the wood from the adjacent hedge. Nonetheless it's well able to withstand the dense shade of the young wood, though it does look lonely amid the bare earth and leaf litter. Behind it the creeping shoots of ivy are spreading out from the hedge. Ivy is typical of new woods. Curiously for such a shade-tolerant plant, it's uncommon on the ground in ancient woods, except on the edges. An abundant growth of ivy covering the ground can be taken as an indicator of a recently formed wood.

Much depends on whether the new wood adjoins an old one or not. Most wildflowers will spread slowly over the boundary into suitable new habitat, though the speed of their advance is very variable. I know an ancient bluebell wood which has a recent extension on one side. The age of the recent part can be gauged by looking at old editions of the Ordnance Survey map. The map of 1904 shows it as an open field and that of 1930 as partly wooded, so it's probably around seventy-five years old. The bluebells have spilled over the small bank which marks the boundary and spread a metre or two into the new. By contrast, I know another place where a small plantation of native trees was made some hundred metres away from an ancient wood with bluebells in it. Little more than a decade after it was planted I saw a clump of bluebells growing in the new wood. Could it be that a bluebell bulb was brought here by a squirrel in the way that jays spread acorns? Or was it a perhaps a two-legged mammal who wanted to enhance the new planting? Maybe the bluebells were already there before the trees were planted, growing under a canopy of bracken as they sometimes do. There is bracken in the little plantation.

Over the centuries different plants and animals colonise till there's very little difference between a wood that's always been woodland since the days of the wildwood and one which grew up during the Middle Ages. But in tracing the course of succession from young woods to old we're bedevilled by the two problems: the process takes too long to observe directly, and all our woods have been managed for centuries. We can't observe a natural succession. We can only infer it from the glimpses of the process which we can observe, such as this one from my notebook.

Bach y Gwydel, 30th April 2004

The wood has a canopy of oak with a few beech. Under the canopy the regeneration is entirely beech. In the ungrazed pasture outside the wood there are oaks coming up through the matted mulch of last year's grass.

You can see this same pattern again and again wherever both oak and beech are present. Even under a pure canopy of oak, if there's a mature beech nearby to provide the seed, the next generation is all beech. Oak can't reproduce in the shade but beech can, while beech casts such heavy shade that no other tree can grow beneath it. This suggests that in the end any wood will end up being dominated by beech, the ultimate stayer among European trees. A simple theory was built on this observation. It stated that all successions, whatever their earlier stages, ended up with pure beech woodland. This end point was called the climax. Where the soil or climate are unsuitable for beech there would be other climax trees, perhaps a mixture of species. But the vision for lowland Britain under a purely natural regime would be a narrowing down of diversity to this single tree species. Herbaceous plants and shrubs would also be rare under a pure beech canopy, although there might be a great diversity of fungi and invertebrates living on the dead wood and leaf litter.

That's more or less what ecologists used to think. But the glimpses we can catch of the later stages of succession suggest that it's really not as simple as that. One of the best places to catch such glimpses is Lady Park Wood, a large ancient wood which lies in the gorge of the lower River Wye. Since 1944 it's been kept as a non-intervention nature reserve, untouched by human hand, just observed. So it's reasonable to suppose that what happens there reflects what might happen in a truly natural wood. (See photo 23.)

It's one of the few places in Britain where you can almost imagine yourself back in the wildwood. Many of the trees do still show signs of its past as a coppice wood by their multi-stemmed form, but the wood is also beginning to acquire some of the characteristics of a wild wood. The most obvious of these is the accumulation of dead trees, which lie where they fall and gently decompose. The opposite bank of the gorge is also covered with native woodland and if you look across and see this you can for a moment believe you're in a primeval landscape, wooded from coast to coast. Then you catch a glimpse of the meadows at the foot of the opposite slope, or hear the voice of a holiday-maker wafting up from the cycle path below, and the illusion is broken.

Typically for an ancient wood on limestone, there's a wide range of tree species in the wood, and these include beech. Parts of the wood haven't been touched since 1870 and these are the most natural-looking, full of big, old trees. Other parts were felled in the 1940s shortly before it became a nature reserve and these have regrown with a strong coppice structure. The topography is also varied. The lower part of the wood, beside the river, is very steep, with patches of vertical cliff. The middle and upper slopes are gentler and on the upper slopes the soil becomes thinner.

The lower slopes make a dramatic landscape. The tallest ash trees I've ever seen reach up to the sky, drawn up by the cliffs which deny them light on one side. With such a length of clean, branchless bole they look like trees from a tropical rain forest. On such a steep slope they stand on shaky foundations and their crowns are one-sided, growing away from the cliff, which makes them even less stable. Every now and then they come crashing down and you can see them lying there like giants slain in a battle with the gods. So this part of the wood is open-canopied, dynamic and full of shrubs and young trees. It's not the sort of place where beech would ever dominate. A stayer like beech needs stability above all. Frequent disruption gives the advantage to pioneers.

By contrast, the rest of the wood is just the kind of place you would expect beech to dominate in the long run. At least that's how it looked during the first three decades after the wood became a nature reserve. It seemed to be the sort of stable environment where the contest of shading could be played out to the very end. Then came the drought of 1976. In the old-growth part of the wood, the part which wasn't felled in the 1940s, the big, old beeches were hard hit, especially on the thin soil of the upper slopes. Some died, others weakened and died over the next few years and the survivors grew very slowly. Moderate-sized gaps appeared which allowed other trees to grow. Meanwhile on the thicker soil of the middle slopes the canopy stayed intact.

Of course beech was still there on the upper slopes, still ready and able to produce seed for a new generation which would one day make the drought of '76 seem like no more than a blip in the steady march towards its destiny. Beech reproduction comes in pulses. At intervals of five to fifteen years it produces a bumper crop of nuts, enough to leave a substantial surplus after all the nut-eaters in the wood have had their fill. Some ten years after the drought one of these pulses of regeneration had grown into saplings about a metre tall and showed every sign of being the canopy trees of the future. Then there was a population explosion of voles, which love to eat the bark of young trees. All the young beeches were killed. Although other species were hard hit too, this episode put off the dominance of beech by a generation. The 'inevitable' destiny of beech to dominate began to seem more doubtful.

Meanwhile the part of the wood which had been felled in the 1940s had developed a mixed structure. The felled trees had regrown in multi-stemmed form. Some had not been felled and these now formed an over-storey of big single-stemmed trees. There were also young single-stemmed trees which had got started at the time of the felling. By the 1980s beech and ash were becoming the most important trees. But 1983 was a particularly bad year for grey squirrels. Beech is one of their favourite trees and they stripped the bark of most of them in this part of the wood, killing their leading shoots. The beeches were reduced from potential dominants to understorey bushes. This left the way clear for ash to become the dominant species in the current generation of trees.

So Lady Park Wood today is a patchwork of different stands with a variety of structure and species which shows no sign of becoming more uniform with time. It's not a natural wood. Some of the influences on its present condition are human-made, such as the fellings of the 1940s and the introduction of grey squirrels. But others are wholly natural and they serve to illustrate how the simple vision of a steady march towards climax is in practice disrupted by a whole series of accidents. We can imagine a purely natural wood as being a dynamic ecosystem with frequent change as the only constant.

Backwards and Forwards

Succession can go into reverse. The events in Lady Park Wood which I've just recounted show how succession can be set back by just one step, from stayers to pioneer trees. But sometimes the whole process can be put into reverse gear and woodland can turn into grassland. This is always as a result of human action and the immediate cause is almost always grazing. It can happen either by accident or by design.

It's happening by accident in many semi-natural woods, especially in the uplands of Wales, northern England and Scotland. In the past these woods were valuable as sources of firewood, timber, or oak bark for tanning and they were carefully preserved. They would also occasionally be used as winter shelter for cattle and sheep but this would be kept to a level where it didn't interfere with the ability of the wood to reproduce itself. Now the only function which has any economic value at all is that of shelter for the animals. Even this is now less important as more animals are kept indoors in winter or even trucked off to lowland areas for the colder months. It's no longer worth repairing the fences, hedges or walls that once restricted the animals' access. Now the woods are open to the surrounding grassland and grazed along with it. In time this grazing turns woodland into grassland.

The first part of the wood to go is the ground layer. Woodland herbs aren't tolerant of grazing or of trampling. If the wood has a dense canopy they will be replaced by bare soil. If more light reaches the ground they will be replaced by grasses, which do tolerate both grazing and trampling. The shrub layer is next. Shrubs are actually very difficult to kill by browsing alone but in a wood they have the added stress of shade and this combination is lethal in the long run. Any saplings will be swept away with the shrub layer and once this happens the wood is dying. The existing trees can live out their lives but as long as grazing continues they can't reproduce. By this stage the value of the wood as shelter is greatly reduced. The wind whistles through the bare trunks of the trees, funnelled between the canopy and the ground. The wood will still be a warmer place on a still, frosty night, as the canopy acts like a blanket which reflects back the warmth of the ground and of the sheltering animals themselves. But in windy weather it will give no shelter at all.

The trees may survive for a long time like this. A sadly common sight in the uplands is a windswept wood of gnarled oaks standing in a carpet of short grass which merges seamlessly with the surrounding pastures. All vestige of a hedge or fence which once protected them has long since vanished, or maybe a low, grassy bank marks where once there was a dry-stone wall.

Sometimes reverse succession is carried out intentionally by nature conservationists. This may seem strange. If woodland is the natural vegetation of this country surely letting trees grow wherever they will is good for biodiversity? But it's not necessarily so. As we've already seen, a new wood is very low in biodiversity. Grassland, on the other hand, can sometimes be home to a whole range of sun-loving wildflowers and insects. These are semi-natural grasslands, ones which have escaped the agricultural improvements of modern times. Like ancient woodland they're the end point of a succession but in this case an arrested one. Succession has been arrested at the grassland stage by constant grazing but not altogether stopped. Just as woods become more diverse with time so do grasslands. There are several factors affecting the diversity of a grassland but longevity is certainly one of them. If you see a meadow full of wild flowers you can be sure that it's been a meadow for a long time.

Semi-natural grasslands are very rare now and unfortunately they're often just the ones which are most often allowed to succeed to woodland, as they're usually in places which for one reason or another are difficult to farm intensively. In some places conifers have been planted on semi-natural grassland for much the same economic reasons. There's a clear gain to biodiversity from removing the young woodland or plantation and allowing the diverse grassland to reassert itself. Seeds and bulbs of many wild flowers can remain viable in the soil for decades and sprout again when the sunlight returns. Sometimes there are also odd corners of grassland left among the scrub, or along the rides in a plantation, and these can recolonise the newly exposed ground when the trees are taken away.

Removing the trees is straightforward enough but getting back from the scrub stage to grassland can be more difficult. Without the added stress of shade the shrubs are hard to kill by grazing alone. Also, the soil may be full of the seeds and vegetative parts of recently removed shrubs, and there's plenty of bare ground for these to germinate in. A combination of grazing and topping can do the trick but it's not always as simple as that. For one thing, on a cleared plantation site the ground may be too rough for topping, especially if the land was 'ploughed' before planting (see page 190) or if the trees were grubbed out rather than merely felled.

Another problem is that grazing which is intensive enough to prevent shrubs from growing may be too much for the wild plants which are the object of the whole exercise. Both of these factors operate at Powerstock Common in Dorset, where nature conservationists are trying to bring back former semi-natural grasslands. It's a large and complex nature reserve with a great diversity of habitats, including patches of cleared conifers and ancient woodland. The pride and joy of the grassland which is re-establishing in the wake of the conifers is the devilsbit scabious. It's quite a rare wild flower and it's also the sole food plant of the marsh fritillary butterfly. Since the plant can't stand heavy grazing only a few cattle are kept on the reserve, not enough to prevent the growth of gorse, which is coming in on those parts where the conifers have been grubbed out. (See photo 24.)

Some of the other patches of conifer were cleared by volunteers when the trees were still small enough to cut by hand. Every now and then they would come upon a self-sown oak sapling among the dark evergreens, and insist on leaving it to grow. The oak has an almost sacred status in the hearts of many people who care about nature. The professional conservationists would have much preferred to clear away all the trees but it's hard to argue with people who are doing the job for free out of their own good will. Those little oaks, released from the competition of the surrounding conifers, have grown tall and wide. Already after only a few years their canopies are reaching out towards each other and in some parts they will surely meet before long.

Perhaps the shade that these oaks will one day cast will help to suppress the shrubs, but whether it will be good for the devilsbit scabious is another matter. It seems to me that the reserve is slowly heading towards a future as a wood pasture. The cattle have access to the woodland as well as the cleared ground. In the long run they'll open out the lower layers of the former coppice woods, leaving the mature trees more isolated. Meanwhile the supposedly open areas will become more wooded as the young oaks increase the spread of their boughs and the occasional new tree gets established among the gorse. This would make for a varied mosaic of more and less wooded patches, probably very rich in wildlife. But it will surely be quite different from what was envisioned in the management plan which was drawn up before the first conifer was cut. Nature conservation is not a precise science and landscapes can have a will of their own.

In this chapter I've looked at how individual ecosystems change and mature. Most landscapes are made up of lots of different ecosystems in varying stages of maturity. Very rarely do we get the opportunity to see how a whole landscape matures simultaneously. For that, a wide area of land would have to start from bare soil at the same time. One place where this has happened is on the polders of the Netherlands. A polder is land which was once under the sea but has been turned into dry land by making a dyke around it and pumping out the water. Hence the traditional Dutch windmills and the old saying, 'God created the Earth but the Dutch made the Netherlands.' On one of the trips I made to the Netherlands to teach I made this observation in my notebook:

Warmanderhof, 11th August 1996

There's an interesting comparison between Beemster, the polder where Fransje lives, and this one, Flevoland. Beemster was the first polder to be made, I think in the early seventeenth century. There aren't many old buildings there but the banks of the rhynes* and the roadside verges are full of wild flowers. The rhynes are home to ducks and other water birds. The only trees are roadside avenues and the trees around houses.

Flevoland is just thirty years old. There are lots of plantations, mainly willow and poplar but also ash and a little cherry and oak. The herbaceous vegetation is immature, with nettles and thistles very common in the plantations and most of the ditch banks nothing but couch grass. The willowherbs and coltsfoot are also common. There's much less open water, with no rhynes and all the land under-drained. There's a network of deep ditches but these are pretty well dry at this time of year. The only water birds I've seen here are herons flying over to the canal and some ducks on the canal itself.

* Somerset word for a drainage channel.

So even such heavily humanised landscapes mature with age. Flevoland looked raw. In some of the plantations the understorey was nothing but a mass of spear thistle, in others there was a shrub layer of pure elder. Beemster had a mellowness, despite its rigid rectangular layout. In its narrow uncultivated places it had that diversity of wildflowers which only comes with time. You could even describe the road verges of Beemster as semi-natural grassland. Yet it wasn't only that which made the two landscapes different. The changes in human technology over the three and a half centuries that separated them meant that they were made differently in the first place. By the time Flevoland was made underground drainage had made the open channels of Beemster redundant. Also by then it was considered appropriate to devote some of the land on a new polder to forestry. Perhaps the notorious grain and butter mountains of EEC surpluses were with us by then and the idea that every valuable acre must produce food was already slipping out of fashion.

CASE STUDY

Josh's Wood

The Study Area

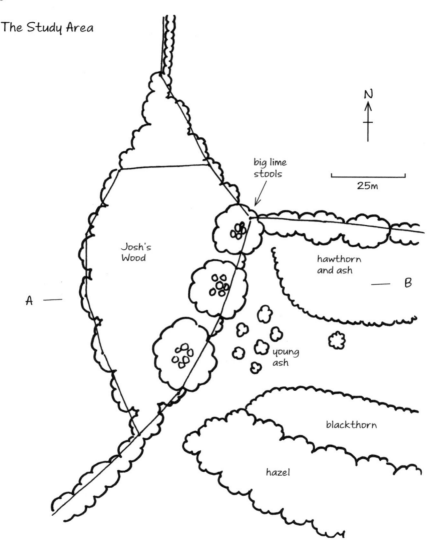

Although we can't live long enough to see a single place pass through all the stages of succession, sometimes we can see several different stages in one place simultaneously. Over the past few years I've got to know just such a place at Ragmans Lane Farm. It centres on a little wood, only some hundred metres long by fifty wide, known as Josh's Wood. In the wood itself and its immediate surroundings you can see various types of woodland, scrub and grassland, all representing different processes of succession.

The wood itself is clearly ancient. Its trees are ash, maple, hazel, hawthorn, spindle, elder and wych elm. There are a few weak brambles and the wild flowers include wood anemone, lesser celandine, dogs mercury and bluebell. Such diversity of trees and the presence of the woodland wildflowers are sure signs of great age, as is the absence of ivy in the interior of the wood. You can clearly see why this little patch has long been left as woodland: it's the steepest part of the slope and also very stony.

The hedge on the eastern boundary of the wood shows signs of being even older than the wood itself. It contains three massive coppice stools of small-leaved lime, a tree which is absent from the rest of the wood. Lime is an extreme stayer with very little ability to colonise new woodland. It's most often found in woods which have never been cleared from the days of the wildwood to this. Its presence in the hedge suggests that the hedge is a relic of the wildwood rather than one which was planted on open ground at some time in the past. The absence of lime in the rest of the wood suggests that at some point in history the wood disappeared for a while. Given the stoniness of the ground this would have been due to grazing rather than deliberate clearing for cultivation. The diversity of trees and wildflowers in it suggests that it re-established itself a long time ago, in the Middle Ages or before.

The hazel coppice to the south-east of Josh's Wood is also on a stony slope. It has a similar range of wild flowers to the wood but the trees are much less diverse, being almost entirely hazel with a few wild cherry. It's likely that hazel has been encouraged here at the expense of other trees. Historically hazel was an important tree in the rural economy. The slender coppiced wands were ideal for making hurdles and the wooden staples known as spars which are used in thatching. It's many decades now since the coppice was cut and the stems have grown thick and crooked. It's a magical place, floored with mossy stones. The occasional curvy-trunked wild cherry tree, bright in its shiny striped bark, stands out among the gnarled hazel like a princess among peasants.

Growing out from the north edge of the hazel coppice is a dense thicket of blackthorn. Even if you didn't know that blackthorn is a pioneer shrub you could tell this thicket is a recent development. The boundary between the hazel and

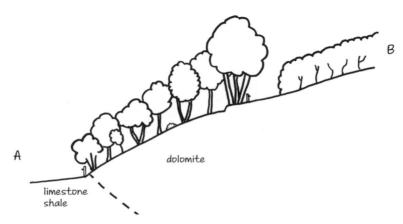

the blackthorn gives the clue. While the hazel stems in the middle of the coppice grow up more or less perpendicular, those on the margin with the blackthorn grow out at a low angle. The perpendicular ones grew up surrounded by other trees on all sides and needed to go straight up to get at the sunlight.  The low-angled ones were on an edge and reached out into the field where there was no competition.

Opposite the blackthorn a similar thicket has grown out from the hedge to the north, but this one is composed of hawthorn and ash. Chance must surely have played a part in this segregation between blackthorn on one side and hawthorn and ash on the other. There's no logical reason for it that I can see. In the decade I've been working at Ragmans Lane I've seen the thicket of hawthorn and ash develop. My earliest note describes it as "hawthorn with a few ash" but today it looks like a young ash wood with a hawthorn understorey. Over the short time I've known it you could say it's stepped over the boundary from scrub to young woodland.

These thickets, both blackthorn and hawthorn-ash, are very young. Their lives are hardly to be measured in decades, compared to the centuries which differentiate the ancient woodland from the even older hedge. Their youth is evident in the small number of woody species and the lack of herbaceous plants beneath. The grassland plants have all been shaded out and the wildflowers of the ancient woodland have not yet started to move in.

Between the two thickets, on the eastern side of the wood, is an area that has succeeded from grassland even more recently. I've also seen this area change. At first it was a fairly thick stand of bramble and bracken with a few contorted ash saplings just raising their heads above it. Their twisted stems showed how each autumn the bracken had slumped down on them and nearly crushed them till eventually they were tall enough to grow away from it. Then one year some cattle were put in the field instead of the usual sheep and they severely trampled the bracken and brambles. By now the ash trees were stout individ-uals some three or four metres tall. They were not only strong enough to withstand the attentions of cattle but perhaps starting to cast enough shade to affect the brambles and bracken, their erstwhile protectors. This area is now best described as young ash trees with a few wisps of bracken and bramble beneath.

These three patches of young woodland – the blackthorn, the hawthorn and ash, and the pure ash – have established themselves on three sides of an old pasture. The grassland which is left in the middle of them is very diverse, with a full suite of typical limestone wildflowers. It's a classic example of the most diverse grassland being the first to scrub up. It contrasts strongly with the grass field on the downhill side, to the west of the wood, which is much less diverse, much more productive and lacks any scrub. The main reason for this is simply that the two fields are in different ownership. The ancient hedge is the boundary between Ragmans Lane Farm to the west and Glasp Farm to the east. Ragmans, although an organic farm, is in the business of food production while

Glasp is more of an informal nature reserve. The Ragmans field has been grazed more intensively than the Glasp field, with regular manuring and topping.

So succession is arrested at the grassland stage on one of these fields while on the other it's going ahead. Meanwhile, in part of Josh's wood succession has gone into reverse. This is the northern part, which is unfenced and open to the cattle which graze the Ragmans field to the west. It's a very small area of woodland so when the cattle come and take shelter in it they have a heavy impact. This is most obvious on the ground layer, which is mostly bare earth. The only survivor among the woodland wildflowers is the lesser celandine, which is particularly resistant to compaction and trampling. In the spring it makes a cheerful splash of yellow against the dull brown of bare soil and stones. At present this small piece of unfenced woodland is a benefit to the welfare of the animals in the field but in the long term, if it stays unfenced, it will disappear.

Josh's wood and its surroundings are a microcosm not only of succession but of the eternal dance between human and natural factors which goes to make up the landscape. The underlying rocks have fixed the location of the wood. The steep bank it stands on is caused by the outcrop of a harder rock. All this landscape lies on limestone of one kind or another but the wood stands on the boundary between limestone shale to the west and the harder dolomite to the east. The harder rock makes for a steeper slope. (See profile on page 159.) Within this framework determined by the rocks, the idiosyncrasies of human behaviour have had free play. The two landowners have very different approaches to the land, productive on the Ragmans side and laissez-faire on the Glasp side. All the examples of forward succession are on Glasp and the single example of reverse succession is on Ragmans.

In that sense the most important feature in this little landscape is the ancient hedge which marks the boundary between the two farms. I wonder how long it has been a property boundary. Boundaries are often the longest-lived of all human features in the landscape. Could it even have been the boundary between two farms down all the long ages since the fields were first carved out of the wildwood?

Trees
as Individuals

Trees are almost infinitely varied in shape and size but only some of the difference between one tree and another is down to their genetic potential. External influences can have at least as much effect on the form of a tree, if not more. For instance, coppicing makes a tree grow in a multi-stemmed form, so there's far more difference between a maiden ash and a coppiced one than there is between a maiden ash and a maiden maple. Coppicing is the most extreme case, but there are other ways a tree's form can be moulded by its environment including browsing, windflagging, competition from other trees and pollarding.

Free-Grown Trees

Trees which grow far enough from other trees so that their branches don't touch are known as free-grown. This doesn't mean that they aren't affected by any external influences at all. In fact trees which have grown completely undisturbed are rare, rare enough for a colleague to tell me in some detail about some examples he found.

June 1994

> Phil Corbett tells me he knows three apple trees, grown from discarded cores, in a place with full light, no grazing and no water stress. They're very wide and low, with branches at the edge of the crown going up vertically "like osiers". The crowns are very dense. The only competition is from herbaceous plants and he sees this shape as a competitive response. Some nearby oaks are growing in a similar shape.
>
> A thought: there's no such thing as the natural shape of a plant. A tree grown entirely without competition or browsing and with no climatic stress will adopt a certain shape. But that's not a natural situation.

Even those apple trees were not entirely free of external influence, as Phil reckoned the form was at least in part a response to competition from the herbaceous plants beneath the trees. A low, wide canopy would shade out the maximum area of grass, thus reducing competition for water. On the other hand it may just be that a tree with nothing to interfere with the growth of its branches will assume a more or less hemispherical shape with its branches reaching the ground on every side. I once saw an oak in a churchyard which had grown without any significant influence on its shape and it was almost hemispherical, though slightly taller than it was wide. It was a fairly young oak and the more upright shape compared to Phil's apples may be a sign of immaturity.

Younger trees tend to have a more upright shape than older ones. Many species have a pronounced conical shape when young which rounds out to a broad, spreading crown when mature. It's as though they point up to where they want to go till they get there, and then relax. Most conifers do this. The familiar Christmas tree shape indicates that they're still growing. Because timber trees are harvested before they stop growing you rarely see a mature one, except for the native Scots pine. The broad crown of a mature pine, shaped like passing clouds,

contrasts with the tight cone of the young tree. Among the broadleaves, alder and ash are two which follow this pattern. The drawing shows three ashes which I sketched one day at Ragmans Lane Farm. They grow alongside the old green lane which gives the farm its name. Notice how the shape changes with age.

Another sign of age is a stag-headed tree, that is a tree which has dead branches among the living ones. Stag-headedness is particularly characteristic of oaks. It gives the tree a sickly look and some people think a stag-headed tree is dying. But oaks can live happily for hundreds of years like this. In most cases it probably results from the tree's response to a serious drought in years gone by. Trees constantly lose water from the surface of their leaves and every branch bears thousands of leaves. Shutting down some of its branches by allowing them to die reduces a tree's need for water, enabling it to survive a drought which otherwise might be fatal. Oaks aren't the only trees to do this. Ashes do it too but the evidence quickly disappears because ash wood soon rots away when it dies. When an oak branch dies the bark and the outer layer of wood rot away fairly quickly but the heartwood at the centre of the branch will last for a hundred years. Stag-antlers on an oak are a record of a past victory over drought. They don't mean that the tree's in terminal decline any more than the scars of past fights on the nose of a tough old terrier mean that the dog's about to die.

The outer layer of wood in a branch or trunk is the sapwood, the living, growing part of the wood which moves water up from the roots to the rest of the tree. The heartwood consists of dead wood and the tree uses it as a depository for the unwanted waste products of metabolism. When the tree, or a single branch, dies these waste products act as a preservative and the heartwood is very slow to decompose. But while the tree's alive exactly the opposite happens. The living sapwood is equipped to defend itself from rot while the heartwood eventually succumbs to the rots and moulds of time. Once an opening is made to the outer air, perhaps through an old knothole, the rotting organisms can enter.

The process of decomposition takes a long time but in some cases the tree can end up hollow. Being hollow is no disadvantage to an old oak. A healing layer is formed on the inside of the shell of sapwood and this hard layer, though thin, is actually stronger than the solid mass of the heartwood. I've heard that hollow trees withstood the great storm of 1987 better than solid ones.

Oaks are normally felled for timber at around a hundred and fifty years old, when the vigorous growth of their youth starts to slow down. But they can live for hundreds of years more, just as we can live for several decades after we stop growing in our late teens. It's during these centuries of middle age and decline that oaks become stag-antlered, rotten and hollow. Although by now they're useless for timber, from a biodiversity point of view they're priceless. There are many insects and other invertebrates which are completely dependent on this decaying wood and many of these creatures are now rare and endangered. The occasional old tree may still have a value for forestry, because such a richness of invertebrates will support a good population of woodpeckers and other predators, which can then move on to the crop trees and control any pests which are there. (See photo 25.)

The oldest trees are not necessarily the biggest ones, though they do usually have very thick trunks. As a rule of thumb you can estimate the age of a free-grown tree by measuring the circumference of its trunk in inches and taking each inch to represent one year. For smaller trees, like crab apple and hawthorn half an inch represents a year, as it does for woodland trees, whose growth is slowed by competition with each other. But this is only a rough guide. For big old oaks it more often underestimates the age than overestimates it as there are more things which can slow down a tree's growth than there are things which can speed it up. These two pollard oaks, both growing at Middlemarsh in Dorset, have trunks of almost exactly the same circumference. But, by Oliver Rackham's estimation, the larger one is about five hundred years old and the smaller one a thousand. The large one, with its tall, wide crown can clearly put on plenty of growth each year and probably has done so throughout its life. The small one has a much smaller crown and there's no sign that it ever had bigger branches, so its annual growth rate has probably always been much less.

It's the same with young trees: you can't always assume that the smallest ones are younger than their larger neighbours. A tree in a wood only has to fall slightly behind its neighbours for them to start suppressing it. Usually this results in the death of the smaller tree, but if it still receives enough light it can stay alive as a virtual bonsai. This can happen both in semi-natural woods and in plantations but it's much more obvious in a plantation because you know that all the crop trees are the same age. Any differences in size must be due to differences in the speed of growth. I once saw a Norway spruce which had the misfortune to be planted in the compacted soil of a tractor rut. It hardly came up to my knee while the surrounding trees were a good five metres tall.

Severe microclimates can also keep trees small. Those contorted little oaks you often see on hilltops and mountainsides are often far older than you might think, especially where the soil they're growing on is thin and poor. Salty sea winds are the worst climatic conditions for trees. They not only stunt them but can also prune them into fantastic shapes which hug the ground. The most extreme windflagged tree I ever saw was an apple.

Oswald Beach, Lulworth, 4th May 1999

On the path down to the beach I saw an apple tree, probably grown from a discarded core.

The seaward side was a hedgehog of dead vertical twigs with a few leaves growing under it. The lee side was full of leaves and blossom. Nearby was an elder, very similar but more upright.

The 'hedgehogs' on these two trees, although made up of dead material, were important functional parts of both plants. It was the shelter they gave that enabled the rest of the tree to bear leaves and blossom. The fact that the elder was more upright may have reflected some subtle difference in microclimate which I couldn't see, or possibly it was because elder is a tougher tree than the domestic apple.

These were extreme examples of windflagging, which is a great aid to reading the microclimate and has been described earlier on. (See page 80.) Most free-grown trees show some sign of windflagging, however slight. Another influence which is almost universal on trees grown out in the open is browsing. The classic parkland tree has a rigid browse line at the base of its crown. The height of the browse line depends on the kind of animals grazing the park. Horses reach higher than cattle, which in turn reach higher than fallow deer, the usual deer of parks. A tree which has only been browsed by sheep is rare. Above the browse line parkland trees have no restriction on their shape and the influence of the tree's species shows up more clearly than it does in a wood. The towering column of common lime contrasts strongly with the broad sweep of beech, while the gangling ash can often be distinguished at a distance from the more compact walnut. But these species differences are less than that between free-grown and woodland trees, regardless of species. While the woodland tree

typically has a tall, branchless trunk, the open-grown tree has a shorter, thicker trunk and big branches growing both high and low.

Compare the parkland tree on the previous page with the trees in the sketch on the right, which I came upon one spring day in a field in Wiltshire. It was obvious that they hadn't grown up as the isolated parkland trees which they now are. Their form shows clearly that they spent most of their lives in a wood which had been cleared in recent years. I asked some people who live nearby and they said that was just what had happened.

A completely different browsing effect can sometimes be seen on evergreens. Where the deer population is very high young conifers can be restricted to a cylindrical shape by constant browsing. Interestingly, this is much the same shape that a holly develops if it's repeatedly cut for Christmas decorations. But when a free-grown holly is browsed by ponies it develops a curious skittle shape.

A skittle-shaped holly

The unbrowsed tuft at the top is presumably out of reach and below that browsing is more intense at head level than lower down. In woods, browsed hollies develop a browse line like any other tree, as the combined effect of browsing and shade is too much for the lower branches to survive. In the New Forest, where holly is a very common tree, I once saw one growing on a woodland edge which had the skittle shape on the sunny side and a browse line on the shady side.

I've seen something similar on gorse bushes, also in the New Forest. The rounded part of the bush is rather more spherical than in the case of the holly, perhaps reflecting the natural shape of gorse bushes. Overall they're much shorter than the skittle-shaped hollies but they still have the topknot. This is puzzling because the topknot is well within the reach of ponies. Perhaps my theory about their formation is wrong. I've never actually watched the process in action so I can't be sure about it. I once saw blackthorn bushes which had been browsed in the same way. It was only a fleeting glimpse from the window of a train as it sped down the Avon valley towards Bath. In this case the unbrowsed topknots were taller than the browsed part at the bottom. It was blossom time but though the topknots were white with flowers the rounded bits below were green. This is because blackthorn flowers on second year twigs, so a part of the plant which is constantly bitten back never gets to flower.

Sometimes a tree is shaped both by browsing and by pruning. This tree has been subject to browsing for most of its life but now the field it stands in is used for arable crops and it's been high pruned to enable tall farm machinery to work right up to its trunk. The short branches below the canopy have regrown since it was pruned, but only

above the old browse line. Below that, on the part of the trunk which has been browsed for almost all the tree's life, there was no regrowth when I made the sketch. (Now, a decade later, dormant buds have woken on the lower part of the trunk and formed young branches, though shorter than the ones above.)

A more traditional reason for pruning a free-grown tree is to improve its form for timber. Removing the lower branches gives it a longer length of straight, knot-free timber which can be sawn up into planks. In fact it's the only way of producing usable timber on a free-grown tree. Pruning needs to be started when the tree is young and done regularly every few years. But people have rather lost the habit of pruning and unskilled attempts can have comical results, like the two trees below. They're sycamores which grow in a hedgerow near where I live. When they were pruned a few years ago they had far too much taken off at once, leaving crowns which were too small for the amount of energy in their roots

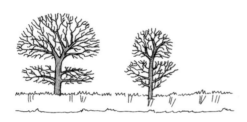

and trunks. All that excess energy burst out in regrowth of the side branches. The tree on the right is the more extreme case. The crown that was left was ridiculously small and the regrown branches are now actually longer than the uncut ones. So much for getting a good timber form!

Woodland Trees

Much the more common way to grow timber is to put the trees together in a wood so they draw each other up. The job of the pruner's saw is done by the shade each tree casts on its neighbours. They all

reach up to the light above and their lower branches are

progressively suppressed as the trees grow. The closer they are together the more each tree gets drawn up. These sketches shows the contrast between conifers grown at wide and close spacings. The wide-spaced ones (above) were planted as a shelterbelt and will never make usable timber because they're too branchy. The close-grown ones (left) are possibly too drawn up, with excessively thin trunks. The plantation should have been thinned before this stage. (See pages 190-191.)

Drawing up doesn't necessarily mean that all the trees in the wood will be totally uniform. It may do in a well-managed single-species plantation but in an unmanaged and mixed woodland there will be a variety of tree shapes. This is partly due to irregular spacing and occasional gaps but also to the differences between species.

Yew is a very shade-tolerant tree that hardly gets drawn up at all. However dense the trees around it, it stays obstinately low and rounded in shape. The only other tree which comes close to it is the holly, which sometimes gets drawn up and sometimes doesn't. In some parts of the New Forest you can see yew and holly forming an understorey, indifferent to the shade, while the deciduous trees all around them reach up to the light above. Their shade-tolerance may be partly due to being evergreen, which means they can make use of spring sunshine before the deciduous trees come into leaf. But on the other hand the Scots pine, another native evergreen, is one of the least shade-tolerant of trees.

As a general rule the most shade-tolerant trees are also the ones that cast the densest shade, and that's certainly true of the yew. This sketch (right) shows a group of trees in Lady Park Wood. The remorseless outward growth of the yew has caused the young oaks and beeches to lean away from the deep shade of its boughs till now they form a cup around it. But there doesn't need to be a difference in species for one tree to be deflected in its growth by another. It's  only necessary for one to grow more vigorously than the other for the weaker tree to be deflected. The most extreme case I've seen of a tree being bent by competition was an oak, so crowded by a yew and a couple of hollies that it grew

 almost horizontally (left). It looked perfectly healthy despite its bizarre shape. On the upper side of the elbow where its trunk changes direction there's a small scar like a branch scar. This marks the position of the original trunk. When the dense shade cast by its neighbours suppressed the trunk a branch took over and eventually became the horizontal trunk we see today.

Even where trees are all drawn up more or less vertically there can be differences between the form of different species. The least shade-tolerant ones grow straightest and are less likely to have any lower branches. The Top Wood at Ragmans Lane Farm contains ash, wych elm and wild cherry. The ash and wych elm both have smooth boles leading up to a crown of upward-reaching branches. The only lower branches which remain are occasional big ones. Sometimes it's more of a case of a divided trunk than a trunk and a branch. These lower branches are more frequent on the wych elms, which are more shade-tolerant trees than the ash. Wild cherry has a different branching habit altogether. Its branches grow at a more horizontal angle and they're more persistent. When wych elm and ash branches die they rot away leaving little or no trace on the face of the trunk. But cherry branches stay there, sticking straight out from the stem, more like the branches of a conifer than those of other broadleaves. Wild cherry can make valuable timber but to get the

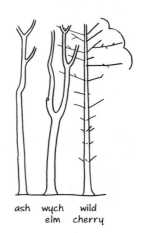

ash   wych   wild
elm   cherry

highest quality the trees need to be pruned even if they're grown close together. Otherwise the wood of the trunk grows round those dead branches, leaving deadwood knots which will fall out and leave a hole in the timber.

An old 'bender'

Trees can get excessively drawn up by their neighbours. Sometimes a young tree is so drawn up that it only stays standing due to the support of its neighbours. If any of them are lost, by felling or some accident, it will bend over into a curve with its crown brushing the ground. If it survives like this, vertical branches will sprout up from the curved trunk and it may go on living in this shape for many years. I've seen this happen to pussy willows and ash but no doubt it can happen to other trees too.

Trees which grow on the edge of a wood are hybrids between free-grown and woodland trees: branchy on the sunny side and clean-stemmed on the shady side. Sometimes you come across a tree with this form standing on its own with space all around it. Then you can deduce that the trees which made it that shape have been felled. It must have spent most of its life on the edge of a ride or clearing, or even a small gap. Sometimes you can find the tell-tale stumps among the undergrowth. The tree in the sketch below stands on the side of a ride in a large plantation. The compartment behind it has been felled. Apparently the forester decided to leave this one to grow on and become a veteran. It's a good idea because an edge tree like this is too branchy to be any use for timber. In fact these days foresters sometimes plant up the edge with a row of mixed native trees for wildlife habitat. There's little point in planting crop trees in a position where they'll never make marketable timber.

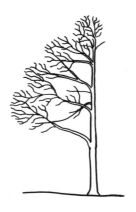

A tree can grow one-sided for other reasons apart from being on an edge. In a wood on a steep slope almost every tree will have more branches on its downhill side, as the tree above it shades its stem and the one below doesn't.

(See illustration on page 81.) In a mixed wood a tree can grow one-sided if there's a big difference in the size of the trees on either side of it. This sketch (left) shows an example of this from Josh's Wood. If the trees either side of the large central ash were felled it would look very much as though it used to be on the edge. You could only find out what really happened by examining the ground on either side and finding the stumps. But the stumps of these small trees would soon be covered by undergrowth and rot away so it could be hard to piece together the story.

Two of the trees in this sketch have multiple stems instead of a single trunk: on the left a hazel and on the right a maple. A multiple stem is a sign of coppicing. In fact hazel is unusual in that it grows in a multi-stemmed form naturally, so this tree may or may not have been coppiced. But the maple certainly has been. The shape of a maiden tree, especially a young one which is growing vigorously, is ruled by the leading shoot. It produces hormones which flows down by gravity and slightly inhibit the growth of lower shoots. This is known as apical dominance and it's the reason why many trees have a conical shape while they're still growing in height. When the tree is cut down there's no longer a leading shoot. Instead there's a ring of dormant buds around the circumference of the stump, between the wood and the bark. Now that these are the highest buds in the tree they're all equally free to sprout and grow. The resulting multi-stemmed tree is known as a coppice stool.

Almost all broadleaved trees respond to felling like this. Among native trees beech is a bit of an exception. Some beeches regrow well enough after coppicing but others grow weakly or even die. This means that coppicing gives a competitive advantage to other tree species, so it can maintain diversity in a wood which would otherwise become dominated by beech. Suckering trees, such as wild cherry and aspen, rarely coppice. Felling the tree stimulates a mass of suckers to sprout from the roots but the stump usually dies. Very few conifers will coppice and the one or two exceptions aren't important in the landscape. The produce of coppicing is a crop of poles, which can be used as fuel or for various crafts. This is in contrast with single-stemmed trees which produce timber, a single log which can be sawn up to yield planks, beams and so on.

The stems on a coppice stool follow nature's typical pattern of regeneration, a large number of offspring followed by a high death rate. A cut stump usually produces a mass of little shoots in the first year but by the second year as many as half of them may have died by mutual competition. The rate of loss slows down after that but a tree which started out with a hundred shoots may end up with a dozen by the time it's ready to be felled again. The number of stems varies, depending on the species of tree and the length of the coppice rotation, that is the number of years between fellings. Hazel is both naturally multi-stemmed and cut on a short rotation of about seven years. When coppiced it produces a large number of thin wands which are just right for making hurdles and thatching spars. All other woodland trees are naturally single-stemmed and are cut on a longer rotation than hazel so they have many fewer stems by the time felling time comes round.

In fact these days very few woods are still coppiced and the stools you're most likely to see are old, overgrown ones. This big ash stool is typical. Note that one of the stems has recently died and is rotting away. Eventually old stools can be thinned down to a single stem, but when this happens there's usually a

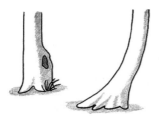

tell-tale sign of its origin as a coppice stem. This may be a bulge at the base of the tree, the vestigial stump of another stem or a curved base to the trunk, known as a swept butt (left). Oak was often coppiced right down to ground level and the stems grow up from the roots rather than from an above-ground stump, so the trees can look deceptively like maidens. But if you look carefully you may be able to see that they stand in small groups and that all the members of each group strongly resemble each other.

A swept butt on its own doesn't necessarily indicate that the tree was once coppiced. Trees can also acquire this form in a wood on an unstable hillside where the soil creeps downhill. Young trees slip down with the soil and as they do so they lean over. New growth of the main stem is always vertical. So, with the lower part of the stem at an angle and the upper part vertical, they develop a kink. Slipping may go on till the roots grow big enough to hold the soil in place, so the young trees develop a whole series of kinks in the stem which soon smooth out into a curve. You can get much the same effect on a very windy site with an unstable soil, where the young trees get bent over by the wind. In both these cases it's fairly easy to deduce the cause because all the butts are swept in the same direction. I did once see a plantation on the Flevoland polder where many of the trees had swept butts, all curving in different directions. As the land was only thirty years old this was clearly not the result of coppicing in some bygone age. It must have been due to subsidence of the soil, which was still drying out and shrinking when the trees were young.

Nor do all coppice stems have swept butts. Whether they do or not mainly depends on how close together the stools are. If they're far apart the young stems on a freshly-coppiced stool grow away from each other and only start to grow upwards when they meet the stems from an adjacent stool. This gives them a curved butt. If they're crowded by other stools right from the start they grow straighter. Swept butts are in fact a sign that the wood is understocked. The yield of wood would be higher if there were more stools per hectare and the poles would be more useful for craft purposes if they were straight.

In its lifetime a coppice stool will be cut many times. In fact coppicing can greatly increase a tree's lifespan because every time it's cut down its above-ground parts are rejuvenated. An ash tree can live for up to two hundred years as a maiden but as a coppice stool it can live for a thousand. A very rough guide to the age of an old stool can be had by measuring its diameter. An ash stool four feet wide may be four hundred years old and one eight feet wide, eight hundred. On a waterlogged or infertile soil they will be much smaller for their age. This rule of thumb probably holds true for lime, oak and hazel as well as ash. Maple stools grow a bit faster and sycamore and chestnut perhaps twice as fast. The age of the oldest stool in a wood gives the minimum age of the wood itself.

There are three distinct time cycles in the life of a coppice stool: the annual cycle, the coppice rotation and the lifespan of the stool. In an actively coppiced wood you may come across a stool whose roots and stump are several hundred years old while its stems and branches are nothing but slender young wands. As the centuries go by the stump grows wider and wider, producing more and more stems. Eventually the centre of the stump breaks up and the individual stems become isolated, often with little clue to their common origin.

This big oak stool which I saw in a field in Pembrokeshire gives some clues to the history of the little valley where it grows. It almost certainly started life in a wood because trees in fields are never coppiced as the regrowth would be mercilessly browsed by grazing animals. It could possibly have grown in a hedge, but there's no sign of a former hedgebank under it and in this area hedges are grown on substantial banks. The wood it grew in must have been an old one because the width of the stump shows that the tree was coppiced many times in its life. But the wide spread of its branches shows that for most of the time since its last coppicing it has grown out in the open, so the rest of the wood must have disappeared not long after it was last cut. The size of its stems suggest that this was over a hundred years ago.

In a traditional coppice wood there are almost always some trees which haven't been coppiced: scattered among the coppice stools are a number of single-stemmed trees, originally grown for timber and known as standards. Their form is quite distinct from that of the trees in a wood which is grown purely for timber because they're not close enough together to draw each other up. They're drawn up by the coppiced trees but as these never reach the same height as a timber tree only the lower part of a standard's trunk is free of branches. The height of branch-free trunk depends on the length of the coppice rotation. The longer it is the taller the stools grow and the higher the clean trunk.

Above that the branches grow out, often at a low angle, unimpeded by any competition. The angle of the branches depends more on the genetics of the tree than any environmental influence. This standard is growing in a mainly hazel wood which is still coppiced. Note the short length of clean trunk. This is an outsize tree and would have been harvested for timber long ago in a traditional coppice wood. (See drawing on page 181.)

In a neglected coppice wood the coppice will have grown up taller than it used to be. The lower branches of the standard trees, which were formed when the coppice was shorter, are now shaded by this new growth and they often die. As the standards are usually oaks the dead branches can stay on the tree for a long time, preserving something of the history of the wood in the form of the old trees. This example is from Lady Park Wood.

Pollards

Pollarding a tree is like coppicing up in the air. The tree is cut not at ground level but some two or three metres up the trunk. This makes it possible to get regular crops of poles in a place where there are grazing animals. It's harder work than coppicing and the yield of wood is slightly less, but the tender regrowth is up out of the way of hungry mouths. A large proportion of trees growing outside of woodland, whether in hedgerows or free-standing in pasture, used to be pollarded. So were many of the trees in wood pastures. But pollards are not normally found in coppice woodland, except on the bank which encloses the wood.

Those brush-headed willows which you see along river banks are pollards. They used to be pollarded every three years to get thin sticks for making thatching spars. This is less than half the normal hazel rotation of seven years but the sticks grow to the same size. The difference is partly because willow's a much faster growing tree than hazel and partly because the pollards are grown further apart than coppice stools so there's less competition. Crack and white willows are the species used for pollarding. These are larger trees than the rather shrubby pussy willows. They have narrower leaves and lack the catkins which turn from silver to gold in the spring and give the pussy willows their name. But when the crown of a white willow pollard gently bends to a summer breeze, the silver undersides of its leaves sparkle in the sunshine like a shoal of little fish in the current of the wind.

A maiden tree

Not much pollarding is done these days. Most of the old pollard trees are left to quietly grow on, their branches becoming fewer and larger over the years just like the stems on an old coppice stool. Eventually they reach a stable condition with just a few large branches and will live on like this with little change. Most of the pollards you see now are old grown-out ones of this kind. Although they're not as obvious as a freshly

An old pollard

pollarded tree they're quite easy to recognise once you get your eye in. Usually all the branches start at the same height above the ground and there's often a ring of bosses at that point where previous generations of branches have grown and been cut. (See photo 48.)

Sometimes this basic pattern is confused because a tree has been pollarded at two different heights. Lower down you can see the point where it was first pollarded and up above you can see where the big old branches have been pollarded individually at a later date. This may have happened where pollarding was stopped and then restarted. Repollarding the thick old branches at their base would have been too much of a shock to the tree and might have killed it. The ash tree in the photo 27 has obviously had a complex pollarding history.

Although the branches are irregularly arranged on the tree, you can see it's a pollard by the clear contrast between the thick old trunk and the much younger branches which are all the same age.

A variant on the pollard is the stub. This is a tree which has been pollarded low down, usually at about chest height. Stubs are often boundary markers, especially inside a wood. Where all the other trees are either coppice stools or standards a stub makes an unambiguous landmark. In a wood there's no need to make it as high as a pollard, just high enough to stand out among all the coppice stools.

The tree which is most often still regularly pollarded is willow. There's still a small demand for thatching spars. Although thatched hay and corn ricks are a thing of the past there are still houses which need thatching. Some years ago I made part of my living pollarding willows and making spars. Pollarded willows are very much part of the landscape here in central Somerset. With so much low-lying wet land, the water-loving willow grows well and easily out-yields hazel. There are hazel coppices locally, but hazel was traditionally reserved for making hurdles, as it lasts longer than willow when exposed to the weather. Thatching spars are mostly kept snug and dry under the thatch and don't need to be durable.

Willows are also sometimes pollarded just to keep them in good condition. Most other neglected pollards steadily lose branches over time, healing the wounds of each dead branch as it goes. Willows sometimes do this but often the weight of the crown becomes too much for the tree and the branches break off. This not only looks untidy and gets in the way of farm work but the resulting wound lets in rot and eventually this can kill the tree. Willow pollards are prone to break up like this because they're fast-growing trees with weak timber, so the strength to weight ratio isn't favourable. Oak lies at the opposite end of the scale in this respect. It grows slowly, its timber is strong and when a branch dies it turns slowly into an 'antler' and stays on the tree for a long time.

Pollarding, like coppicing, increases the life of trees. If you see a large maiden tree and a smaller pollard, don't assume the maiden is older. A maiden oak may live for five hundred years but a pollarded one can reach a thousand or more. Pollarding, like coppicing, constantly rejuvenates a tree. Another reason for longevity is that pollards are usually grown in the open, away from other trees. This means they can go through a period of retrenchment without being suppressed by younger, more vigorous trees, as they surely would be if they grew in a wood.

Old pollards often grow into fantastic shapes and they can be among the most beautiful of trees. They're also usually rich in the dead wood habitats which are so valuable to invertebrate life. The bole frequently rots and becomes hollow while the tree is still being pollarded and producing regular crops of fresh young poles. In the crown, at the top of the bole, a miniature ecosystem develops. Organic debris collects there and turns into a humus-rich soil. Ferns often grow in the crown, but this isn't unique to pollards as they also grow directly on the branches of old maiden trees in the wet west of Britain. What is unique to pollards is that other trees and shrubs can grow in them.

Wild roses and brambles are common in willows, which is a nuisance when you come to pollard the tree. But elder, hawthorn and even ash are fairly frequent too.

The most extraordinary example of this kind of relationship that I've seen was a sycamore growing in an ash. Many years ago the ash had been a pollard and the sycamore grew in its crown. Eventually the sycamore's roots found their way down, through the rotten wood in the heart of the pollard, to the ground. Now the pollard's trunk is hollow and the sycamore fills it. The sycamore is growing much more vigorously than the old ash and one day it will surely supersede it.

The Different Kinds
of Woodland

When you walk into a wood you enter a little world where the rest of the landscape can seem remote. This is not just an illusion created by the fact that you can't see the open country; woods really are different. They're much more three-dimensional and this multiplies the number of niches which can be filled. Beneath the trees there's space for shrubs and saplings and below them for herbaceous plants, ranging from tall bracken to tiny violets. Lichens, mosses and even ferns can find a home on the trunks and branches of the trees, along with climbers such as old man's beard and ivy. The occasional bracket fungus does no more than hint at the hidden mass of fungal threads pervading the soil and the trees, both living and dead. A wealth of birds find their homes in the different layers of vegetation, from the acrobatic blue tit high up in the branches to the wren bustling in the undergrowth. Insects make a living at anything from harvesting the huge resource of tree leaves to the quiet hidden life of a dead-wood-eater.

People are part of the woodland ecosystem too. The importance of wood to the people of previous ages is hard to imagine now. Almost everything they used was made of it and in most parts of Britain it was the only fuel. In short, wood was essential to survival, as it still is in other parts of the world and may be here again in the future. Although today we may value woodland more for its beauty and biodiversity than its physical produce, we still have a functional role in the ecosystem. Either by changing the way we manage a wood or by not managing it at all we change the woodland community itself. Traditional small-scale harvesting of trees used to create a unique microclimate, a combination of sunlight and shelter, which is needed by many woodland creatures such as butterflies. Modern large-scale felling lets in sunlight but takes away the shelter, whereas leaving the wood unharvested maintains the shelter but keeps out the light. Creatures which thrive on dead wood and humidity will benefit from a lack of harvesting but the butterflies will disappear. Woods are dynamic. They don't stand still because we do. Leaving a wood alone is in effect launching it onto a new path which will eventually turn it into a different kind of wood.

Woods are very rarely homogenous. With the exception of the large-scale plantations of the uplands it's unusual to come across a tract of treed land that's the same all the way through. It may be partly semi-natural and partly plantation. If it's all plantation it will probably be made up of various compartments containing trees of different species or different ages. If it's semi-natural the tree species will probably vary from one part of the wood to another and there may be differences in the structure of the trees too, as in Lady Park Wood. (See pages 153-154.) Often there's more variation within a single wood than there is between one wood and another, so the unit of observation is not so much the wood as the stand. A stand is a distinct part of a wood which differs from other parts in species, age or structure. It can be any size from a few square metres to many hectares. It can be the whole wood, but this is unusual. One of the first questions to ask about any wood is how many stands there are and how they differ.

The wood is, of course, part of the wider landscape and another key question is why it's located where it is. Why is this particular piece of land covered in trees and not put to some other use? If it's an ancient wood the answer is usually that the land is unsuitable for farming. It's not that woodland was less valuable than farmland in the past. In fact medieval records show that it was often worth more than arable land. It's that trees, being the native vegetation, are much less demanding than other crops. So in a farmed landscape woods are not sited where trees grow well but where farming is difficult. In hilly parts of the country woods are usually sited on the steepest slopes. There are exceptions to this. In chalk country, for instance, the escarpments are usually grassy sheep walks while the intractable clay-with-flints, which lies on flat hill tops, is often wooded. (See photo 28.) But from many a vantage point where you have a wide view of the landscape you can actually see how woodland marks out the steepest slopes with surprisingly few exceptions. There's also a tendency for woods to be more frequent on the colder, north-facing slopes than the warmer, south-facing ones.

In flatter country ancient woods are often sited on poorly-drained land. This doesn't mean alongside the rivers, where the land was valuable meadow in medieval times, but on flat clay land, often a low plateau well away from any river. There's also a tendency for woods to be relatively far from villages, near the parish boundary. Woodland needs less frequent attention than fields so this placement made good sense in terms of people's time and energy. It's likely that woods were relegated to the boundary not as a deliberate plan but because land clearance spread outwards from the village and only stopped when the remaining woodland was beginning to acquire scarcity value. By no means every parish had woodland. Those without it could get some firewood from hedges but there was also a constant trade in wood.

In recent times woods have sprung up in places where other activity has declined. As grazing has lapsed on commons many of them have succeeded to woodland over the past couple of centuries, while on enclosed farmland steep fields are usually the first to be let go. Conifer plantations tend to be on land which was cheap to buy. In lowland areas this means they've usually replaced heath and ancient woodland, both of which had lost their traditional functions by the mid-twentieth century. In upland Britain, where all the land is unfavourable for farming, it can mean anywhere and often the choice of sites seems quite arbitrary.

A view which brought this home to me was that from the ancient road known as Sarn Helen as it crosses the Cambrian Mountains north of Ffarmers in Carmarthenshire. Where the valley of the River Twrch veers away from the road towards the north-east it opens up a long view into the heart of the mountains, a view which includes various plots of treed land. The one ancient wood you can see clings faithfully to a steep hillside, while an equally steep hillside on the other side of the valley is succeeding to new woodland. But the dark slugs of conifer plantation are scattered at random, on flat land and sloping, from fertile valleys to blasted moorlands. The plantations seem to be sited not according to the physical conditions of soil, slope and microclimate but according to the whim of individual landowners. It appears that the key factor in the siting of these plantations was the location of farmers who were prepared to sell off a bit of land for forestry.

(Since I wrote that paragraph I've revisited the spot. A couple of the plantations have been clear felled since my earlier visit and this has changed their appearance. One has been replanted and the light green of the young conifers makes it quite unlike the 'dark slugs' of the unfelled plantations. The other, which happens to be the one in the lowest, most fertile situation, is further away and harder to see clearly. I couldn't make out whether it's been replanted or put to another use. The unity of the view, which spoke so clearly of random placement when all the plantations stood out against the lighter background, is gone. The change in the view between my two visits is a good example of rotational change in the landscape. See page 14.)

In recent decades planting decisions have sometimes gone from the arbitrary to the downright contrary. Tree planting has been used as a way of reducing agricultural surpluses and higher rates of grant have been paid for planting on the most productive farmland. If this had really caught on it would have turned the landscape upside down, at least in some places. But I can't say I've noticed much effect of these grants in my travels around the country. The instincts of farmers are too deep-seated for such sacrilege.

In farmed landscapes woods are discrete parcels of land with definite, permanent boundaries, but in the Highlands of Scotland they can be much more mobile. Also, in these harsh conditions they are found in places which are good for trees rather than ones which are bad for other things. The alternative land use here is not farmland but moorland, which is less demanding than trees rather than more so.

North Side of Loch Leven, August 1992

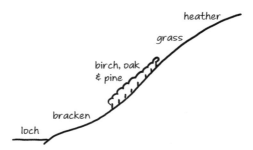

There's much more regeneration of young trees outside the wood than within. Does this mean that woods change location cyclically? But the woods here seem to be sited predictably rather than randomly: a) on land too steep for intensive grazing, b) on the land which is best for trees, the burnsides and lower slopes.

In posing my question I assumed that Highland woods must behave in one of two ways: either they wander round the landscape at random or their location is determined by slope, soil and microclimate. In fact the answer is that both are true. The country is big enough and the woods few enough that they can migrate from one site to another while staying within a general area that favours trees. Birch and pine are both extreme pioneers which rarely regenerate in their own shade and woods made up of them are usually mobile. Oak, since the coming of the mildew, has started to behave in the same way. (See page 127.)

Grazing pressure follows much the same pattern here in the Highlands as it does in other places. It's less intense on the slopes than on the flat. What is different from the south is that the harsh climate and poor soil are a limitation on tree growth. Rainfall is very high so soil drainage makes a significant difference. Slopes are better drained than flat ground and, contrary to expectation, the banks of a burn are usually well drained because the burn itself acts as a drain. The soil by a burn is also better supplied with bases. (See page 254.) In terms of microclimate, lower slopes are usually less exposed than upper ones and, although the burns may not carve deep valleys, even a little dip in the ground is a help to a seedling. In a land which is so marginal for tree growth the woods do indeed tend to grow in those places which are best for trees.

Ancient Woods

In the beginning was the wildwood. Although there's some debate about whether it was solid woodland or not, pollen analysis gives us a good idea of what tree species it contained. In its most developed phase at the dawn of the Neolithic there were three distinct regions, which for convenience we can call the Lowlands, the Uplands and the Highlands. A glance at the map will show that these names aren't completely accurate geographically but they do reflect the broad divisions of tree communities in prehistoric Britain. In fact they have some relevance for woodland right down to the present day and I'll use them throughout this chapter.

■ highland
▨ upland
☐ lowland

The three regions of the wildwood

In the Lowland region the commonest tree was lime. Beech had arrived in Britain but still hadn't spread very far, so lime was the most extreme stayer, the tree which could out-compete all comers on the most favourable sites. Oak, hazel, ash, elm and alder were also common and could be dominant in some places. In the Upland region oak and hazel were the commonest trees, probably with oak dominant on the poorer soils and hazel on the richer. The central Highlands were the stronghold of pine. To the north of this birch was the main tree, dwindling down to treeless tundra on the north coast and the Outer Isles. This was the general picture but within it there was much more complexity, with many different species and mixes occupying all sorts of different soils and microclimates.

In general there was, as there is now, a wider range of tree species in the softer south than in the harsher conditions of the north.

We can't be nearly so sure about the structure of the wildwood as we can be about its tree species. Certainly there would have been a lot of dead wood and some very big trees indeed. But there must also have been young trees in all stages of growth, giving a variety of structure. How did regeneration happen? Young trees need a break in the canopy in order to grow. Was this mainly a matter of individual trees dying of old age and leaving small gaps or were there more general disturbances? The great storm which struck the south-east of England in 1987 may give us a glimpse, however obscure, of the dynamics of the prehistoric woods.

When the storm came most people thought it was a disaster. Hundreds of years of tree growth destroyed in one night! In some places they rushed in with bulldozers to clear away the fallen trees and replant. This caused soil compaction and erosion and in some woods completely destroyed the ground vegetation, the woodland wildflowers. It turned ancient semi-natural woods into impoverished plantations, and it was completely unnecessary. In the woods which were left alone most of the uprooted trees survived quite happily in a horizontal position. Regrowth, both by seed and vegetatively from the fallen trees, was vigorous. The main exception was beech. In some places the dominance of beech was thrown back for a generation while in others the new mix of trees was very similar to the old. The patches of windthrow varied greatly in size from half a hectare to a hundred, but in all they only amounted to a tenth of the ancient woodland in the path of the storm.

Storms as strong as that of 1987 come on average every two or three hundred years. Less violent ones which blow down fewer trees happen about once a decade and local gales more frequently. The effect is always patchy. Some places may not be touched by windthrow for a thousand years while others be hit twice in three years. Nevertheless the average time interval isn't so different from the lifespan of trees. The wildwood may have been an irregular mosaic of patches, each blown down at a different time and each at a different stage of regrowth.

The woods first emerge into the light of history during the Middle Ages. By this time the wildwood had been gone for millennia. The typical wood in the Lowland region was an isolated coppice surrounded by farmland. The coppices had a distinct three-layer structure. Topmost were the standard trees, which were almost always oaks and are grown for timber. Beneath them were the coppiced trees, harvested on a cycle of anything from five to twenty years. At ground level there were woodland wildflowers, thriving on the alternating light and shade conditions of the coppice cycle.

actively coppiced wood   neglected coppiced wood

These coppice woods were formed out of the wildwood simply by regular harvesting of the trees which were there. No trees were planted nor were less favoured species grubbed out. Many of these woods are still with us, though few of them are still coppiced.

Until quite recently these woods were thought of as oakwoods and people believed that the natural vegetation of the Lowland region was 'mixed oakwood'. Indeed the standard trees do make the most visual impact and they can make up a high proportion of the biomass in a wood. But they're a minority of the trees. In the days of the wildwood oak was even more of a minority. Since then it has been deliberately favoured by generations of woodsmen. Being a naturally durable tree it was the best species for building, which was always the main use for the standard trees. So oak seedlings were selected as standards while everything else was coppiced.

Though the rise of oak is easy to explain the decline of lime is not. This is the native small-leaved lime, not the common lime, which is a modern hybrid. Once the commonest tree in the wildwood of the Lowland region, lime is now only found in a few limited areas. Where it does occur it's tough and competitive and often forms a pure stand in the coppice layer. Why such a tree has disappeared over most of its former range is a mystery. Ash, maple and hazel are now the commonest coppice trees in most of the Lowland region, while hornbeam is specially common in woods around London. Like lime, hornbeam is both shade-tolerant and casts a heavy shade so it often forms a pure stand beneath the standard oaks.

Another puzzle is the origin of the woodland wildflowers, the violets, primroses, wood anemones and bluebells. They're so well adapted to the light-and-shade cycle of coppicing that it's hard to imagine what place they had under a radically different regime. The wildwood, with a blow-down once in a couple of hundred years, certainly had a completely different light regime from a coppice wood with its cycle of five to twenty years. The pollen record shows that at least some of these plants were here in wildwood times. Perhaps they managed to eke out a living under light-shading trees such as ash. But they can hardly have been as abundant in the wildwood as they were in the heyday of coppicing.

This is the glory of semi-natural ecosystems: humans don't just co-exist with wildlife, they actually benefit it. The destruction of the wildwood must have exterminated many species which couldn't adapt to the new conditions. But others have done better with humans as an active part of the ecosystem than they could ever have done before. It's not just the wildflowers but song birds, butterflies and many other animals which have thrived in the conditions created by coppicing, and suffered as it has fallen into disuse. Coppicing also leads to a greater diversity of trees. At every felling both competitive and uncompetitive kinds get cut down together and have an equal chance to regrow. Beech is the one broad-leaved tree which doesn't regrow reliably after being cut, so regular coppicing can stop it suppressing other trees.

These old coppices are the ancient woods. Some of them have been woodland continuously since the end of the last ice age, in which case they're known as primary woods. Others have sprung up by natural succession on former farmland, heath or moor. These are called ancient secondary woods. They've been woodland long enough to have been colonised by most of the plants and animals characteristic of primary woods. It can be hard to tell the difference between primary and secondary ancient woods but it's not a very important distinction because ecologically both are very rich. There is a big difference, though, between ancient and recent woodland, both in ecological diversity and in appearance. An ancient wood is defined as one which was there in 1600 in England and Wales and 1750 in Scotland. These are fairly arbitrary dates, chosen mainly because they're the earliest that you can get direct evidence for the existence of a wood from maps.

On the ground, one clue which suggests a wood is ancient is a wood bank round all or part of the perimeter. (See page 95.) Big, old coppice stools are another sign. But the real test is a suite of plants known as ancient woodland indicators. These are extreme stayers, slow colonisers which are rarely found in recent woods. They were first identified in a famous study of the woods of Lincolnshire by George Peterken and Meg Game in the 1970s. They took a large sample of woods whose ages were known from historical documents and surveyed the herbaceous plants in them. They found a big difference in the herbaceous communities of ancient and recent woods and compiled the first list of ancient woodland indicators. Since then other people have made similar studies in other parts of England, with slightly different results in each region. (As I'll shortly explain, the concept of ancient woodland indicators is less relevant in Scotland and Wales.) In the box overleaf, under 'All England' I've listed plants which occur in more than half the lists and are also easy to identify.

Most of the regional variations are too complex to list here but I have included a regional list for Eastern England, from Essex to Lincolnshire. The east stands out from other regions because of its dry climate. Woodland herbs are adapted to the moist, shady conditions of woods and need a certain level of humidity to survive. In places with wetter climates some characteristic woodland herbs can live outside woods, in hedgerows or even in the open, and this makes it easier for them to migrate to new woods. The plants in the Eastern England list are ones which can't survive outside of woods in that region. Oxlip is confined to the Eastern England list because it doesn't grow in other parts of England at all. The false oxlip, a hybrid between the primrose and cowslip, grows all over the country but it's usually easy to tell the one from the other because the true oxlip carpets the ground while the hybrid is solitary.

Most of the plants in the Eastern England list are typical of ancient woodland everywhere. If you see them in a wood in another part of the country you can take it as a clue, if not as firm evidence, that the wood is ancient. In fact some of them are included in other regional lists. Ramsons, for example, is listed for the north and south-west of England and almost occurs on enough lists to make it onto the All England one.

A single ancient woodland indicator species on its own is not conclusive. Any one of them can occasionally be found in a recent wood but two or more would be unlikely. As with soil indicator plants, what you're really looking for is a community.

ANCIENT WOODLAND INDICATOR PLANTS [*]	
All England	**Eastern England**
Trees wild service small-leaved lime *Wildflowers* woodruff wood anemone common cow wheat moschatel or town hall clock wood spurge herb paris	*All the above plus:* bluebell dogs mercury primrose ramsons or wild garlic oxlip

[*] This list is based on the comprehensive survey of all the lists by Oliver Rackham in his *Woodlands* (2006) in the New Naturalist series published by Harper Collins.

While some parts of the wildwood were being turned into coppice woods other parts were being transformed in a different way. Rather than being surrounded by a big bank to keep animals out they were deliberately grazed. In many places this resulted in grassland or moor but where some trees were retained it resulted in wood pasture. Wood pasture varies enormously, from the widely-spaced oaks of a deer park like Moccas to the dense woodlands of the New Forest. But one thing all wood pastures have in common is that coppice is impossible because the animals would eat the regrowth, so the trees are either maidens or pollards. (See photo 29.)

In terms of biodiversity wood pasture is complementary to coppice. While coppice has retained the diversity of trees and herbaceous plants from the wildwood, wood pasture has lost both to grazing. Woodland herbs are unadapted to grazing and trampling and were soon replaced by grass or bracken. Over the centuries the more palatable trees have been browsed out. The ones which are left are predominantly unpalatable ones, like oak and beech, or ones which can protect themselves, like holly with its thorns or the poisonous yew. On the other hand wood pasture does have one thing which coppice doesn't and that's old trees. Constant coppicing and felling of standards meant that trees could never grow old in a coppice wood, whereas primary wood pasture has a continuous succession of old trees going right back to the wildwood. Many of them have had their lives prolonged by pollarding. Those beetles and other creatures which depend on dead and decaying wood and which have survived the demise of the wildwood have mainly done so in wood pasture. Another group which is dependent on a continuity of old trees is the lichens, which grow very slowly and are even slower to colonise a new tree. Wood pasture is the stronghold of lichens, especially rare ones.

Very little wood pasture is left now. The modern trend towards monoculture has seen it converted either to plantation or to farmland. Here and there a very old pollard or a small scatter of them in farmland shows where once there was a wood-pasture common. The one place in Britain where a great deal of it does survive is in the New Forest. My notebook records a visit to an ancient wood in the Forest.

Ridley Wood, Winter Solstice 1999

The wood has a feel of wildwood about it, especially in the amount of dead wood lying around. There's a contrast between the parts which have a closed canopy and the parts where there's windthrow and regeneration of young trees. The former feel quiet, the latter full of event and movement. I saw four or five treecreepers together, working up neighbouring trees. I've never seen more than one at a time before. Maybe it's because they flock in winter and previously I've only seen them in summertime.

On the other hand, two things make it feel quite unlike a wildwood, the overgrazing and the low number of tree species. Within the wood I saw only beech, oak and holly. There's a severe browse line and virtually no ground vegetation. Even bramble is severely browsed. There are tiny holly seedlings, but nothing above ankle height. The poor ponies have nothing much to eat. There's no grass longer than a billiard table, either in the wood or on the heath outside. They were eating beech and oak leaves in the wood – there can't be much food value there – and browsing gorse and a rare small holly outside.

In the Upland region, as shown on the map on page 180, there's less of a hard-and-fast distinction between wood pasture and coppice. Many upland woods have had a patchy history which includes both coppicing and grazing, or at least being used as a winter refuge for cattle and sheep. There's also much less distinction between ancient and recent woodland than there is in the lowlands.

The typical ancient woodland of East Anglia is an island of semi-natural vegetation in a sea of intensive cultivation. When you step outside the wood you go into another country, with a different climate, a transformed soil and hardly an inch of land which hasn't been sown or planted. The further west and north you go the more this contrast softens. In the uplands the woods merge gently into the surrounding landscape, much of which is semi-natural vegetation. Often there's no hedge or fence between the wood and the surrounding pasture or moor. The damp climate makes it easy for many woodland plants to survive outside of woods, so they're there in situ if the land becomes wooded again. There are also alternative habitats which can give shade much like that of a wood. Bracken, for example, can simulate the light-and-shade regime of deciduous woodland: it dies down in the autumn and comes up again in early summer, just when the trees would be leafing. As long as it's not too dense it can make a permanent home for bluebells, violets and other woodland flowers. Trees can often hang on in the face of grazing on steep slopes and in rocky places, like the ravines at Kinlochleven. (See page 150.)

The oak-hazel woods which dominated the uplands in prehistory are now mainly oak. Hazel, which favours the more fertile soils, has mostly been cleared for farmland. The oakwoods are probably much purer than they were. In many upland woods browsing has reduced the more palatable trees while in others oak was favoured for tan bark production in the nineteenth century. Of course there are mixed woods in the uplands and regional variations too, such as the ashwoods on the limestone of the Pennines.

I live in the Lowland region, but from the window of my house I can see the curving profile of the Quantock Hills, the first rampart of the West Country and thus of the Upland region. The highest bump on this profile is a hill called Dowsborough.

Dowsborough, 9th July 2000

The strange thing is that the map shows considerable areas as open moor which are in fact wooded. The woods may have increased since the survey was done – the map is dated 1979 – but some of the stands which are shown as open county are old coppice.

The trees are very stunted, some four to five metres tall, with no large ones except in a steep valley. They're almost all oak, with very occasional birch and rowan inside the woods and a fringe of birch where the wood is advancing. The advance is controlled where small patches of moor have been burned here and there.

The ground vegetation inside the woods is mainly whortleberries*. I found a few ripe berries, though the canopy is dense. There's also cow wheat, a few very small brambles and occasional bracken. Wherever there's a break in the canopy, such as under a dead tree, the bracken thickens up. There's some rather poor heather growing in the woods, possibly in spots where there was formerly more light.

As soon as you move out into the open the relative proportions of the plants change. Bracken and heather greatly increase and whortleberry decreases in proportion. Cow wheat is replaced by tormentil. I didn't see brambles in the open but I did see gorse.

* Somerset word for bilberries or blaeberries.

In the Highland region the concept of an ancient woodland more or less melts away, at least in the sense of a definite site with fixed boundaries. Pine, birch and oak all migrate around the landscape. Although a particular wood may stay within the same area for centuries, it will occupy different parts of that area at different times. You can sometimes see this process in action, where a pinewood is dying away on one edge and spreading onto new ground on the other – if the deer permit it. The spread often happens in two stages. First a few scattered trees get established in more favoured places, such as where the peat is worn away by erosion, exposing the richer mineral soil, or in a rocky place that discourages browsing. As these mature they flood the area with seed and a second generation grows up. The young trees are close together and drawn up while the older generation have the broad, spreading crowns of trees which grew in the open.

The acid litter of pine needles combines with the underlying rocks and the climate to produce classic podsol soils. In this the pine woods are more like heather moorland than birch woods, whose sweet leaf litter tends to turn a podsol into a brown earth. The ground vegetation of Highland woods is rarely very different from the surrounding moorland, perhaps because they have a long history of grazing. Neither the cattle and goats of the times before the Clearances nor the more recent sheep and deer have ever been fenced out of the woods. Now that the deer population has grown higher than ever, regeneration of the woods is impossible in most parts of the Highlands without fencing. This is extremely expensive but the charity Trees for Life have put up miles of deer fences in Glen Affric and the surrounding area. Where once there was moorland dotted with the odd surviving tree now there are wide expanses of young woodland.

WOODLAND TYPES		
Ancient Semi-Natural	**Recent Semi-Natural**	**Plantation**
Primary has always been woodland *Secondary* became woodland before 1600/1750 these may be either: *Coppice* trees coppiced on regular rotation *Wood Pasture* maiden or pollard trees with grazing animals *Upland and Highland woods* less distinction between coppice and wood pasture than in the Lowland Region	*In Lowland Region* has self-seeded since 1600/1750 *Upland and Highland woods* less distinction between ancient and recent woods than in the Lowland region	*Timber* single stemmed trees, usually grown on the clear-fell system *Coppice* rare, mostly chestnut in SE England *Amenity or wildlife* Usually mixed native species

Recent Woods and Plantations

A wood which has arisen by natural succession since the end of the Middle Ages is different from an ancient secondary wood only in degree. A brand new wood will be composed entirely of pioneer species and the diversity of both plants and animals will be low. The stayers, the species which are typical of ancient woodland, will colonise it slowly and gradually diversity will increase. But it's not just a matter of time. Being adjacent to an ancient woodland speeds the process up considerably.

At Hayley Wood in Cambridgeshire Oliver Rackham has observed mercury, bluebell and oxlip advance from the ancient wood into a recent extension at a rate of a metre a year. Anemone and sanicle have spread more slowly. This is just one example and shouldn't be used to calculate the age of a new extension in another place but it gives an idea of what's possible. By contrast, a Lincolnshire plantation of mainly native trees, three and a half miles from the nearest ancient wood, still "had the air of a grassland that had acquired trees" after a hundred and twenty years. A few woodland wildflowers had colonised but not enough to change the character of the place.

The herbaceous plants of recent woods are more typical of hedgerows or even grassland than of ancient woods: ivy, lords and ladies, goosegrass, nettles, Jack-by-the-hedge, cow parsley, hogweed and creeping buttercup. They're all to a greater or lesser degree lovers of nutrient-rich soils. The soil of a recent wood is essentially a farmland soil, enriched by years of manuring. How different these plants are from the carpets of wildflowers in an ancient wood! As for the trees, much depends on which species are growing near enough to act as seed parents. Very often a new wood starts off as a pure stand of birch or some other pioneer. Sycamore is common in recent woods in many parts of the country. It's a tree which has great powers both as a pioneer and as a stayer. Its seeds are borne on the wind by their little 'helicopters' and the seedlings are quite resistant to browsing. It's also heavy-shading and tolerant of shade, so it can persist into the second generation more effectively than light-demanding pioneers like hawthorn, birch and pussy willow.

Recent woods have no old stools in them and usually no sign of coppicing at all. Sometimes they contain the occasional veteran tree, maiden or pollard. These will have the broad spreading branches which show that they were there before the wood grew up. A recent wood never has a big woodbank round it, though there may be the ditch and bank of an ordinary hedgerow along one or more boundaries. This isn't diagnostic, though, because some ancient woods haven't got banks either.

All these features combine to give new woods a raw feel compared to the mellowness of an ancient wood. Personally I don't find them attractive places. Recent extensions to ancient woods are less raw than new woods which stand alone, and often more interesting. Sometimes it's hard to be sure exactly what you're looking at.

Rew Copse, Isle of Wight, 29th March 2003

This is an ash-hazel wood on steep chalk. In the southern part of the wood there are two distinct stands, separated by a small bank with the remains of a wire fence running along it. On the uphill side there's hawthorn and elder along with the dominant ash and hazel, and the ground cover is mainly ivy with some dogs mercury and bluebells. In the rest of the wood, below the little bank, there are a few spindle trees and maples among the ash and hazel but no hawthorn or elder. Here the ground vegetation is bluebells, dogs mercury, wood anemone and moschatel, with ivy only in patches. This must be an ancient wood but the upper part shows every sign of being recent. Did the upper part go through a period of grazing?

Well that's a possible explanation. The main part of the wood is certainly ancient. The presence of anemone and moschatel indicate that. The upper part could have lost its woodland ground layer during a period of grazing that didn't last long enough to destroy it as a woodland altogether. When grazing ended it was recolonised by ivy, perhaps the most distinctive ground-layer plant of recent woods. Mercury and bluebell, both more mobile than anemone and moschatel, are slowly following.

However, if that had been the story you'd expect the mix of tree species to be the same in both parts of the wood. The upper part lacks maple and spindle which, while not exactly ancient woodland indicators, are characteristic of older rather than newer woods. Instead it has the pioneers hawthorn and elder. So the upper part could be a recent extension to the wood rather than a part which has been modified by grazing. The lack of any discernable woodbank or lynchett on the upper boundary of the wood and the presence of a little bank between the two stands supports the idea that the upper part of the wood is new. But the presence of coppice stools and a few big old ash trees in the upper part supports the opposite story. An investigation of the old maps could settle the matter.

Mount Sylva, 22nd March 2005

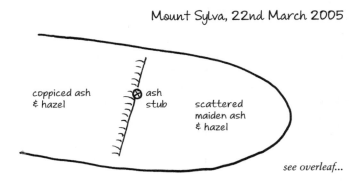

see overleaf...

189

> Both parts of the wood are ash and hazel with a thick ground cover of ramsons all over. Neither the bank nor the difference between the structure of the ash in the two parts were immediately obvious.

Mount Sylva in Dorset is another ancient wood with what looks like a recent extension, but it's almost a mirror image of Rew Copse. Whereas in Rew Copse the structure of the two parts of the wood is the same but the plants are different, in Mount Sylva the difference is all in the structure of the trees. The species of trees and ground layer plants are the same on both sides of the bank but on one side the trees are coppiced and on the other side they aren't. This showed that the part of the wood on the right hand side is recent. Although ramsons grows abundantly in both stands and is an ancient woodland indicator in this part of England, it just goes to show that you can't come to any firm conclusions on the basis of one plant.

Self-sown woodland like this, whether it's adjacent to an ancient woodland or not, is still semi-natural. No-one has planted it. A plantation is something else altogether. Even though it may look like a wood it's really a crop, like wheat or barley. The main difference between plantations and arable crops is that plantations take many years to grow and in that time many things can happen.

If you look at a road atlas of Britain, one which colours woods and plantations in green, you'll see that by far the biggest areas of green are in the Upland and Highland regions. These are mostly large-scale monocultures of exotic conifers. In sheer size they dwarf all other tree-covered land in the country, both lowland plantation and semi-natural woodland. They're grown on a short rotation of just a few decades, often not thinned, then clear-felled and replanted with the same species. Much of the land where they're grown is too wet for good tree growth. To dry it out it's 'ploughed'. This is really more of a ditching operation than ploughing. A huge machine draws out a ditch for each row of trees and the trees are planted on the ridge of spoil left alongside it. If the land later reverts to moorland these furrows and ridges leave an unmistakable record of its former life as a plantation. Many of these soils are peaty. As the peat dries out it starts to oxidise, releasing far more carbon dioxide into the atmosphere than the trees will ever remove from it as they grow.

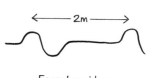

Forestry ridges

Broadleaved trees and the more demanding conifer species can't produce an economic crop on these difficult sites so they're grown at lower altitudes where microclimate and soil are more forgiving. These are higher-value crops and they receive more care and attention. The ground is planted with thousands of trees per hectare and these are progressively thinned over the years till perhaps only ten percent of them are left as the final crop. Growing trees this way improves the quality of the timber. Initially the close spacing draws the trees up into good timber form and suppresses lower branches. Later the increasing space allows the trees to expand into well-proportioned individuals with a good girth.

Thinning also allows the forester to chose the best trees for the final crop. A plantation which is neglected will develop thin, spindly trees which are not worth much as timber and are prone to windthrow.

Although most plantations are monocultures there are mixed plantings too. Mixtures require more careful management because one species will always be that bit more vigorous than another and threaten to suppress it. But they also mean that more of the available niches are filled and this can make the plantation more productive. Perhaps the commonest kind of mixture is that of a nurse crop along with the main crop trees. The nurse trees are of a tough species and they help the main crop trees by providing shelter when they're young. If the nurse crop is coniferous and the main crop is broadleaved the nurse also helps by drawing up the broadleaves with its dense shade. This combination has a commercial advantage too because all the nurse trees come out as thinnings and conifers are more valuable than broadleaves when they're young. They can be used for fencing stakes and so on whereas young broadleaves can only go for pulp or firewood. When it comes to the mature crop the tables are turned and broadleaves are more valuable. At least they are if they're well grown, and getting off to a good start with a nurse crop contributes to that.

A typical layout is to alternate three rows of the main crop with three rows of the nurse. Early thinning can then simply be a matter of removing whole rows of conifers. In hilly country these alternate strips can stand out clearly on a hillside. Some people disparagingly call this 'pyjama forestry' and complain that it doesn't look natural. Personally I like it. It shows that someone is doing something a little more thoughtful than simply planting monocultures. I agree it doesn't look natural – but neither do hedges, which are held to be one of the glories of the countryside.

In other mixed crops two or more species are grown to maturity side by side. This kind of mixture is often planted in small groups rather than strips, with each group just the right size to be thinned down to a single tree by the time the plantation is mature. Here again the niches of the component species can interact to their mutual benefit. Broadleaf trees can enhance the growth of conifers by improving the soil, while conifers give year-round shelter to the broadleaves and provide winter roosts for birds which feed on pests. There are similar benefits in all-conifer or all-broadleaf mixes but they may not be so marked.

Virtually all plantations are harvested by clear-felling. That's to say the whole plantation or compartment is felled together and then replanted, giving large blocks of even-aged trees. This is very disruptive to wildlife and to the visual appearance of the landscape, and exposes the soil to erosion. Just occasionally you may see a completely different kind of plantation in which there are trees of various ages growing side by side. This is known as continuous cover forestry because the trees are felled a few at a time so the land is never completely denuded of trees. It's an alternative approach which is beginning to gain ground among more progressive foresters. The most usual form it takes is group selection, in which trees are felled in groups of half a hectare or so. Regrowth is often by natural regeneration rather than planting. The plantation takes on an uneven structure, with different groups at varying stages of growth. There are never large areas of bare soil exposed at one time and wild plants and animals are more able to cope with the trauma of felling when there's untouched habitat nearby. This pattern of regeneration in groups may in fact be the closest analogy we can make to the original structure of the wildwood. The groups are not totally dissimilar in size and time interval to the windthrow patches which may have been the pattern of the wildwood.

Continuous cover is a commercial form of forestry. Although it lacks the economies of scale of the clear-fell system, its advocates say it's actually more profitable because small stands of trees can be treated individually to get the best from them. It's rather like bespoke tailoring compared to mass-produced clothes. But there are also plantations which have no commercial purpose and have been made entirely for amenity or wildlife purposes. They're usually very easy to recognise.

Slapton Ley, 6th May 1993

We passed by a typical amenity/wildlife planting of trees, perhaps ten or fifteen years old. It was oak, ash and wild cherry, regularly spaced and planted so far apart that their branches aren't yet touching. There's a high proportion of cherry and the occasional shrub of impeccable nativeness, such as dogwood and spindle. The whole effect is only slightly more natural than a closely mown lawn.

A few hundred yards further on we passed some natural regeneration of about the same age. It was a hundred percent sycamore, with the trees very close together, drawn up and already killing each other with hot competition.

Perhaps I shouldn't have been quite so sarcastic about the plantation. I've planted something quite similar myself in the past. There's always the temptation with wildlife planting to make it unrealistically diverse and include more of the rarer species than you'd find in nature. Why not? If we wanted a semi-natural wood all we'd do is leave the land alone and see what happened. A plantation is always artificial.

Almost all plantations, whether they're for timber production or amenity, are made up of single-stemmed trees. Coppiced plantations are very rare. One exception is the pure chestnut coppices of Kent and Sussex. Although some of these coppices are probably semi-natural many were planted in modern times. It may be hard to tell the difference between planted and semi-natural ones because chestnut grows so vigorously as a coppice stool that it suppresses other trees, so a semi-natural stand which starts out with a preponderance of chestnut may become pure chestnut in time. Chestnut is an excellent fencing wood as it's durable without the use of preservatives. A major output is those temporary fences made of cleft stakes bound together with wire which are sometimes used for crowd control, and many chestnut coppices are still in commercial production.

Some pure hazel coppices are still worked for hurdles and thatching spars, mainly in Hampshire and Dorset, but most of them are believed to be semi-natural. Nevertheless hazel has been planted for coppice production, often with oak standards. These plantations date from the eighteenth to early nineteenth centuries, when the age of coppice wasn't yet over but the age of plantations was already under way. I chanced upon one in Chase Wood, just south of Ross-on-Wye in Herefordshire. It must have been a very late example of the genre because neither coppice nor standards looked as though they'd ever been harvested. The plantation felt surreal, almost ghostly. The oaks and hazels stood in solemn, uniform rows, like people waiting patiently for something that's never going to happen.

Plantations have a bad press. Many of them have replaced diverse ancient woods with dark blankets of conifers. The irony of it is that many of these plantations now stand derelict and unharvested. With the fall of the former Soviet Union vast areas of forest became available for exploitation and the price of timber plummeted. On large-scale upland plantations the economies of scale have kept the industry alive. But in scattered compartments on the sites of former woods, many with poor access, the cost of harvesting is often more than the value of the timber.

These plantations have a sad feel to them. They usually haven't been thinned, which means a high-value final crop is now unlikely or impossible, and high value is the only way to make small-scale forestry economic. On some sites coniferisation has been total and the only clue to their past as woodland may be an irregular outline and the presence of a woodbank. Sometimes there's a 'dishonesty belt', a screen of the original trees left on the edge to give the casual passer-by the impression that nothing has changed. On other sites some of the woodland trees survive inside the wood. How many survive varies enormously, from the occasional one or two to a resurgence of the original vegetation which is clearly winning the competition with the conifers. Which gets the upper hand on a particular site depends on various factors. One is the choice of the conifer species. The better it's matched to the soil and climate of the site the better the conifers will do. Another is the kind of trees in the original wood. The usual procedure was to kill off the trees with Agent Orange, the herbicide once used by the Americans to defoliate the forests of Vietnam. Oak, beech and birch succumbed to it more completely than ash, hazel and maple. But lime, the small-leaved lime that was once the dominant tree in the lowland wildwood, has proved almost impossible to kill. Many limewoods have survived more or less intact, including the ground vegetation, and some are now nature reserves.

CASE STUDY

TOP WOOD

Top Wood, Ragmans Lane Farm

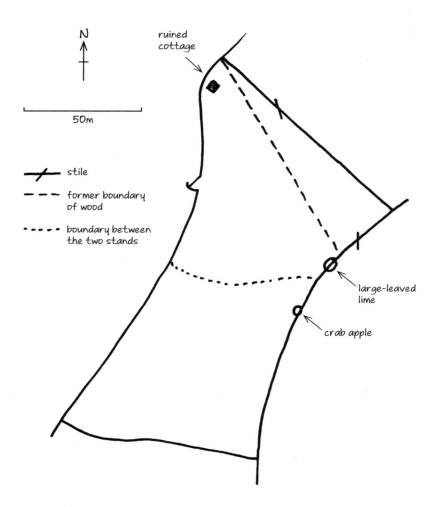

The Top Wood at Ragmans Lane Farm is located on a steep, stony slope where the hard dolomite rock outcrops. If you go there in springtime its ancient origin is immediately obvious from the abundance of anemones, violets, ramsons, bluebells and dogs mercury which cover the ground. There's also quite a bit of goosegrass, perhaps encouraged by the extra fertility left by the sheep which often wander in through the gappy fence of the neighbouring farm. But soon you notice a scattering of larch trees among the native broadleaves and you know that some planting has gone on. Larch is the only deciduous conifer, an easy tree to recognise in winter and spring, and one often used for forestry on relatively fertile sites.

The northern part of the wood, where all the access points are, is somewhat chaotic. There are trees of all different shapes and sizes, some with signs of past coppicing, some without. On the whole they're widely spaced and branchy, so they have little or no value for timber. In one place there's a big gap in the canopy with a dense understorey below it, partly of elder and partly of bramble. The trees are mainly ash, hazel, hawthorn and yew. By the ruins of the old cottage there's a clone of suckering damson and a single domestic apple. After making your way through this jumble of trees and shrubs you come out into the southern stand. Here the trees are closely spaced and drawn up, with a closed canopy and no shrub layer. Ash is the commonest species, followed by wych elm, and there's a scattering of wild cherry. The larches are being suppressed by the other trees and most of them are dead. Here and there an old oak stump, rotted down to the heartwood, gives a clue to a previous generation of trees. (See page 169.)

This stand is clearly a neglected plantation. I suspect the planted trees were a mix of ash and larch. Some of the ash looks older than the larch and may have already been growing when the planting was done. The wych elm and some of the younger ash are probably self-seeded while the cherry may have survived from the former woodland as root suckers. The stand has never been thinned. Perhaps the plan was to take the larch out first and leave the ash to grow on. Much the same effect is now being achieved by competition between the trees rather than the woodman's saw, and the produce is being consumed by wood-boring insects rather than by the timber trade. There are a few ash trees which are well enough grown to make good timber, but only a few. With the lack of thinning most of them are excessively drawn up and too spindly to ever make a decent sawlog. There's no sign of seedlings coming up anywhere and only the occasional wisp of bramble. Sheep certainly have access to the wood and, though there are no explicit signs of them, the lack of any natural regeneration even under the prolific ash suggests that rabbits and deer graze here as well. This means that coppicing wouldn't be possible here without some serious investment in fencing, as the regrowth would be browsed mercilessly.

Nevertheless, the wood which stood here before the present generation of trees was probably coppice with oak standards. The oak stumps are about the right distance apart for standards. The lack of any remains of the coppice stools is not surprising as oak is the only species that lasts long as a stump. The jumbled northern stand may still look something like the southern stand did before replanting. Although there are a few larches in it, it wasn't cleared for replanting the way the southern stand was. It's not as though the trees which were left standing at planting time were useful for timber as they include coppiced ash and non-timber species, such as hazel, hawthorn and holly. It seems that the clearing of the wood was a job which was started but never finished, and here in the northern part of the wood the larch were fitted into whatever gaps there were.

The boundary of the wood has changed in two places in recent times. The large-scale ordnance survey map still shows the western boundary a bit further in than it is on the ground. Now it has expanded to include the ruined cottage and its garden. More recently the northern boundary shifted to take in a bit of grassland with scattered trees and this area is now beginning to acquire an understorey of bramble and young ash. A few years ago you could still see the remains of a rabbit fence along the line of the former boundary, clearly a relic of the tree planting episode. Along one part of the line you can see a row of hollies, close set and with signs of having been laid as a hedge. (See page 294.)

Wild service leaf

There's no wood bank on any part of the perimeter, which is not unusual for ancient woods in the locality, but some interesting trees do survive on the eastern boundary. Halfway along it is a crab apple and near the northern end is the wood's pride and joy, a large-leaved lime stub. This is a very rare tree and an ancient woodland indicator in this part of the country. One winter's day I was showing this tree to a group of students. As I scanned the ground for a sample leaf to show them, my eye lit on the unmistakable leaf of the wild service tree. Almost as rare as the large-leaved lime, this is another ancient woodland indicator. I've looked for the tree itself, but I haven't found it yet.

How Woods Work

There's nowhere you can see the workings of an ecosystem more clearly than in a wood, both in terms of structure and of cycles in time. The three layers of vegetation, trees, shrubs and herbaceous plants, not only use different parts of the available space but also have different annual cycles. These are specially noticeable in spring, when first the herbs come into leaf, then the shrubs and then the trees. There are also the cycles which operate on longer timescales than a single year. Coppice, wood pasture, recent self-sown woods and plantations all have their distinct rotational cycles and many plants and animals find their niches in this constantly moving kaleidoscope.

Trees, Shrubs and Herbs

Go to an ancient wood in late January. Though the trees and shrubs are still deep in their mid-winter slumbers things are already beginning to happen at ground level. Some of the wildflowers which spend their dormant season below the ground are already beginning to peep. Here a bluebell spears the soil, there a dogs mercury leaf unfolds. Come back in February and you'll see many herbaceous plants in leaf and the first celandines may already be flowering. In March and April the shrubs come into leaf, but the trees don't green up till late April at the earliest and many of them not till May. By the time the canopy closes most of the wildflowers have bloomed and some of those which were up so early in the year start to die down again. Ramsons is gone by the end of June. The deep shade makes high summer a quiet and uneventful time on the woodland floor. (See photo 30.)

This annual sequence is the result of a trade-off between the two niche factors, light and temperature. The plants of the ground layer leaf early in order to have at least some time in direct sunlight. It means they have to invest some of their energy in making themselves frost-hardy, but plants which live below deciduous trees have little option. The trees can save themselves this expense by leafing late and it's because they do this that the herbs have their time in the sun. The shrubs, being intermediate in height between herbs and trees, occupy an intermediate niche in this respect. How long has it taken for this beautiful piece of dovetailing to evolve? To me it epitomises what ecosystems are all about. An ecosystem isn't just a collection of plants and animals but an integrated community in which each species fits in with the others around it like the parts of an intricate machine.

As well as differences between the three main groups of plants there is, of course, plenty of variation within the groups themselves. In the case of the trees the most obvious distinction is between broadleaves and conifers. In fact conifers don't so much occupy different niches from broadleaves as come from different environments. The broad leaf is designed to intercept the maximum of solar energy. It's the ideal adaptation to a mild, well-watered climate like ours. The narrow but thick leaves of conifers are adapted to harsher conditions, to the drought of the Mediterranean or the cold of the far north, where frozen soil often creates a virtual drought. Conifer needles lose less water than broad leaves but at the cost of being less efficient as solar collectors. The yew, one of only three native British conifers, is most competitive in very dry situations, such as the exposed southern bank of Burrington Combe. (See page 74.)

Why, then, do conifers totally dominate British forestry? If broadleaves are better at converting sunlight into biomass surely they must grow faster than conifers? The answer is that they do but it's the wrong kind of biomass. Firstly, it's denser than conifer wood, so you get less volume of timber for the same weight of tree. Secondly, a higher proportion of the biomass of a broadleaf is in the branches, which aren't marketable. This means that conifers produce a greater volume of marketable timber than broadleaves. The toughness of many conifer species also means that they do better than broadleaves on the harsh upland sites where most forestry goes on.

The greater biological productivity of broadleaves explains why the many species of conifer which have been introduced to this country for forestry have on the whole failed to become naturalised. The main exception is Scots pine, which is not a native tree south of the Highland Line but has become a pioneer on the light sandy soils of lowland heaths. On less drought-prone soils it can't compete with native trees and if it's grown in a plantation it must be kept well weeded to stop it being taken over by native, self-seeded trees. "I remember a landowner telling me his father had planted pine on a field and he'd just harvested a good crop of ash from it," says Oliver Rackham.

The key to successful forestry is to match the tree species as closely as possible to the soil and climatic conditions. In semi-natural woods this has of course been done by nature but the match between trees and soil isn't always very close because chance can also play a part in the distribution of trees. One way this may happen is through the phenomenon of mast years. Mast is tree seed and many trees don't produce the same amount of it each year. They take it easy for a number of years and then have a bumper crop, known as a mast year. Oak typically has mast years once in two to seven years, beech once in four to fifteen. If a particular species is having a mast year just when space for regeneration becomes available it can become the dominant tree in the next generation even if it's not the tree most perfectly fitted to the site.

When looking at the composition of an ancient coppice wood the first thing to do is to forget about the standard trees. In most woods they'll all be oak and they're there because they've been deliberately encouraged by past generations of woodsmen. The coppice trees are the ones which differ from one wood to another and which give a wood its character. In the Lowland region oak was only coppiced where there was more than enough of it to provide the standards and it only grows in such abundance on the least fertile, acid, sandy soils. Birch, hornbeam and chestnut are the trees of relatively base-poor soils, while maple, ash and elms favour the richer soils. Alder is confined to wet soils but always where the water is moving, while aspen will grow where the water is stagnant. Beech is found on any well-drained soil, from light, acid sands to thin soils over chalk, though it does occasionally grow on clay. Hazel will grow on any soil except the least fertile.

In woods on base-poor soils the landform can have a strong effect on the distribution of trees. You can often see this in woods in the Upland and Highland regions, as this sketch from my notebook illustrates.

Kinlochleven, 7th April 1995

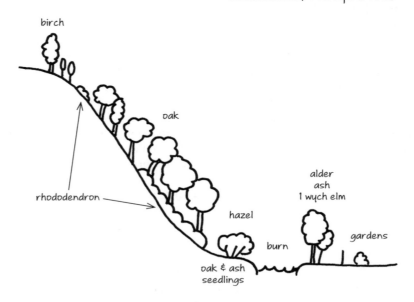

In this region of high rainfall and old, base-poor rocks the soil is easily leached. There's a strong contrast between the leached, acid soil of the slope and the soil on the flat ground beside the burn, which catches some of the leached minerals. The more demanding trees are confined to the burnside while oak rules on the slope. Rhododendron has seeded from the gardens on the other side of the burn and seems to be spreading up the slope. It's a strong indicator of acid soil and has leapfrogged the base-rich soil at the bottom. The change from oak to birch at the top of the slope may have more to do with succession than soil fertility or microclimate. In other words the wood may be extending.

Herbaceous plants often indicate the woodland soil more accurately than the trees do. Why this should be so isn't fully understood but their smaller size and shorter lifespans would tend to even out the effects of chance events. This means there's often a poor match between the distribution patterns of the trees and the herbaceous plants beneath them. The least fertile soils are often indicated by bracken, bilberry and cow-wheat, moderately infertile by honeysuckle and bluebell, moderately fertile by dogs mercury and primrose, and the most fertile by nettles.

The two plants at the opposite ends of this list, bracken and nettles, can some-times grow so vigorously that they prevent the regeneration of trees. They're both tall, densely-shading plants and on the most extreme soils they can outcompete all comers. In the case of bracken this means sand. With its deep rooting system it thrives on a very sandy soil where other plants, including tree seedlings, are disadvantaged by the lack of water in summer time. Any seedlings which do survive the summer are unlikely to be big enough to survive being swamped by the bracken as it dies down in autumn. Where a patch of very sandy soil occurs in a wood a bracken glade can form. This is an area which is devoid of trees, except perhaps the occasional oak, simply because the bracken is so competitive.

Nettle glades, by contrast, can form in flat-bottomed gullies on heavy soils. This is the sort of place where nutrients collect naturally but many nettle glades are enhanced by runoff from fertiliser-rich fields or dungy farmyards. A tall, dense stand of nettles, with their broad, horizontal leaves, can shade out almost any plant which tries to grow within it. At most there may be the occasional hogweed or elder poking its head above the nettles.

In many woods the only difference in soil from one part of the wood to another is its moisture content. The sketch below is from a wood in the Lowland region, where leaching is not such a major factor as it is in the Highlands. The soil is much the same on both the steep bank and the flat bottom except that the bank is well-drained and the flat ground is wet, fed by a line of springs at the base of the slope.

Park Wood, Chaffcombe, 10th May 2003

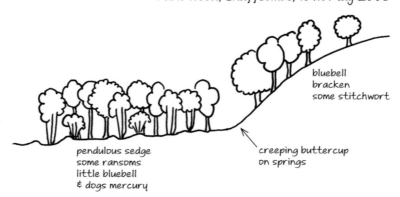

bluebell
bracken
some stitchwort

pendulous sedge
some ransoms
little bluebell
& dogs mercury

creeping buttercup
on springs

There's a dramatic contrast between the electric blue of the bank and the deep green, enlivened by the odd patch of white ramsons, of the flatter ground below. The difference, so visible at this time of year, appears to be due solely to moisture content, which in turn is due to the landform. I did the finger test on soil from both parts of the wood but wasn't able to detect any difference in the sand, silt and clay content.

Bluebells, bracken and greater stitchwort, with its flowers like little white stars, are all dry-soil plants. Pendulous sedge, that big grass-like plant with flowerheads like long hazel catkins, likes it wet. Ramsons, whose globes of white flowers light up the woods in May, has a complex relationship with soil water but in most places it favours damper soils. In the wet area there are a few plants of bluebell and dogs mercury, which also likes dry soil. This is because the ground is uneven and there are some dry patches. By contrast, on the steep bank there are no wet spots and no moisture-loving plants.

The distribution of herbaceous plants can also be influenced by the distribution of the trees. Some trees cast a deeper shade than others and the herbs vary in how much shade they tolerate. I know a place where a single common lime, which is a heavy-shading tree, stands among an open canopy mainly of the light-shading ash. Under the lime is a patch of pure ramsons, which is particularly shade-tolerant. The outline of the patch matches the canopy of the lime pretty closely. Under the surrounding ash is a mix of bluebells, wood anemone and dogs mercury. You can get the reverse effect under canopy gaps. In an oak plantation with a mixed ground layer of dogs mercury and bramble the dogs mercury dominates the majority of the wood where the canopy's closed but the brambles take over under the occasional gap.

Leaf fall can have an effect too. Beechwoods are famous for having little or no ground layer but this isn't only because they cast a heavy shade. It's also because beech leaves are the slowest of all to decompose and they build up into a layer so deep that the little roots of germinating seeds can't get to the mineral soil below, which they must do in order to grow. Ash is at the opposite extreme. It casts a light shade and its leaves are the first to be taken by earthworms. In a plantation with compartments of both beech and ash there will be a sharp contrast in the ground cover.

Mosses tend to do better under conifers than broadleaves, because they're sensitive to the smothering effect of the autumn leaf fall. Conifers do lose their leaves but it's a constant trickle through the year rather than a sudden shower. The deep shade of the conifers helps mosses too because it reduces competition from more light-demanding ground-layer plants. Mosses also have the advantage of not needing soil to grow in, so in a deciduous wood they're often found on stones and logs, where they're free from competition and usually above the level of leaf litter.

The shrub layer can be an elusive element of the woodland community. Under a closed canopy it's often missing altogether because there's not enough light. Where there is one it's mainly defined simply by being lower than the tree layer. In an actively coppiced wood the coppice stools can be regarded as the shrub layer, in comparison with the standard trees, though if coppicing stops they'll eventually join the standards in the canopy. Hazel, whether coppiced or not, can make a distinct understorey to taller trees, although there are pure hazel woods where it makes the canopy. Saplings of the canopy species can be part of the shrub layer and in some woods they're all of it. Then there are the undershrubs like bramble and bilberry, which may be closer in size to the larger herbaceous plants.

The shrub layer can also be a passing phase. In a plantation there may be shrubs in the early stage when the crop trees are still small. They often need to be cut back to stop them swamping the crop but once the canopy closes the shade becomes too much for them, especially in a conifer plantation, and they disappear. They may come back towards the end of the rotation when the process of thinning opens out the canopy and lets in more light. In a recently self-seeded wood there may be a shrub layer which consists of the remains of the scrub stage of succession. But as the trees start to cast more shade it gradually thins out and only survives on the edge of the wood. Here there's light from the side and shrubs can persist indefinitely. There may be some shrubs in the interior of a wood but they're usually small, drawn-up specimens which never blossom. Meanwhile on the edge of the wood, by a ride or glade, or under a gap they grow strongly, flower and fruit, providing nectar for insects and berries for birds. The woodland edge can be a place of great productivity and diversity.

There's not always a concentration of shrubs on the woodland edge. Ancient woods usually have a hedge, either on top of a bank or on the flat, and this is often composed of much the same species as the coppiced trees of the wood itself. These days a combination of neglect and increasing shade from the grown-out coppice will have made the hedge useless as a barrier in most woods but you can usually recognise its remains. In ancient woods the edge is often characterised by ivy. In the interior of the wood you may sometimes see it on the trees but rarely on the ground, whereas within some ten metres of the edge it's common on the trees and may grow on the ground too. In recent woods ivy grows throughout on both the trees and the ground.

An actively coppiced wood is full of edges and full of light. The dark interior of an undisturbed wood may seem a dull place by comparison but it may contain just as much diversity, though the species will be different. The creatures which shun the edge are often drab and boring compared to the bright butterflies and song birds of coppice. The stag beetle, which I described towards the end of the chapter on niche, is top of the range for visual interest. We often wax lyrical about the wildlife of coppices but who are we to say that fritillaries and chiffchaffs have more right to exist than some greyish grub that lives in dead logs and few of us has ever seen? Everything has a right to life and wild creatures are not there purely to please us.

Edge vegetation can sometimes be found fossilised in the interior of a wood which has expanded. Shrubs may need an edge to get established and indeed to reproduce but they don't necessarily need it to survive. In ecology, they say, possession is ninety percent of the law. Once plants get established they can often hold their ground even though the conditions which enabled them to colonise that place have changed, as this entry from my notebook illustrates.

Brockley Combe, 19th July 2000

This is a recent wood or possibly a plantation, mostly of ash and sycamore. There's little ground cover except for occasional patches of dogs mercury, some ivy and tree seedlings. In some places there are saplings, about a metre or two high. Ash seedlings are very much more common than sycamore – they form a carpet in some parts – but almost all the saplings are sycamore. Perhaps this is due to sycamore being more shade-tolerant, or to preferential browsing. There's an indistinct but definite browse line visible.

As I got near to the adjacent conifer plantation, hawthorn and spindle suddenly became abundant. I hadn't noticed either of them elsewhere in the wood. This suggests that this was once the edge of the wood and the conifers are younger than the broadleaved stand, planted on fields rather than on a felled part of the wood.

The former edge, with its shrub layer of hawthorn and spindle, makes a strong contrast with the interior, with its shrub layer consisting mainly of sycamore saplings. Sycamore often forms a layer like this. It has a particular ability to grow to a metre or so tall then stop growing and wait for a very long time for a break in the canopy to occur. When a break happens the young sycamores have a head start on other trees and their wide, horizontal leaves cast a heavy shade on any competition. Even if there are only a few mature sycamores in the wood they can produce enough of these saplings to take over the wood next time it's felled. Sycamore is a naturalised tree, introduced some five hundred years ago, and it's increasing at the expense of native trees in both ancient and recent woods. Although a useful timber tree it's often regarded as being poor for biodiversity. Few insect species live on it and it suppresses the ground layer with heavy shade and a dense leaf mulch. Aesthetically it has a dark, lowering presence. I remember how uncomfortable a sycamore wood used to make me feel as a child.

Cycles of Change

Sycamore isn't unique in the way it regenerates. It's just better at it than other trees, able to hang on longer in suspended animation at the young sapling stage. Any tree which is shade-tolerant when young can do this to some extent. It's an important part of the stayer niche. It's called advanced regeneration and foresters who restock their plantations by natural regeneration rather than by planting are always on the lookout for it. Felling a group of trees which has a good stand of seedlings under it is a much surer way of getting the next crop than felling and hoping for the best.

True pioneer trees, not being shade-tolerant, can't do this. They'll only come up where a break in the canopy has already formed. I remember once walking down a ride in a plantation in south Wiltshire. On my left was a stand of mature oak. Underneath it was a dense understorey of sycamore, growing well in the dappled shade of the canopy. On my right was a stand of young pine, with plenty of space between the planted trees and abundant light. The natural regeneration was just as vigorous as it was under the oak but it was all birch. There was no difference in soil between the two compartments. The segregation of sycamore and birch was entirely down to the light levels.

Birch, like most pioneers, succeeds by flooding the area with countless hordes of tiny seeds. Only a minute proportion of them needs to germinate in order to cover the ground with seedlings. Even trees which have large seeds can produce huge quantities in a single season but larger seeds make an attractive meal for seed-eating creatures such as mice, voles, squirrels, pigeons, nuthatches and deer. A key to the reproductive niche of these large-seeded trees is having irregular mast years.

The pattern of mast years is influenced by the weather. A warm summer, or especially two in a row, increases the energy available to the trees and induces a heavy mast. Usually all the trees of one species will have a mast year together. At first glance this seems like a disadvantage. A once-in-two-hundred-years opportunity to reproduce could be lost to a species which didn't have a mast year at the right time to exploit a blow-down. Beech in particular, with its widely-spaced mast years, can easily miss a chance in this way. Ash usually has a mast year every other year and in some woods it can replace beech if a gap or clearing occurs when the beech isn't producing. This is one of the ways in which succession can be knocked back for a generation. But if their mast years weren't synchronised these large-seeded trees would hardly be able to reproduce at all. A steady yield of seed would support a steady population of seed-eaters and they'd eat more or less all of it each year. Instead, the population of these animals is controlled, at least in part, by the amount of seed available in non-mast years. Thus, when a mast year comes there aren't enough seed-eaters to consume all the seed and what remains is enough to produce an abundance of seedlings.

This is one reason why hazels are so vulnerable to grey squirrels. Although nut production does fluctuate from year to year they don't have real mast years. Nevertheless there are exceptional years when there are so many hazel nuts that some of them do ripen and then the trees can reproduce.

Some animals actually help trees to reproduce by eating their seeds, such as the jay which sows the acorn or the thrush which eats a berry and then shits out the seeds. Pigs, both wild and domestic, are also a benefit to tree reproduction. Though they eat a lot of acorns and beech nuts when they get the chance, they always miss some and bury them with their constant rooting. In a good mast year these lost seeds are quite enough to produce a new generation of trees. Foresters sometimes bring in a herd of pigs to help with natural regeneration. They're especially useful in beech woods, where the thick layer of leaf mould needs thoroughly churning up for the seeds to get to the mineral soil below. Badgers can help too. As they dig for worms and other food they make little gaps in the herbaceous vegetation and they may even bury the odd tree seed in the process.

Domestic grazing animals can help as well. So far I've emphasised how they can destroy woods by over-grazing but, as in so many things, it's all a matter of degree. If there's an over-dense herbaceous layer a little grazing and trampling can help to break it up. It can also increase biodiversity by knocking back the more vigorous herbs and shrubs, giving the less vigorous ones more chance to grow. This in turn can create niches for a wider diversity of birds and small mammals. The kind of wood where light grazing may be beneficial to wildlife is usually one which already has a history of grazing. Good fencing is necessary because it's essential to be able to control the intensity and timing of grazing.

The conflict between grazing and regeneration is most acute in traditional wood pastures. In parks, where the trees are widely spaced, any new tree is completely exposed and won't stand a chance of surviving without artificial protection. The oldest trees in medieval deer parks are slow-growing pollards and some of these probably date back to the woodland or hedgerow trees which stood on the site before it was emparked. Maiden trees in parks are probably post-medieval, planted and protected with a timber cage when the deer park evolved into the ornamental park of a country mansion. Planting was very rare in medieval times.

In denser wood pasture there are more chances for seedlings to escape browsing, even when grazing pressure is high, as it is now in the New Forest. Seedlings and saplings can find shelter from hungry mouths among the branches of a fallen tree or in a dense growth of holly. These opportunities are enough to allow a steady trickle of regeneration which can keep the woods going. But over the ages there have also been times when grazing was drastically reduced and a major pulse of regeneration took place. This could happen, for example, when there was an outbreak of cattle disease. It seems to have happened in the New Forest in the years after 1851 when the powers that be made a determined effort to exterminate the deer.

In coppice woods the regeneration of trees is much less of a problem. Every cutting of the coppice is an opportunity for trees to reproduce by seed. Conditions in freshly-cut coppice are so open that even oak can still reproduce in some actively coppiced woods, despite the mildew. (See page 127.) Coppicing also gives an opportunity to wildflowers. Some survive the shady years between each coppicing as buried seed, ready to germinate as soon as the light comes back. Foxglove and wood spurge are examples. Others are mobile and come in on the wind or on the coats of passing animals. The feathery-seeded willowherbs and the sticky-seeded burdock do this. Yet others, such as bluebells, primroses, anemones and violets, survive in situ as plants but only flower after coppicing. Butterflies and other insects feed on the wildflowers and thrive in the sunshine of the felled patch, sheltered by the taller trees in the uncut patches round about. Then as the coppice regrows a whole succession of song birds nests in the dense cover of the regrowing stools.

There are also some woodland wildflowers which are set back by coppicing. Ramsons and dogs mercury are both very shade-tolerant and have the competitive advantage during the dark phase of the cycle. Dogs mercury is also sensitive to trampling so the coppicing operation itself knocks it back. A pure stand of dogs mercury is very much a feature of an abandoned coppice wood.

Only about one in ten ancient woods in the Lowlands is still coppiced. In some of them it's done to provide pheasant cover. A pheasant wood is easily recognised. There's usually a release pen, an area enclosed with two-metre high chicken wire where the young, hand-reared pheasants are introduced to life in the woods. There are almost always pheasant feeders dotted around the rides, dispensing grain to the half-tame birds. In adjacent fields there may be strips of maize, kale or sunflowers to provide food and shelter for the pheasants. In one place I know whole fields of maize are grown for them. Coppicing may be haphazard in pheasant woods and not follow a regular cycle. As long as there's plenty of bushy undergrowth its purpose is fulfilled.

Other woods are coppiced for nature conservation and here coppicing is much more regular. Cutting a small section of the wood each year leads to a mosaic of patches at different stages of regrowth. This pattern is good for wildlife as each species needs coppice at a particular stage of regrowth. The small patches ensure that they never have to go far to find the conditions they need when the patch they're living in becomes unsuitable. Regular coppicing also gives an even spread of both work and produce over the years, which is why it was done that way traditionally.

Commercial coppicing these days is almost completely confined to pure chestnut and hazel coppices grown for craft purposes in the south-east of England. The making of chestnut palings, hazel hurdles and thatching spars involves cleaving the poles, that is splitting them from end to end. Once a pole grows beyond a certain size it becomes difficult or impossible to cleave. This means that when a coppice is neglected it becomes useless to a craft worker. The only economic way to bring it back into a commercial cycle is to make charcoal out of the overgrown coppice and then wait for the regrowth to come up to the right size.

Most ancient coppice woods you'll see these days are completely neglected and have been for a long time. The ground layer becomes less diverse and often rather dull. The plants which survive in an active coppice as buried seed or by migrating around the wood become rare. Those which survive in situ as plants, such as bluebells and anemones, usually survive but may not bloom. Ramsons and dogs mercury increase at the expense of other herbs. The birds and butterflies which depend on coppicing decrease or disappear. The bugs and beetles which thrive in undisturbed woods may not be there to take advantage of the situation. They need old trees and dead wood and will probably have died out during the centuries of coppicing.

As for the trees, in some woods the standards have grown to enormous sizes and reduced the growth of the coppice. In others the coppice has grown vigorously and joined the standards in the canopy. There are some woods where the standards were all felled during one of the World Wars, when there were shortages of imported timber. Some of these woods are still full of pioneers such as birch, hawthorn and pussy willow which came in to fill the vacuum left by the standards. But eventually in all woods the balance swings towards stayers. Hornbeam is a real stayer, similar to beech in both appearance and habit. Coppice woods which were once a mix of hornbeam and hazel have gradually turned into pure hornbeam now that competition has free rein. The standards, being mostly long-lived oak, are still there, but they have no hope of reproducing within a pure hornbeam wood.

These neglected coppice woods are going back towards a more natural cycle, a longer-term one based not on the frequency of felling but on the lifespan of the trees. Recent semi-natural woods follow the same cycle where they're not coppiced. When regeneration happens the seedlings usually come up very thickly. As well as being eaten at the seed or seedling stage, natural regeneration has to contend with competition from herbaceous plants, shade from older trees, drought and fungus diseases. Most trees die as seedlings but the initial numbers are so great that the survivors are usually enough to make a dense thicket of saplings. From now on the main cause of tree death is mutual shading. A hundred saplings may occupy the space which eventually will be taken by one mature tree. Bit by bit, as the trees grow, ninety-nine of them will be overtopped by their neighbours and die. Most dead trees you see in a wood will have died in this way rather than by disease or damage. One consequence of this is that, contrary to expectation, almost all the dead trees in a wood are young rather than old. Once a tree makes it to the mature canopy it will probably stay there for much longer than it took getting there. But one day it will die or be blown over and the cycle can start again.

A modified version of this process goes on in plantations. The baby trees are planted much further apart than nature usually sows them and for the first few years there's space between them. What kind of vegetation fills this space depends on the former land use, whether moor, grassland, woodland or plantation. Often it needs to be cut back or sprayed to enable the trees to grow. This stage is somewhat like the scrub stage in natural succession and some of the wildlife of scrub will inhabit it. On former grassland sites voles multiply in the undisturbed grass between the trees and these in turn support owls and kestrels. When the canopy closes and the young trees start drawing each other up the plantation moves into the thicket stage. The shade is intense, especially if the trees are conifers, and little wildlife survives.

newly planted

thicket stage

high crowns

Thinning takes the place of the mutual suppression which goes on in a natural wood and eventually the plantation opens out into the high crowns stage. Now some light can once more reach the ground. Herbaceous plants and shrubs can move in and with them some animals.

Felling brings a sudden increase in light levels, especially in plantations which aren't thinned and are harvested before they reach the high crowns stage. In plantations on former woodland sites this can be a stimulus to the buried-seed plants which used to respond to coppicing. In some places foxgloves come up in their millions. It will be decades since they last flowered and set seed but they're well adapted to waiting a long time. In the wildwood they might have had to wait much longer. The foxglove is a vigorous biennial and it invests the energy of its first year in prodigious seed production during the second. Each plant produces some three-quarters of a million minute seeds. Only a tiny proportion of these need to survive the long years of dark in order to cover the ground with new plants when the light returns. Then they can make a hillside shine purple in the distance at flowering time. Mobile plants such as rosebay willowherb also make use of clear-fells. These herbaceous plants can flower again and again before the canopy closes once more with the onset of the thicket stage.

Apart from the buried-seed plants, all other sun-loving plants and animals, have to move out when the dark phase comes. While birds and mobile plants can search far and wide for a new habitat, less mobile species need to find it nearby. The clear-felling system makes this very difficult, or impossible if an entire isolated plantation is felled at once. This is why continuous cover forestry is good for biodiversity. Harvesting in small groups gives a mosaic of stands at different stages of regrowth, just like traditional coppicing does.

OBSERVING WOODS

It never does to jump to conclusions based on first impressions. This is true whatever kind of landscape you're looking at but more so of woodland because the trees obscure the view. The limited area you can see at any one time may not be representative and there may be all sorts of surprises in odd corners which are hard to get to. The edge of a wood can be particularly deceptive. I once made a quick visit to a wood I'd not been in before when I was in the area for another purpose. I walked through one end of the wood and concluded it was recent because the ground was covered with ivy. When I visited it again for a more thorough look I saw that it's certainly ancient. All I'd done on the first occasion was to walk through the edge zone. (See page 202.)

Some stands are definitely easier to understand than others, as this passage from my notebook illustrates.

Laurieston Hall, June 23rd 2004

In the woods by the loch there's a very diverse stand with nine species of trees and a ground layer including bluebells, violets, dogs mercury etc. It's on a steep slope. As you go up, the slope flattens out and right on the break of the slope the wood changes to almost a hundred percent young birch with a ground layer of grass and a little bracken.

Further along, on a moderate slope, the picture is less clear. The trees are mostly birch with some old oak and a few others. The ground layer is moderately diverse, typically grass and bluebells.

The two extremes are obvious, but making sense of the intermediate stand would take some careful observation and perhaps some speculation.

There are four kinds of things to look at in a wood: earthworks, ground vegetation, tree shapes and tree species.

Earthworks

A wood bank running round the perimeter of a wood suggests that it's ancient. Woodbanks are usually massive in the east of the Lowland region, smaller in the west and rare in the Upland and Highland regions. The ditch is almost always on the side away from the wood, because it's intended to keep animals out rather than in. If part of the perimeter lacks a bank it's probably because part of the wood has been grubbed out since the end of the Middle Ages. Many ancient woods also have internal banks which marked out the property of different owners.

A woodbank around a plantation suggests the plantation is on the site of an ancient wood. Some recent plantations still have the living hedges of former fields running through them, often in stark contrast to the plantation trees. In an older plantation the hedgerow trees and shrubs may have been completely suppressed, in which case you can't be sure whether a bank inside the wood is a hedgebank or an internal woodbank.

Ancient woods usually have irregular outlines. On a map they often contrast with the more regular shapes of the fields, especially in the planned or champion countryside. (See page 90.) A straight side is usually the result of the wood expanding or contracting in modern times. The outline of a medieval deer park was usually a rectangle with rounded corners and the bank would have its ditch on the inside, to keep the deer in. (See page 96.) A wood whose boundary, or part of it, is like this is probably a former park.

Ridge and furrow inside a wood shows that it's been cultivated at some time and so is secondary. But it may still be an ancient wood. (See pages 91 and 94-95.)

Charcoal-making has left its signs in woods all over Britain. An old charcoal hearth is marked by a circle of black soil a couple of metres or more in diameter. They're hard to spot on flat land but in a sloping wood they form little circular terraces. The black soil distinguishes a former charcoal hearth from a terrace which may have been made for another reason. You can also sometimes find them on moorland and grassland, where they mark the site of a wood which has now disappeared. A forester I know says he uses charcoal hearths as a clue to access. There's always a good path out of the wood from one of them, usually more or less on the contour, even though it might no longer be obvious as a path.

In medieval and early modern times mining and quarrying often went on in woods without necessarily destroying them. Small quarries are quite common and in areas of early coal mining you can sometimes see the remains of bell pits. They're called bell pits because the miner would dig down to a shallow seam then excavate out in all directions, using no pit props, which gave the hole a cross-section somewhat like a bell. When he'd gone as far as he dared he would leave that pit and start another one. Today all you can see is a circular hole with the spoil spread around it in a ring, or on the downhill side if the pit is on sloping ground.

A similar kind of pit and mound, but smaller, is made when a tree blows over. At first the root plate, a great disc of roots and soil, stands up vertically beside the hole it has left in the ground. Over the years as the roots rot away the soil in the root plate slumps down into a low mound beside the pit. Eventually the whole tree rots away leaving only the pit and mound. These are not very common in British woods because under traditional management neither the coppice trees nor the standards ever grew big enough to have much chance of being blown down. But you can see more recent ones with the trees still in situ in abandoned coppice woods. Plantations, especially at high altitude, often blow down en masse. But this usually leaves a generally disturbed soil rather than the distinctive pit and mound pattern.

Ground Vegetation

You get the best view of the ground vegetation between March and June, when most woodland herbs are above ground and many are flowering. Visiting woods at other times of year can be deceptive, especially in early to mid winter, when many woods are bare of all ground plants except for evergreens like ivy. Most woods which have bare soil in winter will burst into life as soon as spring comes. Ones which don't are probably over-grazed.

The ground layer in woods usually indicates the soil conditions more faithfully than the tree layer. (See pages 199-200.) It's also often the best indicator of ancient woodland, as there are more herbaceous plants than trees in the list of ancient woodland indicator plants. (See page 184.) Remember that the simplified list given in this book is far from exhaustive and many characteristic woodland plants like ramsons, violets, wood sorrel and pendulous sedge are ancient woodland indicators in some parts of the country though not in others. If you want a list which is specific to your own area I refer you to the source quoted in the footnote on page 184. The plants which are characteristic of recent woodland are just as distinctive a group as the ancient woodland indicators. (See page 188.) They tend to be taller than the ancient woodland plants and give a completely different character to a wood.

Grasses may indicate that a wood was grassland not long ago but they're also typical of older woods with a history of grazing. In a grazed wood there may also be beds of nettles in places where the cattle congregate, especially by a feeding place. Nettles indicate any spot where nutrients, especially phosphorous, have accumulated. Homesteads are always places where nutrients from the wider landscape are concentrated, especially in the cattle yard, the privy and the compost heap. Phosphorous is incredibly long-lived in the soil, partly because it's resistant to leaching and also because it gets constantly recycled to the soil when the nettles die down each autumn. So a bed of nettles can mark the spot where there was once a habitation in the woods hundreds or even thousands of years ago.

In recent woods nettles can grow simply because the soil has been enriched during its time as farmland. They also occasionally crop up in ancient primary woods on rich soils. These are usually alder woods, because any rich soil which is well-drained will have been cleared for agriculture. In this case the nettles will be scattered rather than concentrated in a bed.

Some characteristic woodland wildflowers may survive in plantations on ancient woodland sites. Wood sorrel, with its white flowers and clover-like leaves with a sharp refreshing taste, is perhaps the commonest example. It's very shade-tolerant and can even flower in shade which would be deep enough to kill most herbaceous plants. Others, such as bluebells, may survive in scattered patches without flowering. In plantations of light-shading broadleaf trees the ground layer may survive almost intact. I know a mixed plantation of beech and ash which is lit up every May by a carpet of bluebells. It's the presence of the ash that allows this. Under pure beech the bluebells would hardly survive let alone flower.

Tree Shapes

Tall, drawn-up timber trees are characteristic of a plantation. But if the trees are broadleaves such as sycamore, ash or oak it may be hard to tell whether you're looking at a plantation or a recent semi-natural wood. In a young stand you can tell the difference more easily because the trees of a plantation will still be in recognisable lines. But as time goes by and trees are thinned as individuals rather than by whole rows the lines become increasingly difficult to detect. Nevertheless older plantations do have a regularity about them. The trees are all much the same size, much the same distance apart and genetically uniform. This last shows up especially in oak plantations. A wild population of oaks is usually an anarchic mixture of individuals, each with its distinctive form. By comparison the oaks in a plantation, all bred from one or two parents of good timber form, stand there like so many soldiers on parade. A mature, well-thinned plantation can have quite a surreal atmosphere if you're used to semi-natural woods. Straight boundaries between the different stands, often demarcated by a ride, are typical of plantations, whereas the stands in semi-natural woods will follow more natural boundaries. If the plantation is a mixture it's likely to be an even mix throughout the stand. In a self-sown wood it's more likely to be varied, with one tree or another dominating in different areas.

There are very few ancient woods consisting entirely of timber trees. Some of the ancient beechwoods on the Chiltern Hills were managed as pure timber and can now look very much like beech plantations, and of course there are the Highland pinewoods. But far and away the most characteristic tree form of ancient woods is coppice. The presence of coppiced trees is not an infallible proof of ancient woodland, though. Recent woods may have been coppiced once or twice and chestnut coppice is usually plantation, though often on an ancient woodland site. But a wood containing big, old stools can't be anything but ancient. A rough guide to estimating the age of stools is given on page 172. Of course the wood may be much older than the oldest stool in it, but it can't be younger.

Pollards are often found on the perimeter bank of an ancient coppice wood but not usually inside it. Old pollards inside a wood may indicate that it was once wood pasture. They could also be field or hedgerow trees which have become incorporated in a recent wood. Stubs may be found on the edge of the wood or on internal property boundaries.

This sketch from my notebook shows a clear story of woodland expanding and contracting simultaneously on different parts of its margin. Most of the clues are in the shapes of the trees, but I did note one ancient woodland indicator plant.

Beacon Centre, Devon, Easter 1991

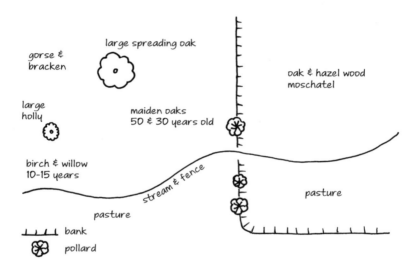

Tree Species

The first rule in looking at tree species in a coppice wood is to ignore the standards, which are almost always oak and tell you next to nothing about the ecology of the wood. (See page 198.) Although the trees may not indicate the soil as faithfully as the herbs there are several which are good indicators. These are listed on pages 64-65. In this chapter the general relationship between tree species and soils is given on pages 198-199. But these are only general tendencies. Ash, for example, is found on a wide range of soils and can never be used as an indicator. A great diversity of trees and shrubs, say a dozen or more species in one wood, is only possible on an alkaline soil.

The other thing which tree species can indicate is the age of a wood. This too is an imprecise science because so many pioneer trees can persist for a very long time in woodland. Oak is the obvious example. Alder is another. It colonises new ground like a true pioneer but it can persist indefinitely on a wet site because other trees can't abide the wetness. Other pioneer trees which readily form new woods are birch, ash, sycamore, hawthorn, pine and pussy willow. But none of them is restricted to new woods. Even birch will jump into an available gap in an ancient wood. A wood is much more likely to be new when these trees are growing in a pure stand. Then you're probably looking at the first generation of trees on that site.

The general tendency for ancient woods to contain a greater diversity of trees than recent ones has some exceptions. Firstly, on favourable sites with undisturbed conditions extreme stayers like beech and hornbeam can eventually form a pure stand. Secondly, on base-poor, acid soils and in challenging climates trees are restricted to the toughest species, such as oak, rowan, birch and pine. Thirdly, a long history of grazing will eliminate the more palatable species.

But in a wood on an alkaline soil, where coppicing only lapsed a few decades ago, and with no history of grazing, a long species list is a sure sign of age. What greater delight can there be than to wander in a wood, totting up the trees as you go, till you realise that even if you've seen no ancient woodland indicator plants you must be in a wood that goes back to the Middle Ages, if not to the end of the Ice Age itself?

Grassland

I have a close personal relationship with grassland. In 1983 an old farmer called Harry Plomley died. He was generally considered an eccentric. When I knew him, towards the end of his life, he milked half a dozen cows by hand on a few fields scattered around the village. He never used any fertilisers or sprays so his fields were havens of biodiversity, filled with a rich mixture of grasses, wildflowers and animal life. His family auctioned off his fields and I was fortunate enough to have inherited some money at that time so I bought one. If he had died a few years later it's most likely that the county Wildlife Trust would have bought them all. But nature conservation hadn't really got into its stride in those days and the true importance of semi-natural grassland wasn't yet appreciated. It was just two years before the publication of that Nature Conservancy Council report which revealed that ninety-seven percent of lowland flower-rich meadows had been lost in the previous forty years. (See page 104.)

For eight years I lived in the field in a tipi. I came to know it like a close friend, experiencing its changes through the seasons and from year to year. I learnt to recognise the grasses and wildflowers which grow there and recorded a total of seventy-three wildflower species and twenty-six grasses. I also had some intimate meetings with wild animals. Each year I arranged for someone to make hay and then graze the aftermath, as we call the grass which regrows after haymaking. I have to confess that I also planted trees on a part of it. As I said, in those days we didn't quite realise just how rare that kind of grassland had become.

Life moved on and it became time for me to live in a house again but I carry that experience of living close to nature inside me. It's as important to my permaculture work as the scientific knowledge I learnt at college. I still visit the field as often as I can and seeing the changes in one piece of land over a quarter of a century gives me an insight into the landscape that perhaps I couldn't have found another way. Meanwhile, just as my life has changed so has the farming scene. As the number of small farmers has dwindled it's become increasingly difficult to find anyone interested in taking the hay off one relatively small field and even more difficult to find someone with a few animals to graze the aftermath. Some years one or the other has been missed and the abundance of wildflowers has diminished as a result.

Two years ago I gave the field to the Somerset Wildlife Trust. My main aim in doing this was to secure its future but the field has benefited in the short term too.

The Trust's network of connections has made it easier to find people to take the hay and the aftergrazing. I'm now the voluntary warden but my relationship with the field doesn't feel any different. I always felt I was its guardian rather than its owner.

Improved and Unimproved

We think of grass as being naturally green but the greenest grass of all is in fact the least natural. It's the product of industrialised farming and consists of a monoculture of ryegrass, fed to bursting point with fertiliser. (See photo 31.) It's the nitrogen in the fertiliser that gives the grass that intense, gloss-paint green. These are usually short-lived grasslands which get ploughed up and reseeded every few years, either to alternate with arable crops or simply because they become less productive as the years go by and need renewing. Ryegrass is the only species of grass used because it responds most vigorously to the very high levels of nutrients provided by the fertiliser. Usually a mix of two varieties is grown, one early and one late, so as to keep production up throughout the growing season. The gross output of animal fodder is extremely high but the net output is another matter altogether. It takes a great deal of energy to produce fertilisers, especially the nitrogen component, which is extracted from the air by means of a gas-guzzling industrial process. If you subtract the total energy input from the energy content of the grass produced, the net output is not nearly so impressive. This kind of farming is only possible while we pay an unrealistically low price for fossil fuels.

Ryegrass
flower

A less intensive approach is to sow clover along with the ryegrass and allow the bacteria which live in the roots of the clover to provide the nitrogen. The overall yield may not be very much less than that of the highly-fertilised monoculture but it does have the disadvantage of being slower to start growing in the spring. Grass can grow at quite a cool temperature but the nitrogen-fixing bacteria in the roots of the clover need a higher temperature before they start work. So growth gets going later in clover-fed fields than in ones which get their nitrogen from the fertiliser bag. It can be quite surreal, the speed at which those highly fertilised fields get going in spring. When the birds are nesting, the cowslips are out and the grass in my field is hardly tall enough to cover the soles of my boots I'm often woken from my reveries by the sound of a forage harvester in the distance taking the first cut of silage. Every year it takes me by surprise. I don't realise how tall my neighbours' fertilised grass has grown till the mower bites into it and reveals its profile. It always seems a little bizarre that someone could already be taking a harvest when the growing season has hardly begun.

A simple ryegrass-clover mix is easy to recognise in the field, even when it's not flowering. The little round leaflets of clover contrast with the lancet blades of the grass and apart from this there's no variety of leaf shapes in the sward. Both conventional and organic farmers may use this simple kind of mixture to establish a ley, as a temporary grassland is known. But some organic farmers use a much more complex mix, especially for a longer-term ley or for establishing a permanent pasture. An example is the classic Clifton Park mix, which contains nine varieties of grass from seven different species plus four varieties of clover and five species of herbs. This kind of diversity means that many different niches are filled. The farmer who grows a monoculture of ryegrass is filling two niches by sowing early and late varieties of one species but these niches vary only in respect of time. The really complex seed mixtures fill a whole complex of niches which differ in many ways.

One of the species used is timothy, a grass with a distinctive cylindrical seed-head. It does especially well on wet, heavy land and produces plenty of leaf in early summer when other grasses are putting their energy into seed rather than leaf. Cocksfoot, whose seed-head has one branch further back than the others like the spur on the foot of a cockerel, is another. It has a deep, well-branched root system which makes it tolerant of dry, sandy soils and occasional droughts and also improves the soil structure. In a field with patchy soil conditions these two will complement each other, each taking a leading role in a different part of the field, though both will probably grow well enough in all parts. They will also complement each other from year to year, with timothy producing more in a wet year and cocksfoot in a dry one.

Cocksfoot flower, left and timothy flower, right

The 'herbs' in this kind of seed mix are plants which are neither grasses nor clovers. Their role is to supply some of the mineral nutrients which are needed in small quantities by plants and animals. They mostly have deep tap roots and each has a special ability to extract certain nutrients from the subsoil. The animals consume these nutrients directly when they eat the herbs and the other plants in the sward get the benefit of them indirectly through the animals' dung. Ribwort plantain is one of them and yarrow, with its feathery leaves and flat, white flowerheads, is another.

A rich mixture like this works in the same way a natural ecosystem does. In the chapter on niche I emphasised how niche differentiation leads to diversity. But what I didn't mention there is that it also enables the ecosystem to have a high overall yield. This is partly because a diverse mix makes fuller use of the available resources than a monoculture can. A mixture of tall thin plants and short bushy ones, and of deep- and shallow-rooting ones, makes fuller use of the resource of space, both above and below ground. A mix of early, mid-season and late species makes better use of the resource of time. Not only that but the plants in the mixture positively help each other. The clovers provide nitrogen, the herbs bring other nutrients up from the subsoil and plants like cocksfoot improve the soil structure, all of which benefits the ecosystem as a whole. In this kind of farming each species of plant is allowed to fulfil its natural role and in doing so it contributes to the life of the community. The result is a high yield without a high energy cost.

Whatever kind of seed mixture has been sown, from pure ryegrass to Clifton Park, the composition of the sward will change in time. In nature nothing stands still and however carefully a farmer may look after the field it will gradually lose some species and gain others. The species and varieties which are deliberately sown are productive ones from an agricultural point of view but they probably aren't the ones most perfectly suited to survival on that site. There will be other plants waiting in the wings which are ever so slightly better adapted to the soil, climate and other niche factors. They may be broad-leaved plants such as buttercups and thistles, or species of grass which are never sown because they're not very productive or even wild varieties of the cultivated grasses and clovers. All of them will reduce the productivity of the sward as they become established in any gaps left by the demise of the sown plants.

The resulting grassland is known as semi-improved. This is a catch-all term for any field of grass that falls somewhere between a productive new ley on the one hand and completely unimproved semi-natural grassland on the other. It can also be formed simply by adding fertiliser to semi-natural grassland. Raising the level of nutrients gives the advantage to a relatively small number of very competetive grasses and herbs which will eventually outcompete their neighbours. In the long run the result is much the same whichever end of the spectrum you start with.

In springtime a semi-improved grassland which is heavily fertilised turns much the same shade of unnatural green as a field of pure ryegrass. At this time of year the grasses are composed mainly of leaf and all species look much the same from a distance. In summer, when the grasses are flowering, improved fields still show green but semi-improved ones turn buff or brown according to the mix of grasses in them. This is because ryegrass flowers are green while those of other grasses are brownish.

When you look at it more closely, a semi-improved sward can seem at first sight to be more diverse than it really is. There may be quite a lot of broadleaved herbs among the grasses and even if the herbs aren't flowering the mix of leaf shapes gives an impression of diversity. But on closer inspection you'll find that they're a pretty limited selection, often not much more than buttercups, dandelions and plantains. In summertime a combination of red and white clover with yellow buttercup, all blooming together, can give a vivid display which brings the phrase 'flower-rich meadow' immediately to mind. But this is really a very modest level of diversity. A field like this may bring joy to the heart and produce a good yield of nutritious herbage, but it has nothing like the rich ecology of a true semi-natural grassland.

Buttercups are perhaps the most characteristic plant of semi-improved grassland. On a suitably heavy or moist soil the abundance of buttercups is largely a measure of how long it is since the field was last ploughed and reseeded. In recent years, with farming in such dire financial straits, bright yellow fields of buttercups have become a more common sight than they used to be. Reseeding a worn-out field is always an expense which can be put off till next year, though it may be a false economy in the long run, so semi-improved grassland has increased in the landscape, and buttercups with it.

Unfortunately there's been no similar increase in semi-natural grassland. It's very easy to 'improve' grassland by adding fertiliser but much more difficult to bring it back to its former diversity. One reason for this is that phosphorous, one of the key nutrients in fertiliser, is very persistent in the soil and remains long after applications of fertiliser have stopped. Another is that potential seed parents have vanished from the surrounding landscape. The drastic decline which was first brought to light in 1985 has continued ever since. These days it's rare to find true semi-natural grassland outside of nature reserves, except on land which is too steep for tractors. Whenever I find myself in a flattish field which supports a wide diversity of wildflowers I start to wonder where the Nature Reserve sign is. But there are some fields which have fallen through the net of improvement. Usually they belong to old farmers, like my predecessor Harry Plomley, who are content to carry on farming the way they've done all their lives. If you ever come across such a field I strongly suggest you let the county Wildlife Trust know about it. Semi-natural grassland has a priceless ecological value and can be destroyed in the twinkling of an eye when the land changes hands. Don't assume that the Trust already knows about it. It's better to take the chance of telling them something they already know than to risk leaving such a jewel of biodiversity to the mercies of someone who thinks it needs ploughing and reseeding – a comment which was once made to me about my own field.

This sketch from my notebook shows how closely the improvement of grassland can be tied to the degree of slope.

Hay Hill, 5th July 1997

There's a gradation from improved grass-land on the lower slope to unimproved on the steepest part, with a narrow strip of semi-improved between the two. The improved sward is dominated by white clover and also contains red clover, ryegrass, other grasses and a few wildflowers such as self-heal. The semi-improved strip has these and some more wildflowers, eg agrimony, in a matrix of mixed grasses. The unimproved grassland above contains all the flowers in the other two parts plus some others such as restharrow and ox-eye daisy. It's succeeding to scrub, mainly of brambles.

There are two very small negative lynchetts. The lower, smaller one is the result of just one ploughing. The upper one, on the fence-line, is a bit bigger and represents several ploughings.

It looks as though this slope was all one field of unimproved pasture till fairly recently. Otherwise the flatter part would be separated from the steeper by a hedge on a big lynchett rather than the existing fence with its small lynchett. It was probably a few decades ago that the bottom part of the field was first ploughed and reseeded. For some reason, last time it was ploughed the ploughman didn't go right up to the fence, leaving the narrow strip of semi-improved grassland between the improved and unimproved.

This site shows the loss of semi-natural grassland in microcosm. It's gone on the flatter land through being ploughed and reseeded and it's going on the steeper land by succeeding to scrub. It's a case of either too much attention or not enough. Both of these mean the end for a field of herb-rich grassland. It's all part of the on-going simplification of the countryside.

GRASSLAND TYPES		
Improved	**Semi-improved**	**Unimproved or Semi-Natural**
Sown grassland, which may be: ryegrass monoculture simple mixture with clover diverse mixture, e.g. Clifton Park	Improved grassland which has deteriorated or unimproved which has been treated with fertiliser	Shows no signs of having been ploughed and reseeded or of being fertilised. May be very diverse. May be either: meadow – tall grasses and herbs, or pasture – short, springy turf

Semi-Natural Grassland

The origins of semi-natural grassland are lost in the mists of time. When discussing the history of the landscape earlier in the book, I mentioned the possibility that grassland is a largely imported ecosystem and that most of the grassland plants came here with the first farmers. (See page 84.) There's also the theory that the primeval landscape of Britain was not continuous wildwood but alternating woodland and grassland. A third possibility is one that occurred to me on a day of brilliant sunshine and blue sea which I spent walking on the south Devon coast path.

Prawle Point, 1st July 1995

Today I counted sixty-four species of herbaceous plants by the sea, not including grasses or ones which aren't flowering at the moment. They were mostly on the clifftop though a few were a mile or two inland. This is surely the highest number of species I've ever counted in a single day.

Most of them were grassland wildflowers. Clifftops are one ecosystem in Britain which has hardly been affected by human activity. Maybe these two facts are connected.

I wasn't the first person to have the idea that natural grassland might have existed on exposed clifftops before the days of farming. Strong salt winds can play much the same role as the munching mouths of cattle. A few trees and shrubs can tolerate salty winds up to a point but none can survive on the most exposed headlands. Here there are only grasses and herbs. In some places you can see a clear gradation from grass on the headlands to scrub further inland to woodland further inland again. The picture below shows the western bank of the Salcombe Estuary, also in south Devon. At the extreme left is the headland which juts out into the sea, with a covering of pure grassland between the jagged exposures of rock. Away from the full force of the salty spray the grass gives way to bracken, dotted with the occasional hawthorn and elder. Then the isolated shrubs merge into continuous scrub and that in turn grades into woodland. (See photo 33.)

This is not an entirely natural landscape. It appears that the whole area, woodland, scrub and grassland, was pasture once upon a time and the graded vegetation we see now only developed once grazing stopped. The clue to this is in the woodland, which is quite clearly recent. Almost all the trees in it are sycamore. Although sycamore is the most salt-tolerant of broadleaved trees and one which is gaining ground at the expense of native species, it's slow to oust other trees from an ancient woodland, even near the sea. I don't know of any ancient wood where it has completely taken over. In the ground layer there are no ancient woodland indicators but there is common sorrel, a grassland plant which is managing to hang on under woodland conditions. So this isn't a primeval landscape which has survived down the ages, but there is every reason to believe that the clifftop itself has been grassland continuously since the end of the last ice age.

One of the marks of a semi-natural grassland is its colour. In spring, when all the fields around it shine with the hard green of artificial nitrogen, an unimproved field may stand out with softer, more patchy hues of green. In April, if the soil is alkaline, the field may well be splashed with the yellow of cowslips. But most of the wildflowers come into bloom in early summer and then it will explode with colour. Though it may still look dowdy from a distance, from close to the shades of green will be lit up by a thousand points of red, blue, yellow, mauve, purple and white. By July, when most of the herbs have finished flowering and the grasses are in full flower, the predominant colours are the buffs and fawns of grass flowers, and then it may be hard to tell an unimproved field from a semi-improved one at first glance.

As yet no-one has prepared a list of plant species which can be taken as indicators of semi-natural grassland on a parallel with the ancient woodland indicator plants. But one to keep an eye out for is the ox-eye daisy, a tall herb whose flower is like a giant version of the common lawn daisy. It's by no means diagnostic of semi-natural grassland. In fact it's quite common on motorway verges. But it's conspicuous and can act as a signal that the field is worth investigating for other flowers which may at first be hidden by the grass. Cowslips are also doing well on some motorway verges but in fields they're a sure sign of unimproved grassland.

Some of the flowers you find may be quite common. Others, like the orchids, may be rare. But what makes this kind of ecosystem special isn't the presence of some rarity so much as the sheer diversity of plants, both grasses and flowers. A single visit in May or June is enough to tell you whether you've found something special and in April the presence of cowslips alone is a promise of floral riches to come. But to get a full picture of what's there you need to visit the field regularly over spring and summer, as many plants have a short flowering season and they're all much easier to identify when they're in flower. Then you'll find yourself delving into the intricate world of grasses with names like sweet vernal, crested dog's tail and Yorkshire fog. You may get to know curious flowers like goatsbeard, also called Jack-go-to-bed-at-noon because it only opens for a short time in the morning then closes for the rest of the day. You'll almost certainly cross paths with butterflies and grasshoppers, and with improbable insects such as the iridescent green beetles which live in the miniature jungle of grass stems, and the steely burnet moth, whose red and black markings warn predators that it's poisonous.

These grassland animals are a vital part of the ecosystem but when you're assessing biodiversity it's a good idea to stick to the plants. For one thing they don't move around or hide so it's less easy to miss them. Also, by and large, where you have the plants there you also have the animals. Animals depend on plants for their food and often for other aspects of their niche such as microclimate and nesting sites. Many grassland animals have a very narrow niche and are restricted to a handful of food plants or even just one. This is specially true of butterflies and moths. Although the adult insects may take nectar from a range of different plants the caterpillars are often restricted to a single food plant. In many cases these are plants which only grow in unimproved grassland. Preserving the plants is the most positive step we can take towards preserving the animals and that means preserving the ecosystem as a whole.

A headcount of plant species is a good rough guide to the ecological value of a grassland but it does need to be seen in the context of the potential diversity on that site. The main factor affecting this potential is the acidity of the soil, as alkaline soils inherently support a wider range of plants than acid ones. So a total species count of a dozen in a field which lies on an acid soil may be equivalent to fifty on a neutral soil. Downland, that special kind of grassland that grows on thin, alkaline soils over chalk or limestone, has the highest potential of all. It may contain fifty species of plants not per field but per square metre.

High alkalinity is not the only reason for the wonderful diversity of downland. It's also due to the way the land has been used in the past. For centuries chalk downs were grazed by sheep, but wool and mutton were not the only outputs, or even the main ones. More important was the dung of the sheep. Sheep have the habit of dunging mostly at night, so people soon learned to graze them on the downland during the day and fold them on the arable land at night. In this way they gathered mineral nutrients from the soil of the wider downs and concentrated them on the arable fields. The 'golden hoof' as it was known allowed cereal crops to be grown on soils which could otherwise never have supported them before the days of modern crop rotations. The effect on the sheepwalks was the opposite. The already thin downland soil became ever poorer in nutrients, the growth of the most vigorous plants was curtailed and a wider range of plants was able to flourish.

Unimproved downland is now mostly confined to steep slopes and military ranges. Before the days of artificial fertilisers the chalk downs of Salisbury Plain were the largest area of unenclosed, uncultivated land in southern England and the army seized upon it for artillery ranges and battle practice areas. These days all the available downland which is flat enough to plough has been converted to productive farmland, but the military area is not available and is still largely unimproved down. Perhaps it's ironic that the biggest force for downland conservation in Britain has been one of the armed forces, but so it is. Most of this land is out of bounds to the public and if you want to stride across springy downland turf your best bet is to seek out the steep scarp slopes, mainly on the northern and western sides of the downs, where the windblown grassland still sweeps down into the clay vales below as it did in times gone by.

In the opposite direction, towards the south and east, river valleys follow the gentle dip slope of the chalk strata. They are home to one of the more curious forms of grassland, watermeadows. This term is sometimes used quite loosely to mean any grassland alongside a river or stream but strictly speaking a watermeadow is an irrigated grassland. In our wet climate they weren't irrigated for the sake of the water itself but for the warmth it brings. Springwater stays at a constant temperature throughout the year and although river water does change temperature with the seasons it does so less than dry land. This means that in late winter and early spring the water is warmer than the soil, so if it's allowed to flood over a field of grass it warms up the soil and gets the grass growing earlier. Although they may have existed in medieval times watermeadows appear to be an invention of the seventeenth century. The first certain record comes from the Golden Valley in Herefordshire but they reached their peak in the valleys fed by the chalk streams of Wessex, where their remains can be seen today.

In their most evolved form the whole meadow was laid out in a pattern not unlike the ridge and furrow of medieval ploughing. Along the top of each ridge ran a carrier ditch and along each furrow there was a drain. The water would flow from carrier to drain in an even sheet. The engineering had to be very precise in order to achieve this and each system was run by a skilled worker known as a drowner. Today much of this detail is lost and watermeadows may appear to be little more than ridge and furrow with a dense growth of rushes in the furrows. In the West Country a simpler version can sometimes be seen, often just a leat which comes from a spring and appears to go nowhere. When it was in use the leat would have overflowed along its last section and flooded the land downhill from it.

Watermeadows were very much a product of the new commercial age of farming, driven by the market rather than the needs of subsistence. The purpose of getting the grass growing earlier in the year was to produce finished lamb earlier than other farmers and thus catch the high prices. The meadows would also yield a crop of hay later in the year but what paid for the investment of constructing a watermeadow was the commercial advantage of that 'early bite'. They fell out of use in the twentieth century because they were labour-intensive and couldn't compete with imported New Zealand lamb.

Another specialist kind of grassland which had a similar rise and fall through time is the traditional orchard. Although the fruit trees may define an orchard, the pasture they grew in was equally important to the farm economy. Trees in an orchard must always be widely spaced. They only produce fruit on those parts of their crown which are exposed to the light, so if you crowd them together like trees in a wood they give very little fruit. This wide spacing means that there's enough light at ground level to grow a crop there. Grass is the ideal crop because it doesn't need cultivation, which would damage the tree roots, and it's more forgiving than corn or vegetables: if the competition of the trees reduces its growth you just get less grass, not a crop failure.

In Medieval times people grew fruit for home consumption in their gardens but it seems that orchards as we know them were another invention of the Modern age. They only became prominent in the eighteenth and nineteenth centuries when fruit became a marketable commodity. The change from a peasant economy to one in which commercial farmers employed paid labour may also have spurred the growth of orchards, because in the real cider counties of Devon, Somerset and Herefordshire the workers were part-paid in cider. Tradition has it that the daily ration was a gallon a day. This must have kept the workers in a permanent state of alcoholic stupor. How they managed the skilled and often dangerous work of old-time farming I can't imagine. On the other hand cider can make a repetitive, uncomfortable task like stooking up corn flow a lot more sweetly, as I can attest from personal experience. Cider trees go especially well with a grassy orchard floor. While desert and cooking fruit needs to be picked by hand, cider apples are harvested by shaking the tree and picking them off the ground and grass is ideal for this.

Almost all the traditional orchards are gone now. The old editions of the large-scale Ordnance Survey maps show a very different landscape in the lowland parts of the cider counties. Each village or hamlet was surrounded by a thick belt of orchards, where now perhaps one or two remain. They needed to be near the farmstead because they're an intensive land use and the harvest is heavy so you didn't want to have to haul it too far. An orchard is more sheltered and shady than an open field, so they were ideal for poorly animals and for cows and mares giving birth. Siting them near the farmstead meant these animals could get the care and attention they needed. Pigs could be let into the orchard to clean up windfalls and chickens could be put there in winter for pest control. Many fruit pests spend winter hibernating in the soil and chickens love to scratch and peck to find them.

Although traditional orchards have little place in the modern economy they're wonderful places for wildlife, especially song birds. The old trees provide the nesting holes which tits need while the mix of trees and grass gives a diversity of food sources which keeps many birds well fed throughout the year. Some trees are thick with mistletoe and others lean over at steep angles. They can live quite happily for decades bent almost to the ground, still producing apples as well as ever. Here and there people plant new trees in old orchards, often more for the sake of tradition than for the produce. In others only a few trees remain in one corner as a reminder that the field was once an orchard.

Modern orchards are quite a different thing. The trees are kept small throughout their lives because they're grafted onto dwarfing rootstocks. Their small size makes them easier to pick, prune and spray. The dwarfing rootstocks also mean that they come into fruit just a couple of years after planting, which is an important advantage when you're paying interest on the cost of buying the trees and planting them. The little trees are planted close together in rows, with alleys wide enough to take a tractor between the rows. A strip of ground directly beneath the trees is kept bare by regular spraying with herbicides because the dwarf trees can't stand the competition of grass. Now these orchards too are becoming increasingly rare as globalisation makes it ever harder to grow apples commercially in Britain.

Meadow and Pasture

Grass does most of its growth in spring, when it produces lots of luscious leaf. In summer it converts this goodness to stems, flowers and seeds and produces little leaf. Intensive farmers do their best to keep it at the young, vegetative stage by repeatedly cutting it or grazing it, so it keeps starting again from the beginning of the annual growth cycle. Rather than allowing the cattle to range over the whole farm they keep them on a small area of land for a short time and then move them on. Each piece of ground gets grazed in this way several times a year. The main winter feed is silage, which is cut from the same field two or three times in the season. It's impossible to stop grass going to seed altogether but this treatment produces the maximum of young, nutritious leaf.

If the winter feed is conserved as hay rather than silage the grass needs to grow to a more mature stage. Silage is grass pickled in its own juice whereas hay is preserved by being dried in the sun. So the first cut of silage is taken in spring, when the grass is young and juicy, whereas hay is cut in summer, when the grass is flowering and the plants are more stemmy. You can only take one cut of hay each year but the aftermath is always grazed.

After cutting, whether for hay or for silage, the grass stubble is a pale, strawlike colour. This can give the countryside a dry look and just after haymaking time people often remark how drought-stricken the land appears. But it's not lack of water which causes the pale colour, it's lack of light. The green colour in all plants is caused by chlorophyll, the pigment which performs the magic of turning sunlight into food. At the base of the grass plants, under the thick shade of the sward, there's no sunlight and the plants don't waste energy producing chlorophyll there. This is the part of the plant which is left after hay- or silage-making and it looks very pale till the growth of new leaf brings the green back into the grass.

All kinds of grassland, improved, semi-improved and unimproved, go through the change from pale yellow to a uniform bright green after cutting. This contrasts with the more mottled green of a grazed field. Mowing takes all the plants in the field back to the young, green-leaf stage, whereas grazing animals are usually somewhat selective and while they eat some of the plants they leave others to mature. In general a grazed field will show less change in appearance as the season goes by than one used for mowing, but it will show more variation at any one time.

Modern improved grassland can be used interchangeably for grazing or mowing but this wasn't always so. Traditionally there was a big difference between pasture, which was grazed, and meadow, which was mown for hay. Meadow needed to be sited on the best land because at haymaking all the mineral nutrients in the hay are removed from the field, whereas in a pasture most of the nutrients stay behind in the animals' dung and urine. In medieval times meadows were confined to the seasonally flooded land beside rivers and streams. The winter floods would bring silt, eroded from further upstream and enriched along the way with the effluent of farmyards and privies. Pasture was poor stuff by comparison. It was usually the land that wasn't good enough to be used for anything else. As farming methods improved the distinction gradually faded but until recently only the best land would be cut for hay year after year.

The difference between meadow and pasture can still be seen in unimproved grassland and in many semi-improved fields too. Land which is constantly used for pasture has a short, springy turf while a meadow sward has a tall, straight-stalked structure. Traditional pastures are not usually grazed on the kind of intensive rotation used on improved grassland. The animals are allowed freer range over a longer time so the grasses and herbs are constantly nibbled. This keeps the plants small and encourages them to branch again and again. This branchiness gives the spring in the sward. Hay, by contrast, is left for two or three months to grow undisturbed before it's cut and the plants reach upwards, like trees in a plantation, competing for the light. Over time this affects the species composition of meadow and pasture. Tall-growing species are favoured in meadows and lower-growing ones in pasture.

The contrast is neatly illustrated by the two common species of clover, red and white. Red clover is tall and is the more frequent one in meadow while the short, creeping white clover is commoner in pasture. In semi-natural grasslands this segregation of the two clovers happens naturally, while in a sown grassland the farmer will choose one or the other according to whether the field is to be mainly mown or grazed.

Most of the plants in a grassland are long-lived perennials, so changes in the sward due to a change in management tend to be slow, as I've seen in my own field. Before I took it on it had been sometimes mown and sometimes grazed but since then I've always treated it as meadow. It was perhaps fifteen years before I noticed an increase in the tall herbs, knapweed and meadow scabious. Some years later I realised that the dandelions had completely died out. They'd always flowered prolifically at the same time as the cowslips, two different yellows clamouring for attention against the short turf of April. To get my fill of pure, undiluted cowslips I had to get up early in the morning, before the dandelion flowers had opened. Dandelions are resistant to grazing, being a very low-growing plant with a strong, sour taste. The tall, shady conditions of a meadow don't suit them. They also like a fairly rich soil and the constant removal of nutrients in the hay may have helped its decline. Cowslips are also low plants but they do very well in hay fields. They do much of their growing in winter and early spring when the grass is still short and seem to cope with summer shade better than dandelions. They also tolerate a poorer soil.

It's no coincidence that dandelion and cowslip both flower in April. Being short plants they need to make maximum use of the sunlight before their taller neighbours grow up and shade them. Some other low-growing plants manage simply by tolerating shade and have much the same annual cycle as the tall ones. Although there may not be an obvious upper and lower layer of grasses in a semi-natural meadow there is a mix of taller and shorter grasses. This gives the sward 'bottom'. In other words it's dense with leaf near the ground, where a sward of pure ryegrass is sparse. Modern farmers who've never farmed a traditional meadow are often surprised by the amount of hay it can yield. They're not used to a field with bottom and when they cast an eye over the crop they only see the upper layer of grasses.

The glory of a semi-natural meadow is its wildflowers. A meadow is shut up for hay during the time when most grassland plants are flowering. By contrast, a semi-natural pasture can look quite drab at the same time of year as the grazing animals munch the flowers along with the leaves and stems. In fact sheep actually have a taste for flowers and will seek them out. On the other hand pastures are better for insect life. Haymaking is a catastrophe for insects and other small creatures, as their whole world comes crashing down around them. Pasture makes a more stable habitat, while its uneven structure gives a range of habitats for different invertebrates which is also favoured by mammals such as hares.

The broadleaved plants which really do well in pasture, both unimproved and semi-improved, are ones which are thorny, poisonous or unpleasant to eat, like thistles, ragwort and buttercups. Ragwort and the tall spear thistle, being biennials, need bare soil to get established and this is most likely to occur when a pasture is overgrazed. The perennial creeping thistle, on the other hand, can reproduce vegetatively and is less dependent on bare soil. Cutting can help to control it but it needs to be done at the right time of year. In spring the plant puts all its energy into vegetative growth. Cutting it then only encourages it to produce more horizontal roots, which just makes the problem worse. But in summer it puts its energy into the flower buds and if you cut it then you remove those flower buds and rob it of that energy. This weakens the plants and if it's done several years running it kills them. Hence the old rhyme:

> Cut them in May and they're here to stay,
> Cut them in June and they'll be back soon
> But cut them in July and they're sure to die.

In practice there's more to it than that, as I once heard from a young farmer.

Talk with Peter Holmes, September 2003

He agrees with the old adage about creeping thistle but on a fertile soil it's hard to eradicate it just by cutting. It's also more difficult if you want the land to look tidy. His father used to top the pasture in front of his house throughout the growing season but it was always longer in front of Pete's house because he liked it untidy. The result was more thistles by his father's house.

If the thistles are growing up thickly in pasture cattle will eat them unintentionally with the grass – and intentionally when the little soft ones come up after cutting. Making hay helps because you cut it at the ideal time to weaken the thistles. But if you're trying to finish beef animals in thirty months without feeding them grain you need the higher feed value which silage has, and that means cutting the thistles at the worst time. The most effective way to control them is probably by grazing with sheep followed by a weed wipe.*

* A herbicide applicator which only treats the tallest plants in the sward.

So understanding the annual cycle of the plant isn't enough. You also have to take into account the soil, the farming system and personal preferences such as tidiness. Even politics enters into it. At the time of our conversation the regulations for controlling the disease BSE were still in force. No cattle over thirty months old could be slaughtered for meat, hence Peter's reference to finishing them by that age.

Buttercups aren't exactly poisonous but they have a nasty, acrid taste and a sap which can bring out a rash on people with sensitive skin. In a field that's always used for pasture they gradually increase over the years as the cattle carefully graze around them. Regular mowing takes away this advantage from buttercups and this is another change I've noted in my field since I've kept it as a full-time meadow. In the lower-lying, moister part there used to be a large patch which would go a brilliant yellow each year at buttercup time. Now, after years of being cut along with everything else, they've been reduced to just another component of the sward and you don't specially notice them.

As well as the differences between meadow and pasture there are differences between the grazing habits of cattle, sheep and horses and these leave clear signs in the sward. By reading these signs you can often tell which kind of animal has been grazing a particular field.

Sheep bite the grass off with their teeth and graze it down much shorter than cattle, which tear if off with their tongues. So sheep pasture is usually shorter and can have a lawn-like appearance. Thus sheep give a competitive advantage to low-growing plants. White clover can take over after very close grazing, as its ground-hugging runners move in and take over any space left vacant by over-grazed grasses. But I have seen this happen after intensive grazing by bullocks, so it's not necessarily an indicator of sheep grazing. Conversely, sheep don't like eating grass which has already grown too long and stemmy. If they're put in a field of mature grass they'll pick away at the shorter stuff below and leave the tall flowering stems. But in terms of the range of plant species they'll eat they're less fussy than cattle and will eat plants like docks, buttercups and thistles, especially when the plants are young. They normally leave rushes but they will graze them if, as a farmer I once knew put it, 'you keep them hungry enough'. One plant they won't eat is nettles and you sometimes see a sheep pasture grazed as close as a billiard table except for the odd clump of tall nettles. They can even eat the poisonous ragwort. Although it's deadly to cattle and horses it's less toxic to sheep, especially in the spring, and judicious sheep grazing can play a part in controlling it.

Cattle are more choosy. As well as leaving docks and thistles they avoid eating where they've dunged. The dark green patches of taller grass which grow up over their dung pats can give a cow pasture a pimply look. The dark colour of the grass is due to the extra nutrients in the dung but it's tall mainly because the cattle won't eat it. A large clump of nettles, or sometimes docks, usually indicates a place where they were fed hay or silage during the winter. Both their dung and the fodder they spill while eating raise the nutrient level and the bare, poached soil leaves space for these nutrient-hungry plants to come in.

Horses are more choosy again and notorious for being the worst grazers of all. They segregate the field into two parts, one for grazing and the other for dunging. They won't take a bite from the dunging area, while in the grazing area they take the grass down even shorter than sheep with their double row of sharp incisors. If horses are kept on the same field continuously the contrast between the two parts becomes stronger and stronger. The dunging area grows tall with rank grasses, docks, thistles and nettles while the grazed area gets nibbled down almost to the bare earth. It also becomes compacted and docks, which tolerate a compaction, often move in wherever spots of bare soil appear. Topping can help a lot and some horse-owners go to the length of removing the dung by hand, but the best solution is to alternate horses with other grazing animals.

Gateways get even more trampling than a pony paddock and they develop a plant community quite different from that in the rest of the field. Plants in a gateway have to contend with both soil compaction and being trodden on. The most intensively trampled area near the gate is usually bare of plants. Moving out a little, or in a less used gateway, you come upon the most tolerant plants of all. These are greater plantain and a funny little flower called pineapple mayweed, a member of the daisy family with an egg-shaped flowerhead and no petals. Where trampling is a little less severe these two may be joined by the tiny annual meadow grass and knotgrass, which isn't a grass at all but a low sprawling plant with spear-shaped leaves and minute flowers. Mayweed, a medium-sized daisy flower, often comes up in the bare soil around a gateway after the trampling is over. Being an annual it benefits from the bare ground and it can tolerate soil compaction. You often see it in the gateway of a hayfield which was used for pasture the previous year. (See photo 34.)

couch grass
& creeping thistle

rest of field:
grasses with
dandelion
& ribwort
plantain

usual gateway plants including:
pineapple mayweed, mayweed,
greater plantain, knotgrass

Further out from the gateway there may be a zone which is less obviously affected by trampling, although the plants are subtly different from the rest of the field. It may have suffered from temporary poaching, perhaps when the field was grazed by dairy cattle during a wet time. As dairy cows go in to be milked twice a day they cause much more trampling round a gateway than beef cattle or sheep. This outer zone may be dominated by couch grass for a few years afterwards. Couch is a pioneer grass which can take advantage of a new patch of bare soil but it rarely survives for long in an established sward. Creeping thistle may come up too. Its rhizomes can stay dormant in the soil for years and then emerge when the soil is bared by poaching. The drawing shows a typical example.

On a heavy soil the compaction caused by the poaching round a gateway can be very persistent. I once saw a field with a sward that looked smooth and uniform across the field. But in a quarter-circle around a corner gateway there was creeping buttercup, which was absent from the rest of the field. In this case the creeping buttercup clearly indicated compaction. You can see something of this effect in most grass fields, especially in the distribution of plantains. Ribwort is almost always the commonest overall but as you get near a gateway it's usually replaced by greater plantain.

As well as meadow and pasture, there's grassland which neither gets cut nor grazed. Sooner or later woody plants will move in and take succession forward but before that happens the makeup of the sward can change markedly. At first this change may be more apparent than real. Broad-leaved plants which have been kept in a green, vegetative state by mowing or topping will suddenly stand out as they flower and set seed. Plants like thistles, docks and yarrow which have gone to seed can make the field look terribly weedy.

But compare this field with a neighbouring one which is still being grazed or cut for silage and you may find just the same range of plants, equally abundant but less obviously visible because they're not flowering. You can get the same effect on a roadside verge, part of which is mown for visibility and part left to grow tall. As you drive past in a car the unmown part looks mixed and the mown part pure grass. But if you get out and have a closer look you may find the two are much the same.

A similar process goes on in churchyards and cemeteries where the rigid regime of lawn-mowing has been relaxed in the interests of nature conservation. A patch is designated as 'meadow' and cut just once a year at the end of summer. The sward is often completely unimproved and reducing mowing to only once a year can be like unlocking a long-forgotten jewel box. The most striking example I've seen is the cemetery at Ventnor on the Isle of Wight, which lies on a chalk down high above the main town. It has a substantial patch of meadow alongside the tightly-mown lawn. In spring the meadow is covered in cowslips and in summer it explodes into the full splendour of chalk grassland. (See photo 32.)

It's remarkable how the wildflowers have managed to survive the long years of close mowing. Yet they have, even the tall species such as knapweed and meadow scabious. They're there in the lawn area, kept to a miniature size and a vegetative state, but surviving. Downland plants are adapted to pasture and under traditional management there must have sometimes been long periods when grazing was too intensive to allow them to flower.

Stopping mowing and grazing altogether is another matter. Rather than increase the wildflowers it can drastically reduce them. I saw this happen in my own field when I fenced off the small corner where I used to pitch my tipi. It had been one of the more diverse parts of the field but when it was no longer cut for hay it soon changed. A small number of tall plants took over and gradually crowded out the wildflowers and the smaller grasses. After a few years the only wildflower left was the tall hogweed, holding its own among just three vigorous species of grass. The contrast between this drab display and the riot of colour in the rest of the field was striking. Although there was a huge loss of biodiversity the change will have benefited small mammals such as voles, which like dense cover, and a few new plants moved in. One of these was the meadow cranesbill, a large wildflower with big, bluish blooms which, despite its name, is often found in uncut grass. Creeping thistle and hedge bindweed also came in and a clump of great willowherb, which sprang up on the site of a campfire, persisted for years.

It's the regular mowing or grazing which allows the smaller, less vigorous plants to co-exist among the taller more vigorous ones. Every year all the plants, both big and small, are cut down to the same size and they all have an equal chance to regrow. But when mowing stops the little plants have to contend both with the competition of their taller neighbours and with the mulch of last year's uncut grass lying over them in spring like a thick thatch. As ever, human participation is a necessary component in a semi-natural ecosystem.

Natural Influences

Improved grassland tends to be uniform right across the field. Every effort is made to do away with the variability of nature. But where the human hand rests a little lighter natural influences show through. These include the soil, landform, wild animals and trees.

This passage from my notebook records how a change from a clay soil to a sandy one is reflected in the grassland vegetation. It comes from a walk on the semi-natural pastures of the Kingcombe nature reserve in Dorset.

Kingcombe, 29th June 2007

There's a sudden change in the grassland vegetation as you pass from the Gault clay to the greensand. The change coincides with the line of a hedge, now grown out and no longer a barrier to animals. On the clay there are rushes while on the sandy soil there's bracken and gorse. Apart from these plants there's no other noticeable difference between the two fields. Marsh thistle*, birdsfoot trefoil and white clover are quite common in both of them. Both fields lie on the same gentle slope, the greensand uphill from the clay.

We recrossed the boundary between clay and sand again, this time at a place where the transition was emphasised by a sudden change in the landform.

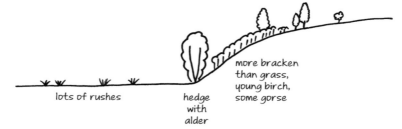

lots of rushes   hedge with alder   more bracken than grass, young birch, some gorse

Here the only noticeable plant common to both soil types is marsh thistle. On the sandy hump tormentil and foxglove grow, in contrast to the orchids and other wildflowers on the clay below. On the better-drained parts of the clay there are clumps of bramble.

*   Despite its name, not confined to wet soils.

Later I checked the geological map and found that the change of vegetation on the ground doesn't quite coincide with the change from clay to greensand shown on the map. In one place the boundary is higher up the slope, in the other lower down. I don't see how the change in soil can be so randomly different to the change in the rocks below and I suspect the truth is that geological maps aren't always very accurate. In both places where we crossed it the boundary was marked by a hedge, which must have been deliberately sited so as to separate fields which would benefit from different management.

Kingcombe is an easy place to read the landscape. The whole farm is a nature reserve and among the wildlife which nature reserves exist to preserve are the very plants which tell the story to a landscape reader. On a more commercial farm the clay fields would probably be under-drained, which would put paid to the rushes, and regular topping would discourage the bracken and gorse. Indicator plants might still be there, such as buttercup in the pastures and foxglove in the hedges, but you'd have to look for them. Kingcombe is rare in that it spent that crucial half-century which followed the Second World War in the care of an extremely laid-back farmer who had no notion whatsoever of 'improving' it. It's also particularly interesting because there are several contrasting soil types on the farm.

That doesn't mean you have to go to a nature reserve to see changes in the soil reflected in the sward. The next extract comes from a much more mundane place, a former pony paddock. I took these notes while teaching the soil observation session on a permaculture course. The field is near my home and had been used for pony grazing for as long as I'd known it, which was nearly thirty years. But three years before our visit it had changed hands and since then it had either been mown or lightly grazed with sheep. From where we entered the field, at its highest point in the south-east corner, you could see three distinct areas. Down the middle was a belt of dark green, to the left was an area of drabber grass and to the right was a flower-spotted sward.

Kia's Field, Butleigh, 12th May 2005

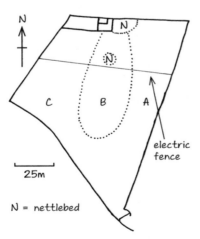

We dug a soil pit in each of the three areas, A, B and C.

A. Low grass, dandelion, both buttercups (mostly creeping), common daisy, ribwort, red clover, black medic. Clay loam, crumb structure, compacted, dark topsoil gradually grading to paler subsoil over 25cm.

B. Tall, dark green grass and not much else. Clay loam, crumb structure, no compaction, very dark topsoil grading to dark subsoil over 30cm.

C. Mostly grass, medium height. Medium loam, crumb structure, no compaction, definite change from topsoil to subsoil at 35cm.

We found earthworms in all three samples.

This patchwork of different swards is the legacy of the ponies. The tall, dark green grass and rich, dark soil in area B is the result of many years of dunging. This central stripe was clearly the ponies' latrine area. The fact that the soil there hasn't been compacted by the pounding of the ponies' hooves is testament to the power of humus to maintain soil structure. The difference between the two areas on either side is due to the difference between a heavy soil and a lighter one. The clay loam of area A is more susceptible to compaction than the medium loam of area C. Compacted soil restricts the root growth of plants and thus denies them access to mineral nutrients. This reduces the growth of the more competitive plants but there are plenty of others whose niche includes tolerating soil compaction. A notable one is dandelion, which is tolerant both of compaction and of close grazing. When the ponies were still in residence it so completely dominated the clay loam part of the field that the sward was almost solid yellow at flowering time. Now that they're gone the earthworms, plant roots and other forms of soil life are gradually healing the compaction.

This, incidentally, is a good example of the importance of timing in landscape reading. My notebook entry ended with the comment:

> The three contrasting areas appeared to go only half way down the field, because sheep were grazing the northern part behind an electric fence and they'd eaten all three swards down to the same level. From a distance A, B & C looked the same in the grazed part of the field and you could only see the difference if you went close up and examined each area one by one. In the ungrazed part they stood out clearly as soon as we entered the field.

If we'd been there a couple of weeks later the whole field might have been grazed down and I probably wouldn't have noticed the difference between the three areas. It's likely that I wouldn't have noticed it at other times of the year, either. In early spring all the grass is short and the wildflowers aren't yet blooming. In high summer the whole field is covered in a buff-coloured canopy of flowering grasses. Late spring and early summer, effectively May and June, is the best time of year to see these differences easily. In fact this year I taught the same course in July and used the same field for the soil observation session. At first glance the difference between the three areas, which had been so vivid the previous year, seemed to have disappeared altogether. Only on a closer look could you see that the canopy of flowering grasses was taller in area B and that under it the grass was dark green, and that in area A the grass grows sparsely and is rich in wildflowers. If I hadn't seen it in a previous May would I have realised that the field divides into three clearly different parts? I doubt it.

The landform often causes differences in the soil. It will be thinner on a slope and deeper on flatter land due to the twin processes of erosion and deposition. Nutrients may also have been washed out of the steeper land and deposited on flat ground at the bottom. These differences can be reflected in the plants which grow on both soils, even if the soils are the same in other respects.

For example you may see bracken on a slope and creeping thistle on flatter ground at its foot. But you can't always assume that the soil is in fact the same in other respects. A difference in the underlying rocks may be the very reason why one part is sloping and the other part flat, as in the example from the Kingcombe nature reserve shown in the sketch on page 232.

The landform also influences the way animals graze. Since they feel more comfortable on flat ground they graze the flat parts of a field more than the steeper parts. You can see this most clearly on strip lynchetts, those large terraces made by medieval cultivation. (See page 93.) Through much of the year the grass on the flat parts is short and green while the steep 'risers' bear the buff colour of stemmy, flowering grasses. Two things reinforce this effect. Firstly, the flat bits get more dung on them so the soil becomes richer. Secondly, the flat parts absorb summer rain while the steep bits shed it. Both these effects make the grass greener on the flatter land. It's also a self-reinforcing spiral because close grazing keeps the grass in its leafy stage of growth which makes it even more attractive to the animals. (See photo 15.)

This is the pattern of grass growth you get in any field with an intimate mixture of steep and flat land. But where the flat area is large and accessible it will almost certainly be improved and this will override any contrast caused by the grazing preferences of the animals. You might wonder why, if lynchetts were originally created by cultivation, they shouldn't be improved now. If you can get on them to plough surely you can spread fertiliser on them? But there's a big difference between what was worthwhile to hungry medieval subsistence farmers and what will pay in the modern age of big machinery and the market economy.

In areas of the country which are largely arable there's a similar contrast between flat and sloping land, though here the flat is not improved grassland but arable. Leys may be grown as part of the arable rotation but almost all the permanent grassland is on land which is too steep to plough. In the chalk country this often produces a sandwich landscape with arable on the hilltops and in the valleys and permanent pasture on the steep hillsides between.

A common feature of grassland on steep slopes are the little horizontal terraces known by the rather inelegant name of terracettes. When grazing animals, especially cattle, do venture onto the steeper slopes they prefer to walk across the slope rather than up and down it. This is much more comfortable for a four-footed animal. It's also easier for them to follow in the footsteps of another animal which has been that way before, so eventually a series of paths gets worn across the slope. Soil is eroded away on the uphill side of each path and deposited as a little ridge on the downhill side and thus a miniature terrace is formed. In wet weather the paths get poached but when there are no cattle in the field they soon green up. Each path is close enough to its neighbours so the slopes between them can be grazed without the cattle having to leave the path. The result is a whole series of terracettes running

across the slope, giving it a corrugated texture. Terracettes are very much a feature of grassland. I did once see them in woodland but I took it as a sign that it must have been a recent wood. (See photo 15.)

Terracettes can influence the distribution of grassland plants. I know a shady, north-facing hillside covered with terracettes where both primrose and lesser celandine grow abundantly. The primroses, being sensitive to trampling, only grow on the steep bits between the paths. The celandine, which is resistant to trampling, grows both on and off the paths. This gives the hillside a stripy appearance in spring, with the pale yellow of primrose lying in swaths across the contour and the darker yellow celandine dotted all over.

The following pair of sketches from my notebook illustrate some of these interactions between slope and grassland.

Wearyall Hill, Glastonbury, 26th Aug 1995

A couple of days after the first rain for a long time:

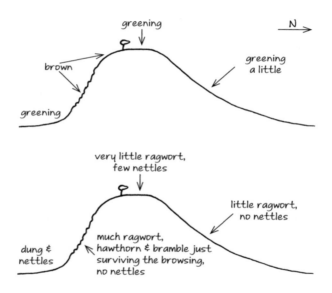

Overall there's a short grass sward, with yarrow, white clover etc. The grazing is mainly by cattle.

The first sketch illustrates how slope influences soil moisture. Summer rain coming after a dry spell gives the best opportunity to observe this but you need to wait at least a couple of days for the grass to green up. In this case the difference is reinforced by aspect as the steeper slope faces south and the more gradual one north. The second sketch shows how slope affects the intensity of grazing and thus influences the vegetation. The lack of ragwort on the more intensively grazed areas is probably due to the tighter sward, which gives it less opportunity to germinate and grow. By contrast, there's plenty of bare soil on a steep slope with terracettes.

The influence of wild animals on grassland vegetation is usually smaller and less obvious than that of farm animals. Herbivores ranging in size from deer to grasshoppers may graze but it's hard to distinguish what they eat from what's eaten by the domestic animals. The main exception to this is the rabbit, which tends to graze a small area very tightly and leave the rest. (See page 116.) The bare earth left by creatures such as badgers, rabbits, moles and earthworms gives an opportunity for seeds to germinate. This is especially important to the few annual plants which grow in grassland. One of the most visible grassland animals is one which actually spends its life underground, the yellow meadow ant. The visible part is its home, the anthill, is a characteristic feature of old pasture on a dry soil. In size and shape anthills are like overlarge molehills with flattened tops but unlike molehills they're covered in a tight sward of grass and herbs. (See photo 10 and 35.)

You don't get anthills on meadow, because they would be flattened by mowing. Even on moderately intensive pasture they'd be destroyed by chain harrowing, topping and even the trampling of the cattle. So they're an indicator of lightly grazed pasture. They're also an indicator of dry soil, because the ants live under the hill as well as in it and they don't tolerate wetness. In some places you can see quite clearly how they've sought out the best-drained part of a field. In a steep field on a clay soil you may see them strung along the little ridges on the outer edge of terracettes, like so many beads on a string. On the Mendips I've seen them seek out the line of a former wall, now just a grassy ridge a few inches higher than the ground on either side. But you'll often see them spread out across a field, usually favouring the warm, sunny south and south-west slopes. The ants love heat and they will abandon their hills if a pasture is left ungrazed and the shade of the tall grass cools them down too much, though they may recolonise an old hill if conditions become suitable again. The hills are said to grow at an average rate of one litre per year, but this can vary widely and can't be used as a reliable way of dating how long the field has been keep under a regime of light grazing.

Rabbits seem to have some affinity with anthills. They often leave their droppings on the top and dig away a chunk of the downhill side. This seems to be their territorial marking scrape transferred to the convenient vertical surface provided by the anthill. Anthills also have their own distinctive vegetation. Plants which like hot, dry conditions seek them out. I once saw a piece of downland on the Mendips where there was marjoram on every anthill but none at all on the ground between them. Wild thyme often grows on them but often only on the dry, sunny south side. Mosses usually only grow on the moister north side. On a cloudy day you could use this pattern as a compass to help find your way.

If anthills mark the warmest points in a field the coolest places in summer are under the trees, and animals congregate there on hot days. The trampling easily leads to bare ground as the plants are already weakened by shade from the tree. Hedgerow trees are more shady than free-standing ones because the hedge adds to the shade. Free-standing trees are more likely to have bare soil under them if they're the only shelter in the field and so get heavily used. Nettles are very common under trees in grassland due to the dung of the sheltering animals. In the wood pasture of Moccas Park that there's hardly a single big tree without a bed of nettles under it. Sometimes you get a 'doughnut effect' with bare soil in the shadiest zone near the trunk and a ring of nettles further out. In orchards which are no longer harvested there may be nettles round the trees because of the accumulated nutrients from fruit which has fallen and rotted in situ. Sometimes nettles grow under trees not so much because there's a specially high level of nutrients there but because tractors can't get in there to mow. Most modern fields are quite rich enough in phosphorous to support nettles. It's only their intolerance of cutting that stops nettles from out-competing the grass.

OBSERVATION SUMMARY

Kinds of Grassland

Improved. This is a sown crop. It's usually pure ryegrass but may be a mixture of ryegrass and clover. In rare cases, usually on organic farms, it may be a rich mixture of grasses, clovers and herbs. Ryegrass is green throughout the year whereas fields composed of other grasses take on a buff hue in summer when the grasses are in flower. Highly fertilised ryegrass is very dark green.

Semi-improved. This is either improved grassland which has deteriorated or unimproved grassland which has been regularly fertilised. It usually contains a limited range of grasses, with or without clovers, plus the commonest grassland herbs, including buttercups, plantains, dandelions and thistles. A high population of buttercups is usually a sign that a field has not been ploughed and reseeded for a long time. In spring, before the grasses flower, a heavily fertilised semi-improved grassland can be the same dark green as pure ryegrass.

Unimproved or semi-natural. This is very rare. It contains a wide variety of grasses and wildflowers. You can expect much more diversity on alkaline soil, e.g. chalk downland, than on acid soil. Plants which may alert you to a semi-natural grassland include cowslips in April and ox-eye daisy in summer. The wildflowers are usually at their peak in June. Unimproved grassland is often a yellowish or brownish green compared to the dark green of fertilised fields.

Orchards. The trees in traditional orchards have tall trunks and spreading crowns with continuous grassland below. Modern orchards consist of small, usually conical trees with a strip of bare soil beneath each row.

Mowing

Silage is cut early, mainly in May, from young, leafy grass and a second or third cut may be taken in summer or autumn. Hay is cut later, usually in June, from grass with more stem in it. After both hay and silage have been cut a field looks pale and yellowish till it greens up with new growth. The grass which regrows after haymaking, the aftermath, is normally grazed.

A traditional meadow, which is usually mown rather than grazed, has a tall, drawn-up plant structure. Tall-growing species, eg red clover, are favoured in the long run.

Grazing

A pasture, which is grazed rather than mown, has a short, springy sward of compact plants. Short-growing species, eg white clover, are favoured in the long run. Many grassland herbs survive in pasture without flowering.

Sheep graze short and leave the sward with a lawn-like appearance. There are few plants they won't eat, apart from nettles.

Cattle leave a longer sward and reject more of the coarser weeds such as docks, thistles and ragwort. They also reject the grass which has grown up in their own dung pats, which often appear as clumps of taller, uneaten grass and may give the field a pimply appearance.

Horses are the worst grazers. They use part of the field as a permanent latrine and never graze there. If horses are not alternated with other grazers the latrine area develops a sward of tall grasses and coarse weeds such as docks, thistles and nettles. The grazed area is often eaten down extremely short and may be compacted. In extreme cases the grazed area may be colonised by docks, which are tolerant of compaction.

Overgrazed pasture in general is characterised by thistles, docks, ragwort, bare ground and poaching.

Plants of gateways have to tolerate both soil compaction and trampling. Greater plantain, pineapple weed, knotgrass and annual meadow grass are the commonest gateway plants. Mayweed is often the first pioneer on the bare, compacted soil when the animals are removed.

Anthills indicate an old, lightly-grazed pasture. They favour the warmest, driest patches of soil.

Unused Grassland

Tall grasses and herbs, e.g. hogweed, which were present in the original sward take over. A few plants which do well in uncut grassland, e.g. meadow cranesbill and hedge bindweed, may colonise.

Slope

Flat ground is almost always improved unless the area is too small for convenient tractor working. The most likely place to find unimproved grassland is on a slope too steep for tractors. Sometimes you will find a patch of semi-natural grassland in a steep corner of an otherwise improved field.

A field which is an intimate mixture of steep and flat parts, e.g. strip lynchetts, is usually used as pasture. For much of the year the flat parts will be green while the steep parts are brown because the flat parts are more intensively grazed.

Time of Year

In spring fertilised fields are dark green. Unfertilised fields, either unimproved or semi-improved, stand out with a more yellowy, mottled colour.

May and June are the best times to see wildflowers. In general, distinctions between different swards stand out most at this time of year.

In July, when the grasses are flowering, improved swards of pure ryegrass will still be green while those of other grasses will turn buff-coloured. Distinctions between unimproved and semi-improved swards are obscured by the tall, flowering grasses and wildflowers are difficult to spot.

Just after a field has been mown or heavily grazed all swards look very much the same. A pale yellow colour in a recently cut field does not indicate dryness.

When the grass of a mown field starts to regrow it is a uniform green, contrasting with the more mottled green of a grazed field.

Heaths and Moors

An open moor has a real feeling of wildness. With nothing in sight but mile upon mile of heather and moorgrass, and no sound but the sigh of the wind and the croak of a raven, you can feel yourself truly immersed in nature and far from any human influence. But then you see sheep quietly grazing and you know that the apparent wildness is an illusion.

Most of the moors in Britain are semi-natural, forged from the wildwood by grazing that was persistent enough to prevent the regeneration of trees. If the trees were felled or the land was cultivated for a while in the process of making the moor, it makes little difference. What prevented the wood from regenerating on that site was grazing. Whether this grazing led to grassland, heath or moor, depends on the soil and the climate. Where they're favourable you get grassland. Where they're not you get that rougher landscape which we call heath when it's in the lowlands and moor when it's up in the hills.

Heath literally means heather and heather is the characteristic plant of both heaths and moors. But they also contain other plants, such as bracken, gorse, grasses and sedges, and can be dominated by them over large areas. In fact there's no hard and fast distinction between a rough, unenclosed grassland and a grassy moor. As always, our definitions are no more than a crude attempt to impose our own order on the infinite variety of nature. The distinction between heath and moor isn't hard and fast either, though there is a broad difference in the factors which have formed them. Heaths are what they are only because their soil is poorer than that of the surrounding farmland while moors are formed by both soil and the harsh climate of high altitude. But there are many similarities and much of the information below under Heathland applies to Moorland and vice versa.

Heathland

Heaths are areas of the lowland landscape which are just too infertile to farm, usually on sands and gravels. These soils are acid, poor in nutrients and prone to drought. The heather itself increases soil acidity and the change from deciduous woodland to heath can mean a change from brown earth to podsol. (See page 50.) Broadleaf trees, with their deep roots, bring up bases from the subsoil and spread them on the surface when their leaves fall. Heather, by contrast, has shallower roots and leaves which produce an acid litter when they fall.

This acid can easily tip the balance in a soil that's prone to acidity and start the process of podsolization. Before the modern age farmers could do very little to remedy the deficiencies of these soils and the land was left as common grazing. The small but steady loss of nutrients from the heathland soil in the meat, milk or wool produced on it would tend to make the soil even poorer over time. Some heaths may have been cultivated during their history but this only speeded up the decline of the soil. It wasn't till the nineteenth century, when the railways could bring in lime from distant hills and steamships could bring guano and Chilean nitrate from the far side of the world that farmers had the power to turn the heaths wholesale into permanent farmland.

Today very little heathland remains. Once-large areas of heath, such as the Suffolk Sandlings, have been reduced to isolated fragments in a sea of fertile fields. Others, like Breckland and the Dorset heaths, have been mostly coniferised. Small areas of heath, having no real place in modern agriculture, are often under-grazed and succeed to woodland unless they're taken in hand by a wildlife trust. The only really large area of heathland left in the lowlands is the New Forest in Hampshire. About half woodland and half heath, the New Forest is both a royal forest and a common. It's the only forest in Britain which still has something close to its medieval legal structure, with a forest court and officials called Verderers and Agisters. Perhaps more importantly from a landscape point of view it still has a large and determined body of people who have rights to pasture animals on it. Their attitude is perhaps reflected in the name of their organisation, the New Forest Commoners' Defence Association. They're mostly part-time farmers and grazing animals in the Forest is perhaps more a part of their lifestyle than an economic activity. Their determination to go on doing it is what keeps the heathland alive.

A walk I took across some New Forest heathland provides an introduction to the relationships and processes that go on in a heath.

Broomy Plain to Slodden Inclosure, New Forest, 12th October 2006

The heath is a mosaic of different vegetation types but none of them consists of just one plant. They're all a mix, though usually with one species predominating. The size of the patches which make up the mosaic varies a lot according to the landform. On the gravel plateau of Broomy Plain the patches are large, measuring hectares. On the central, flat part of the plain heather* predominates, with bell heather, gorse and grasses mixed in with it. On the sloping edges there's bracken.

I went through Broomy Inclosure and came out in the little valley of the Dockens Water. On the south-facing slope of the valley the mosaic is at a much smaller scale, with some patches only a few paces wide. Where the soil is wet there's a mix of heather and cross-leaved heath, the latter increasing where the soil gets wetter. Where it's very wet the heathers give way to purple moorgrass and bog moss. I soon learned that these were

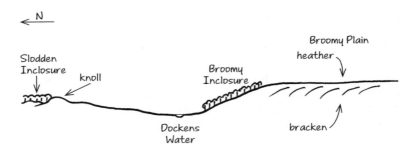

places I had to avoid if I wanted to keep my feet dry! On well-drained loams bracken and grass alternate with each other, with tormentil, sheep's sorrel and the occasional very weak bramble, all growing in the grassy parts. At the top of the slope I found a knoll where heather, bell heather, bracken, bilberry and gorse are mixed. Here the soil is a podsol. On the whole the wetter patches are further down the slope and the drier nearer the top, but not in every case. The boundaries between one patch and another are mostly clear and abrupt.

I made my way along the valley and recrossed the Dockens Water further up. Here the valley floor is a little wider and on the flat ground either side of the stream there's a 'lawn' of close-cropped grasses. The soil, a sandy loam with some silt in it, is well-drained here. The place has a strangely domestic air compared to the chaotic heathland vegetation, almost like a golf course or a garden.

* Here and throughout this chapter 'heather' means common heather or ling, as opposed to its cousins, bell heather and cross-leaved heath.

All the plants I saw that day are typical of poor, acid soils but some plants tolerate poorer, more acid soils than others. That accounts for some of the variation I saw in the vegetation. The other big factor is water, and the wetness or dryness of the soil accounts for the rest.

Broomy Plain is a typical dry heath on a gravelly soil and some of the plants which grow there are specially adapted to resist drought. Plants constantly lose water through tiny pores in their leaves, so drought resistance usually involves some adaptation of the leaf which reduces this flow. In gorse the leaves have been reduced to spikes so as to minimise the surface area exposed to the air. In bell heather the small, narrow leaves are curled in on themselves so their undersides, where all the pores are located, form an enclosed pocket. The air in the pocket stays more humid than the outside air and thus reduces the loss of moisture from the pores. This makes bell heather especially drought-tolerant and perfectly matched to the soil of Broomy Plain. Nonetheless common heather dominates here. It's a very competitive plant and grows more vigorously and reproduces more abundantly than the other two heathers. If common heather didn't exist bell heather would probably dominate on dry sites like Broomy Plain, but thanks to the competitiveness of heather it can only dominate on soils which are actually drier than it would prefer.

As the rainwater drains down through the sandy soil of Broomy Plain it takes with it what little in the way of bases that soil ever contained. The heathers are supremely well adapted to such a poor, acid soil but not so common gorse, which usually grows where the soil is slightly richer. On most heaths it's mainly found by roads, tracks, old boundary banks and gravel workings. In these places the soil has been disturbed and subsoil, which contains some of the bases leached out of the upper layers, has been brought to the surface. The few gorse bushes scattered over Broomy Plain perhaps indicate a soil which is slightly less deficient in bases than others which support only heather. Bracken is a bit more demanding than gorse. It's confined to the sloping edges of the plain. Here it can find more nutrients because where there's a slope there's erosion. This means at least that the subsoil is exposed and at best that the slope cuts into slightly richer strata which lie below the flat cap of gravel.

The other dry area I saw was the knoll on the other side of the little valley and this had very mixed vegetation on it. This is unusual on heathland, where normally one or two plants dominate and the others are strictly subordinate. The soil is clearly a podsol, with a thin layer of pure humus lying over a topsoil so leached that it's almost white. I had no means of digging deeper, but no doubt there's a pan below, where what little iron and clay the topsoil once contained are now lying in a compact layer. What's left is pure sand and I think the extreme dryness of this soil may account for the diversity of plants, as it reduces the vigour of the would-be dominants so other plants can co-exist with them as equals. Bilberry is one of them. Although it's not confined to dry places it's well adapted to them, with a thick, waxy coating on its leaves which reduces water loss.

The water which drains so freely down from the plains and hillocks emerges again in places such as the wet patches I found on the slope. For water to emerge on a slope like this there must be an impervious layer below the sands and gravels, probably of clay. (See page 264.) Here the cross-leaved heath makes its appearance. Just as bell heather specialises in drier soils so cross-leaved heath specialises in wet ones. But the competitive common heather gets in on the act too. Cross-leaved heath, like bell heather, can only dominate on soils which are a bit wetter than it really likes. There's only a narrow band where it's the dominant plant, between the mix of itself and heather on the one hand and the purple moorgrass and bog moss on the other. Purple moorgrass is a tall, tough plant which often grows in tussocks and whose narrow flowering heads are usually, though not always, purple. It's very tolerant of poor, wet soils but it's an unpalatable grass with little feed value and grazing animals will avoid it if they have the choice.

On the patches of well-drained, loamy soil nearby there are sweet grasses and they're a different matter altogether. Like the bracken which also grows on these loamy patches, they need a bit more nutrition and a bit less acidity than most other heathland plants and they get these from the loamy soil, which is a little richer in bases than the typical sandy soils of heathland. The sweet grasses are the most palatable and nutritious of heathland plants and the animals graze them by preference. The grass on these patches is grazed short.

The sweet grasses are in competition with the bracken. If grass and bracken are left to battle it out between them on a deep, well-drained soil the bracken wins every time. This is what has happened on Blawith Common, which I described at the beginning of the book. (See pages 1-2.) But if you throw grazing animals into the balance the contest is much more equal. Since bracken is poisonous, animals hardly eat it but it is vulnerable to being trampled. When bracken is trampled its tissues are crushed and the cells which hold the poison are broken, releasing it inside the plant and thus harming it. Sheep, with their small feet, tend to step between the plants rather than on them. But the main grazers in the New Forest, apart from the deer, are ponies and cattle and they can have a significant effect on bracken. In fact there's a symbiotic relationship between grass and its grazers. The more the animals eat the grass the more they trample any bracken growing along with it and the more the grass prospers, which in turn means more food for the grazers. To some extent this leads to a segregation between bracken and grass, as the animals concentrate on the places where there's more grass and shun the places where there's more bracken, reinforcing any difference which already exists.

The largest patch of sweet grasses I saw that day was the strip of lawn either side of the Dockens Water. The streamside lawns provide the richest grazing of all in the New Forest and though they make up a tiny proportion of the Forest area the animals get a large part of their nutrition from them. The streamside soils are rich and relatively alkaline because this is where the bases which are leached out of the rest of the landscape end up. The poor Forest soils don't contain much in the way of bases in the first place but the area of leaching is large compared to the area where they accumulate. Good drainage is essential for the sweet grasses. Some of the lawns may flood in winter, and this is how they gain much of their fertility, but they all dry out in summer. Where the water can't get away even in summer peat accumulates in the valley and a bog forms. But I didn't see a valley bog that day. (See photo 36.)

What I saw was, of course, a snapshot in time. But heaths, like any other ecosystem, are constantly changing. Much of this change centres around the lifecycles of the dominant plants. Heather has a lifespan of twenty-five to thirty years. It starts out as a little conical plant which doesn't cover much ground and allows plenty of space for other plants to grow beside it. As it grows it develops a strong dome of dense foliage. At this stage it's at the peak of its vigour and excludes most other plants. As it matures it develops a more flat-topped shape and in its old age the plants start to collapse. In the collapsed state each plant is like a horizontal cartwheel, with the spokes of bare branches radiating from the middle and a rim of green growth round the edge. Once again there's space for other plants to grow, this time in the middle of the decaying heather plants.

The collapsed stage may be the time of greatest diversity in the life of the heath, with mosses, lichens, grasses and flowers all taking advantage of the light which is let in by the collapse of the heather. But if bracken is present it may take up the space and become dominant for a time. On some heaths dominance can swing backwards and forwards between heather and bracken according to which of them is in its most vigorous phase of growth, but it doesn't necessarily result in an even see-sawing down the years. All sorts of factors can swing the balance one way or the other.

Fire is one of these. Heather thrives under a regime of controlled burning. After a fire it reseeds copiously and unless the plants are too old or the fire too hot it will sprout from the rootstock too. Fire can be used as a management tool to keep the heather at a young, nutritious stage of growth. But an accidental fire which sweeps uncontrolled across the heath is very much hotter and destroys both plants and seed. Bracken, which has rhizomes deep in the soil where the heat hardly penetrates, is better able to withstand such a fire than other heathland plants and an uncontrolled fire can turn a mixed heath into one of pure bracken.

Grazing is another factor, and it may favour either heather or bracken according to the kind of grazing animals. Many heaths are now grazed only by rabbits and they give the advantage to bracken because it's poisonous and they eat everything else in preference. Cattle and horses, on the other hand, will trample out the bracken and favour the heather. The decline in grazing by domestic animals in recent years has led to the takeover of many heaths by bracken. The result is a loss of biodiversity. Bracken is so competitive that little can live underneath it. Although some spring-blooming wildflowers can survive under a light canopy of bracken, most heathland plants can't and almost nothing lives under a dense stand. Some nature conservationists have resorted to spraying it with herbicides.

Grazers are not the only animals which have a functional role in heathland. Bees are important pollinators of both heather and bilberry. Beekeepers often put their hives out on heaths and moors. The flavour of heather honey is exquisite and the more discriminating beekeepers keep it separate from other kinds and sell it as a single-species speciality. (Unfortunately it's more common in this country to lump all honey together and the only choice we usually get is between set and runny.) Where no hives are put out wild bees and other insects do the job of pollination. Spiders are a major predator on these insects and the bushes of heather and gorse make an ideal framework for their webs. I've seen it suggested that one benefit of farm animals on a heath is that as they barge their way between the bushes they break the webs, thus saving the lives of many insects and improving the pollination of heather.

The bushy structure of heathland is important to other predators too. Stonechats and whinchats, two little birds which are often found on heathland, need shrubs about waist height as hunting perches. It doesn't matter what the plants are as long as they're the right height. Bracken or young trees are equally suitable, though once the trees grow too tall the chats will disappear.

25. Oliver Rackham inspects the thriving ecosystem in the heart of a very old oak.
(See page 165.)

26. This tree was drawn up by a tall, unmanaged hedge which has since been laid.

27. This tree has had a complex pollarding history but the sharp difference in age between trunk and branches confirms that it is a pollard.
(See page 174.)

28. A hilltop wood in chalk country is almost always situated on a layer of clay-with-flints above the chalk. Note the steep slope scrubbing up in the foreground. Salt Hill, Hampshire. (See page 178.)

29. A grazed wood on Dartmoor. The lack of shrubs and the grassy floor are typical of grazed woods. The broad spread of the tree indicates it has never had close neighbours. (See page 184.)

30. A semi-natural wood in April. The herbaceous layer is well in leaf and flowering, the shrub layer just leafing and the tree layer still dormant. Kingcombe Copse, Dorset. (See page 197.)

31. My semi-natural meadow (left) and the heavily fertilised ryegrass mono-culture just over the hedge (right), both in late June.

32. Churchyards sometimes preserve the semi-natural grassland whcich has become so rare on farmland. Ventnor, Isle of Wight.
(See page 231.)

John Adams

33. On this headland the degree of exposure to salt spray determines the vegetation, with grass on the exposed promontory and scrub further inland. South Devon.
(See page 221.)

34. Mayweed growing in the gateway of a hay field.
(See page 230.)

35. A lightly-grazed pasture. Anthills have formed on the south-facing slope (see page 237), while brambles move out from the hedge, taking succession forward.

36. The tightly-grazed turf of a streamside lawn contrasts with the rough vegetation of the heath. Dockens Water, New Forest.
(See pages 243 and 245.)

37. Succession on heathland in Surrey. In the absence of grazing, birch and pine take over from heather and purple moorgrass. (See page 247.)

John Adams

Chris Wormald

38. Bracken on the slopes with heather and grasses on the flatter tops is a common pattern on moorland. (See page 249.) But plantations are located more or less at random. (See pages 178-179.) Dartmoor, Devon.

39. Oilseed rape often grows spontaneously by lowland water-courses. It's palatable to cattle and here it has only survived on the bank which is not being grazed at present. Somerset Levels.

40. Pond vegetation forms concentric rings reflecting the varying depth of the water. (See pages 268-269.)

41. On the Somerset Levels you're never far from the surrounding upland. Part of Wedmore Island seen from Tadham Moor. (See Case Study, pages 370-378.)

42. A typical ditch on the Somerset Levels. Note the distinctive leaves of arrowhead in the middle and the mainly grass-like plants towards the edges. (See page 777.)

43. This Devon hedgebank sports a rich mix of woodland and hedgerow flowers in spring. (See pages 281-282.)

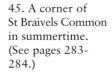

44. An old, mixed hedge in South Devon. The tree with the yellowish fruits is a crab apple, an uncommon tree which indicates an ancient hedge.

45. A corner of St Braivels Common in summertime. (See pages 283-284.)

46. Hedgelaying, Somerset-style, at my field.

47. This line of lime trees in Surrey shows clear signs of once having been a laid hedge. (See page 295.)

John Adams

John Adams

48. Neglected willow pollards like these are rich in wildlife. Suffolk.

In fact whole heaths can disappear as the trees grow, by succeeding to woodland. Scots pine and birch, and also rhododendron where it's present, are avid colonisers of heath and many heaths have become woods as grazing has declined in recent decades. (See photo 37.) Preventing this succession is another reason for burning. The occasional tree is no bad thing, though. Nightjars, those elusive and mysterious birds which haunt the evening gloom with their churring call, need trees as well as open ground in their habitat. They often live on the edge between wood and heath and these days some of them live in forestry plantations, where recently felled compartments take the place of open heath.

Heaths are warm places. Where the shrubs are not too dense they provide shelter without much shade and the dry, sandy soil heats up more quickly than a moister loam or clay. Sandy soils are also looser and their plant cover is more easily worn away by animals' hooves, so there's more bare soil in heaths than in other ecosystems. The heat attracts reptiles. As cold-blooded animals they need external heat to build up enough energy in their bodies to go hunting. Sand lizards need the combination of heat and bare soil as their eggs will only incubate successfully if laid into bare soil which is exposed to the sun. They're one of a number of species which are at the northern edge of their range in Britain and here only live in heaths.

Heather, bracken, bees and lizards are emblematic species which evoke sunny summer days on the heath. But some of the most vital members of the heathland ecosystem are hardly ever seen and not so readily associated with heaths. They're fungi. When a mushroom or toadstool pops its head above ground it's even less than the tip of the proverbial iceberg. It's just the reproductive organ of the fungus, like the flower on the holly bush or the acorn on the oak tree. Almost all of the fungus' body is below ground and it's there all the time, not just for a few days in the autumn. Fungi are everywhere beneath the surface, filling the soil with millions of threads. Some of them live on dead plant material and others are parasites but many of them have a symbiotic relationship with the plants. Without these symbiotic fungi many heathland plants simply couldn't survive.

Most plants can fuse their roots with a fungus which provides them with mineral nutrients in exchange for organic food. (See page 122.) But no plants are more dependent on this relationship than the heather family, which includes not only the heathers themselves but also bilberry, its relatives such as cowberry and cranberry, and even rhododendron. These are all plants which live in the poorest, most nutrient-deficient soils, plants which can live happily on soils which even gorse and bracken would find too poor. But they couldn't do it on their own. No green plant could wring enough nutrition out of such soils without the help of a fungus. Ultimately, every creature in an ecosystem depends on the food that green plants produce by photosynthesis. But plants can't grow without a supply of minerals from the soil, so in heathland the fungi are as essential as the plants. While above ground the drama of competition is played out between plant and plant, helped or hindered by insects and mammals, below ground the quiet co-operation between plant and fungus makes it all possible.

Moorland

The journey from lowland to upland often goes something like this.

Cambrian Mountains, 16th June 2007

Going east from Tregaron you soon leave behind the lowland landscape with its thick hedges full of hazel, their banks brightened with foxgloves and red campion. As you climb a U-shaped valley the hedges give way to fences and the frequent farms and villages to long, lonely stretches with just a rare farmstead. Today, the blue sky and puffy white clouds over the vale have turned into a solid grey mass over the mountains. I went through a big plantation and out the other side onto unfenced country, four to five hundred metres high. It has a harsh, uncomfortable feel after the soft lushness of the lower land. Down there the bracken is chest high, up here hardly knee high and still not fully open. Only the flowering foxgloves were common to both lowland and upland and at the same stage of growth in both places.

This is a common experience. As you go up into the hills a sunny day turns into a cloudy one or a cloudy one becomes a rainy one and when you come back down to the valley you find the weather's the same as it was when you left. There are days when it's sunny on the hills but they're few and precious. The harsher climate that comes so quickly with increasing altitude is an important factor in the formation of moorland. The rocks that hills and mountains are made of do tend to be the old, hard rocks which form acid soils. But the incessant rain leaches bases from the soil so thoroughly that even a rock which would produce quite a fertile soil in the lowlands may support nothing but moorland on the hills.

This sketch from my notebook illustrates how the rock type and the climate can interact.

Blackdown, 19th May 2004

very occasional willow shrub

shrubs, mostly willow & hazel at bottom, mostly rowan at top

N →

hedge, mainly hazel with oak standards

fields of grass

Heather with some bilberry & purple moorgrass, occasional tormentil.
Soil with 25cm peat on top

Bracken & bluebell, with honeysuckle & a little bramble at bottom. Stitchwort right by hedge.
No peat layer on soil.

Hedges, hawthorn with ash & sycamore standards. Ransoms & dogs mercury in shady places

Blackdown is the highest hill on the Mendips. (See page 34.) From it you have a magnificent view down into the intricate landscape of North Somerset, with its small fields and thick hedges, bright green on the sunny May morning I made this sketch. At the foot of Blackdown, but still up in the hills, lies a strip of fields where the land flattens out before it dips down into the gorge of Burrington Combe. (See page 74.) Blackdown itself is made of Old Red sandstone while the fields at its foot are on Carboniferous limestone. Old Red sandstone can make a reasonably fertile soil where the climate's not too harsh, but up here it's another matter and Blackdown is moorland. The limestone which lies under the fields is richer in bases and its soil can support pasture of sweet grasses despite the high rainfall. The plants I noted on the sketch tell the same story as the contrast in land use. On the limestone were ash trees, dogs mercury and ramsons, all typical of alkaline soils, while just over the hedge at the foot of the hillside there were acid-soil plants: stitchwort, honeysuckle, bracken and bluebells.

As well as comparing Blackdown with the surrounding limestone it's interesting to compare it with the other hills of Old Red sandstone which lie along the spine of the Mendips. Blackdown is both the highest of them and the nearest to the sea and these factors combine to give it a much wetter microclimate than the others. It's the only one that's covered in moorland. The others are predominantly enclosed grassland or conifer plantation. One of them includes Priddy Mineries, but this is uncultivated because of lead contamination, not because it won't grow grass. (See page 143.)

Blackdown illustrates a familiar pattern of vegetation you can see repeated on many of the hills of Britain: bracken on the slopes and heather or grass moor on the flat tops. (See photo 38.) Bracken is the most competitive of moorland plants and dominates the vegetation almost wherever it grows but it's limited by soil conditions. It needs more in the way of bases, better drainage and a deeper soil than other moorland plants. On Broomy Plain in the New Forest, I suggested that it was confined to the slopes by the lack of nutrients on the flat plain. That may be a factor on Blackdown but here drainage is certainly important too.

The Old Red is a cemented sandstone which doesn't allow free drainage, so the flat hilltop is significantly wetter than the sloping hillside. The layer of peat which lies over the mineral soil on the hilltop suggests poor drainage. Bilberry favours the dryer parts of the hilltop. It grows on the Bronze Age barrows and on Blackdown's historical curiosity: a regular grid made up of lines of mounds which were built during the Second World War. The idea was to imitate the streets of Bristol in the hope that bombs would be dropped there instead of on the city. I don't think it worked. I've certainly never seen any bomb craters on the hill.

Bracken is an unpopular plant. Farmers don't like it because their animals can't eat it and nature conservationists don't like it because it reduces biodiversity. And it's spreading. Over the past half-century the area covered by bracken has increased enormously. This is partly because of the modern drive towards mono-culture, which in the hills has meant a steady decrease in cattle and an increase in sheep. It's also because bracken is no longer used for anything. In earlier times hill farmers would cut it for animal bedding and the constant cutting kept it under control. I once met a family who had spent a dozen years working a Highland croft as an experiment in sustainable farming. They made the greatest possible use of local resources, including bracken for bedding. When they first arrived there was bracken everywhere and they could cut as much as they needed right beside the byre. But as the years went by the bracken retreated in the face of the constant cutting and by the time they left they had to go to the other end of their land to find it. A third reason for the spread of bracken is global warming. Bracken is not a very cold-tolerant plant and its upper limit on the higher hills is determined by temperature. As the climate warms it's gradually extending upwards.

On Blackdown bracken has decreased. The landowner had it treated with a mulching machine, in effect a large, robust mower, and then grazed by Exmoor ponies. When I saw it on the day I made the sketch I was astonished to see the great whaleback hill shining an electric blue in the May sunshine. I've known it since childhood but I'd never seen so many bluebells on it before. The effect was spectacular but, like so much which goes on in the countryside these days, it was more gardening than farming.

Blackdown is a small, isolated moor in an area with little living tradition of using moorland. In areas where there's more of it moorland is more integrated into the farming system and this has its effect on the vegetation. This example from Mid Wales makes an interesting comparison with Blackdown.

Mynydd Llanllwni, 19th June 2007

On the lower slopes of the common the predominant vegetation is closely-cropped grass, dotted with clumps of gorse and heather. The gorse is more towards the bottom of the slope and the heather higher up. Right at the bottom, against the boundary bank, I can see a patch of pure bracken. So there's a gradation in the vegetation as you go up the slope which, if you discount the grass, can be summarised as bracken to gorse to heather. This seems to be the general pattern round the edges of the moor, though in some places it's more complex.

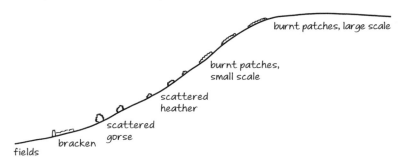

burnt patches, large scale

burnt patches, small scale

scattered heather

scattered gorse

bracken

fields

On the upper slopes, the vegetation at first looked like alternating patches of heather and grass, which I assumed reflected natural variations in the soil. But when I walked over it I saw that the mix of plants is the same in every patch: mainly heather and grasses with some cross-leaved heath and bilberry, and a little heath bedstraw and tormentil. The difference between the mainly grassy patches and the ones where heather predominates is a matter of how long it is since each patch was last burned. In the newly burned places grasses have the advantage. They grow up taller than the little heather plants and from a distance the more recently burned patches appear to be all grass. With the passing of time the heather grows up and suppresses the grass. On the patches which have been left unburnt for longest, dome-shaped heather plants alternate with tightly-grazed grass. Beside a small abandoned quarry some heaps of stony spoil are covered with a pure stand of bilberry.

The top of the moor is a wide plateau. Here the burnt areas are much bigger and generally in a younger stage of growth, indicating that they're more frequently burnt. The general impression is of a grass moor, until you walk out over it and see the heather coming up through the green grass. The sky is full of singing larks and though it's a windy day bumblebees are working the cross-leaved heath, the main plant in flower at the moment. There are sheep all over the place but Sarah (a horsewoman and one of the commoners on Mynydd Llanllwni) says it's undergrazed. She's never seen it looking so green. She says that these days there are no cattle and fewer horses and sheep. The farmers are supposed to leave the moor unburnt for five years, but they're still burning it despite that.

At first the sloping edge of Mynydd Llanllwni seems to be quite different from the hillside of Blackdown. But there is an underlying similarity between the two hills. Both have heather on the top and other, more demanding plants on the slopes. At Llanllwni intensive grazing has banished bracken to little patches on the best soil at the bottom of the hill, thus giving the competitive edge to grass and gorse. There's an interesting gradation in the vegetation from bottom to top: first bracken, then grass with gorse, then grass with heather, and finally heather which would be almost pure if it wasn't for the burning. These plants indicate how the soil gets steadily less fertile as you go up the hill. They reflect the old saying, which I've heard from both Wales and England:

> There's gold under bracken,
> Silver under gorse
> And famine under heather.

The plants were the only clue I had about the soil on the hillside but on the plateau above someone had dug a little pit. I don't know why anyone should want to do that but it gave me a chance to look at the soil. It has fifteen centimetres of peat above a pale yellow mineral soil. It was very wet, though it hadn't rained for a few days, and the bottom of the hole was full of water. This wetness suggests that the soil is either lying on an impervious bedrock or has a hardpan that's so well developed that it's impervious to water. The surface layer of peat, so typical of heather moors, is not just a sign of wetness but also of acidity. Moorland soils are too acid for the of the decomposing organisms which would create the mellow humus of a brown earth soil, and especially too acid for earthworms, which would otherwise mix the organic layer with the topsoil. In other words, the soil is a podsol rather than a gley with a peaty surface. (See pages 50-51.)

You can see from the patches of heather and grass in various stages of recovery from fire that there's a regular cycle of burning. Burning is much more common on moors than heaths and most heather moors are managed this way. The main aim of burning is to rejuvenate the heather, replacing the tough, old plants with tender young ones which are more nutritious. On some moors, like Mynydd Llanllwni, burning is done solely for the benefit of sheep while others are managed both for sheep and grouse. The burning regime is different in either case. Grouse need young heather for food but they also need old heather for shelter and nesting, so the interval between burnings is longer to allow it to grow bigger. As grouse territories are fairly small the burnt patches need to be small too so that each territory contains both young and old heather. The visual effect of these two factors combined is a clear mosaic of contrasting patches on grouse moors.

On moors which are managed for sheep the aim is purely to get the maximum of young heather and grass, so a much shorter cycle is used. Larger areas are burned at a time because it saves labour. The result is a much less obvious mosaic. The top of Mynydd Llanllwni looked almost homogenous to me that day and the only old heather I saw was in odd corners which for some reason have missed being burnt. But grouse are not the only wildlife which appreciate the longer rotation. The shorter cycle favours grass and heather at the expense of other plants and thus reduces overall biodiversity. Hence the ruling that Mynydd Llanllwni should be given a break of five years without burning. Presumably this is a condition laid down by the government as part of the new subsidy regime which links payments to wildlife-friendly practices. The old subsidies were paid on the number of animals each farmer kept and the drop in animal numbers on the moor is a consequence of the ending of those payments. This is why there was more grass on the moor than ever before, though the unusually wet summer could have had something to do with it too. But habits die hard and the farmers were still burning frequently. The urge to 'make two blades of grass grow where one grew before' is deeply engrained in farming culture.

There's an interplay on moorland between the three factors of burning, grazing intensity and soil fertility. If grazing is too hard immediately after burning it can knock the heather back so badly that it fails to regenerate. Then grass, with its structural ability to withstand grazing, out-competes the heather and takes over. Whether the change from heather to grass is of benefit to the farmers or not depends on the soil. If it's very poor the heather will be replaced by purple moorgrass or mat grass, a wiry grass with a grey cast to it. These rough grasses are less nutritious than young heather, so they bring down the productivity of the moor. If the soil's not so poor the heather may be replaced by sweet grasses, which are more nutritious than the heather. Even without burning, hard grazing over a long time can turn a heather moor into a grass moor. This is probably how most grass moors have been formed.

The effect of different levels of grazing can often spring into focus along a fence line where the grazing regime is different on either side. If you see heather abruptly stop and grass begin exactly on the line of the fence you can be quite sure the difference isn't due to natural conditions. This sketch of a hillside I saw in Mid Wales shows a contrast between intense grazing on one side of a fence and no grazing at all on the other. To the left and out of view is an oak wood. The fence has obviously been erected to keep sheep out of the wood, and placed well away from the woodland edge to allow it to spread. The space between woodland and fence is filled with bracken, heather and young birch trees. The land to the right of the fence is part of a large enclosure whose flatter parts have been converted to improved grassland. As you go down the fence line the slope gets steeper and where it becomes more of a cliff than a slope the fence stops. Below this point the steepness is enough to deter the sheep and the lack of grazing has allowed heather and birch to grow on both sides of the line. On such steep ground the thinness of the soil may also favour heather over grass. The bottom of the little valley is inaccessible to sheep and some quite large birch trees grow there.

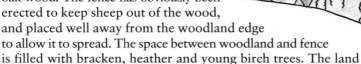

In that example the steepness of the hill had as much effect on the vegetation as the fence did. In fact hillsides are good places to observe the combined effects of soil, microclimate and grazing on moorland vegetation. On smooth slopes, like those of Blackdown or Mynydd Llanllwni, there are broad bands of vegetation which change as you go up the hill. But this pattern can be disrupted on hillsides where the gentle curves are broken by steeper slopes. Here the soil becomes thinner and heather takes over from bracken or even from grass. From a distance the dark heather contrasts with the pale green of bracken in summer and its light brown in winter.

The soil also thins out on the fringes of bare, exposed rock. I made this sketch on the Isle of Erraid in the Hebrides. These massive ledges dip back into the hillside and the deep soil at the back of each one thins out to bare rock at the lip. This is reflected in the repeated crescents of pale bracken, dark heather and light grey rock.

In the western Highlands there are tall mountains that rise up in a series of steps on a much bigger scale. From a distance you can see how heather and birch trees cling to the steep 'risers' while the 'treads' appear to be clothed in grass. In another part of the country you might put this down to grazing pressure being higher on the flat ground, and indeed that may be a factor. But the rainfall here is very high and it's more likely that the flat areas are too wet for heather and trees. What looks from a distance like grassland may in fact be peat bog covered in grass-like plants called sedges.

The mountain in the next sketch doesn't have a complex profile but it's nearly a thousand metres tall and there's plenty of scope for different environments between top and bottom. The distribution of heather and bracken may have more to do with the depth of the soil than any other factor, while the contrast between the birch wood halfway up the mountain and the mixed wood at its foot has to do with both microclimate and soil. Birch is a particularly cold-tolerant tree but, perhaps more importantly in this case, it can tolerate a low level of nutrients. On a steep slope like this erosion is always going on, and rock debris of all sizes, loosened by ice and frost and brought down by gravity or the rushing burns, is deposited at the bottom.

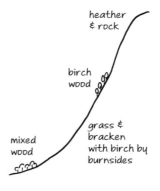

Nourished by the bases in the newly-eroded rock, a range of different trees can grow among the debris. But higher up on the old mountain soils, heavily leached by the constant rain of the western Highlands, only the thrifty birch can survive. This is a general rule in highly-leached landscapes: the more demanding trees are only found where new rock is exposed by active erosion, either at the base of a cliff or where a stream cuts down through the soil to the fresh rock below. The gullies near Kinlochleven, not a mile from the foot of Gharb Bheinn, contain the richest woodland I ever saw in that part of the country, and this is due in no small part to the supply of bases released from the rock by the water that rushes down them. (See page 150.)

On Gharb Bheinn itself even the undemanding birch grows more by the burnsides than away from them, a pattern you can see in many upland and mountain areas. Shelter plays a part in this, as a stream will always carve at least a small valley, but the extra supply of bases is more important. You can see a similar effect in upland oak woods, where the more demanding ash is usually only found in the gullies made by streams. Compared to farm crops trees are not demanding plants and in the lowlands their growth is rarely limited by the supply of nutrients. But compared to moorland plants trees are demanding. Even species like birch, rowan and eared willow, which can grow on many moorland soils, have more chance of survival on a streamside. They grow faster there and can recoup their growth more quickly after the browsing of sheep and deer. Thus they spend fewer years of their life at the small vulnerable size and have much more chance of escaping death by browsing.

At the beginning of this chapter I said that most of the moors in Britain are semi-natural, formed from the original woodland by grazing. But there are some moors which were formed by a purely natural process, the growth of blanket bogs. A bog is an accumulation of peat. Peat consists of partly-decomposed plant material which forms where conditions are too wet for it to decompose fully. Peat itself absorbs and holds a lot of water, so once started the process of peat formation is self-reinforcing and a deep layer of it can form where conditions are continually wet. Trees can't survive in the waterlogged environment of a bog and large areas of wildwood were overwhelmed by spreading blanket bogs during the more rainy episodes of prehistoric times.

You might expect bogs to form in low-lying places where the water can't drain away easily, and indeed they do. But in the wettest hill and mountain areas the rainfall is so high that bogs can form on any flat or gently sloping land where the rock is impermeable. They're called blanket bogs because they cover whole landscapes with a blanket of peat which can be several metres deep. This is on a completely different scale to the thin layer of peat on the surface of a typical heather-moor soil. Most heather moors are found in areas of modest rainfall whereas blanket bogs are the product of rainfall which is so high that saturation is the normal condition of the soil. They're concentrated in the rainy north and west of Britain. On the Pennines they occur mainly on the western side of the range and at higher altitude, while heather moors are more common on the east side and lower down. In Scotland, the heather moors of the Cairngorms contrast with the boggy brown hills of the north-west Highlands. In general, most of the moors in the Grampians have been formed from the wildwood by grazing, while most of those north-west of the Great Glen have been formed by the spread of blanket bog.

Bogs can be wonderful and complex ecosystems, with a rich diversity of plants. The most characteristic plant is bog moss or sphagnum, which not only tolerates these conditions but is partly responsible for creating them as it absorbs many times its own weight of water. Pick up a fat handful of bog moss and squeeze it. Water pours out and you're left with a wisp of moss as thin as a feather in your hand. It needs wet conditions in order to get established but once it's there it makes them all the wetter. Cotton grass grows in both bogs and shallow water. With its distinctive seed-heads like blobs of cotton wool it's a useful visual warning of where it's not safe to walk. Another common bog plant is the short, wiry deer grass. It dominates wide areas of blanket peat and in the western Highlands it's mainly responsible for the golden brown colour of those great, rounded hills in late summer. Despite their names, both these plants are sedges rather than grasses. You can tell a sedge from a grass because their stems are solid and more or less triangular in section while those of grasses are hollow and round.

The extreme poverty of bog soils has given rise to some curious plants, and some interesting folklore. Sundew is a little plant with unremarkable white flowers but very distinctive leaves. The leaves have a sweet scent and a covering of prominent red hairs, both of which attract insects. No sooner does an insect land on a leaf than the hairs, which are covered in a sticky glue, curve in on it and trap it. The plant then proceeds to digest the insect, not for the organic food it contains but for the precious minerals, especially nitrogen. Bog asphodel, with its spike of yellow star-flowers, used to be called break-bone. Farmers knew that sheep which graze where it grows often suffer from brittle bones. They had no way of knowing that it was the deficiency of calcium and phosphorous in the peat that was the culprit rather than the plant itself.

Blanket bogs aren't uniformly wet and in many places they become dry enough for heather to grow and sheep to graze. In time the peat may be broken up by drainage channels. These often start out as sheep tracks or small-scale peat diggings and can end up eroding down to the bedrock. The blocks of peat which remain are called haggs and are much drier than intact bog. Much the same range of plants grows on them as before but the balance of species changes. Only the bog moss dies out entirely and heather is usually the main beneficiary. Pine and birch trees are not limited by the poverty of the soil and may colonise if it dries out enough and the grazing pressure is sufficiently low. But at high altitude or near the coast succession may be very slow to start. Trees find it hard to get established in a cold, wet and windy climate, especially if the wind is salty. In some places there's also a lack of seed parents.

Railway lines can reveal much about succession on moorland. Where a line crosses a moor with a soil and climate that are favourable for tree growth a belt of trees may spring up between the fences which run either side of the track. You can see this, for example, in Glen Ogle in the Trossachs. From the main road you can look across the glen and see where the old railway ran on the opposite slope. It's now a solid ribbon of birch trees, cutting across the bare heather and grass of the hillside. But where a railway line crosses high country altitude also becomes a factor, as the following extract from my notebook illustrates. It records how the vegetation changed over part of a train journey, from its high point on Rannoch Moor down to Crainlarich, which lies in the upper valley of the river Tay.

The West Highland Line, late October 1992

The railway line is fenced against sheep on either side but on Rannoch Moor there's little contrast between the vegetation inside the fences and on the open moor outside them. There's just the occasional birch tree growing inside.

On the section between Bridge of Orchy and Tyndrum there's more contrast. Inside the fences are: birch, willow, rowan, gorse, heather, bracken and grasses. Outside there's only grass and some thin bracken.

Between Tyndrum and Crainlarich the trees inside the fences are denser and broom, bramble and rosebay willowherb appear, followed by male fern and bog myrtle.

It was a bit strange for bog myrtle to make its appearance so late in the journey, when the high moors were far behind. It's a typical bog plant which grows in very wet, acid places. Although there was nothing surprising about finding it down in the glen it must be a hundred times more common up on the moor, where I didn't see it. It just shows how you can never read the landscape on the basis of a single plant.

Another thing I saw from the train that day, as it snaked its slow way through the mountains, was an abandoned farmhouse on the edge of Rannoch Moor. I didn't make a written note of it and I don't remember very clearly what the building looked like but I still have an image of the vegetation. Surrounding the house and its outbuildings, in complete contrast to the wild moorland on every side, was an area of grass, short as a lawn and soft green against the harshness of heather and browning bracken. Clearly the house was on mineral soil because you couldn't build on blanket bog, but so was much of the surrounding moorland. This little area, perhaps less than an acre, must have become grassy when the farm was inhabited. There would always have been some animals by the farmstead, whether sheep, cattle or horses, and their constant nibbling would have favoured the grass. Likewise their dunging would have increased the fertility of the soil, favouring the nutritious sweet grasses rather than rough kinds like purple moorgrass. Although the farmhouse has long been abandoned sheep still graze the moor and they must be attracted to the one place for miles around where sweet grasses grow. So the grazing and dunging continue in a virtuous spiral which maintains this little island of nourishment in a hard, grudging landscape. In a moment the train had passed the spot by but the image stays in my mind.

VEGETATION SUMMARY

The Heather Family

Heather or ling. Poor, acid soil, moist to dry, usually a podsol; tolerates thin soil.

Bell heather. Mixed with heather on drier soils, dominant on very dry ones.

Cross-leaved heath. Mixed with heather on wetter soils, dominant where it's too wet for heather.

Bilberry. Often on drier soil or boulders; shade-tolerant and often grows in woods.

Plants of More Fertile Soils

Bracken. Deep, well-drained soil, richer and less acid than typical heather soil, usually a brown earth, usually on a slope; can out-compete other plants on a favourable soil but is sensitive to trampling; altitude limited by cold.

Sweet grasses (bents and fescues). A mix of fine-leaved grasses, usually grazed down short because the animals like them; they grow on similar soils to bracken.

Gorse, common. Slightly poorer soils than bracken but sometimes more alkaline, well-drained; not cold-tolerant.

Gorse, dwarf and western. Much smaller plants than common gorse, with similar soil preferences to heather.

Plants of Wet Soils

Matgrass. Similar soil to heather but often a bit wetter; both it and purple moorgrass may take over where heather is overgrazed.

Purple moorgrass. Similar soil to matgrass but usually a bit wetter still; often grows in big tussocks, especially when in a pure stand – very hard going for walkers.

Bog myrtle. Wet to very wet soil, often with purple moorgrass; like the alder tree, it's a nitrogen-fixer.

Deergrass. Deep peat and bogs, very wet and acid.

Cottongrass. Bogs, even wetter than deergrass and including open water.

Bog moss. The ultimate peat-forming bog plant, only found in extremely wet conditions.

Water
in the Landscape

All the water features we can see in the landscape – springs, streams, rivers, lakes, ponds, bogs and fens – are joined together. Even if there's no visible overground connection between one and another they're all part of the seamless web of water that pervades the soil and the permeable rocks below. Open water is no more than the visible part of a hidden whole that is much bigger. Water features are accents of extreme wetness in a landscape which is nowhere totally dry. They reflect the character of the landscape where they occur, from high granite moorland to soft, fertile valley. The character is revealed not just by the plants growing in the water but also by the colour of the water itself.

The clear, clean water of rocky mountain pools and streams is a sign of acidity and a low level of nutrients. Up here the rocks are hard and mostly low in bases. The water is clear because there aren't enough bases to support the growth of the microscopic plants and animals which cloud more fertile waters. In short it lacks life. It's usually too low in nutrients for reeds and bulrushes though there may be smaller emergent plants such as water horsetail and spike rush. But in many places the water's too deep for emergents because mountain pools tend to be steep-sided. As you go down into the lower hills the water gradually becomes more opaque. Even if the rocks are still base-poor, the water has travelled further and has had more opportunity to pick up both nutrients and inorganic silt. Streams which flow through chalk have the special quality of being rich in calcium but, because chalk is such a pure form of limestone, clear of silt. The combination of clear, fast-flowing water and an abundance of creatures which depend on calcium for their structure makes them ideal for trout. Trout and their cousins the grayling are indicators of clean, oxygenated water.

Not all upland streams and lakes are clear. Ones which get their water from peat bogs can range from a pale yellow to a mysterious black. Looking down into a dark, peaty pool it can be hard to tell if it's just knee deep or virtually bottomless. If you gaze into the dark depths you don't need much imagination to see where the tales of the monster from the deep came from. Rivers are often peat-stained too. A river which rushes down the hills between woods and fields, Guinness-brown and topped off with a creamy foam, reveals its origins in a distant peat bog.

These dark stains are quite different from the muddy colours of lowland waters. Rivers which meander slowly through lowland vales may be quite clear in dry weather but after rain they turn anything from milk-chocolate brown to ruby red, according to the colour of the local soil. This is an indicator of the high levels of soil erosion caused by agriculture. Although soil does erode under natural conditions the rate is tiny compared to that on farmland. Ponds may contain less eroded material than rivers but more nutrients and these feed a thickening soup of green algae. The resulting colour can range from muddy grey to blue green. The high level of nutrients is partly a natural effect. Throughout the ages bases have been leached and washed from the wider landscape and ended up in the low places from which further movement is slow. But human activity has greatly added to the flow of nutrients.

Sewage is one source. Although sewage treatment is supposed to remove nitrogen and phosphorous from the treated water, in the real world things don't always go exactly according to plan. Even if they do there are still the residues of the past, before the importance of removing nutrients had been realised. Though nitrogen is soon leached out of the landscape, phosphorous is much less soluble and is easily fixed in the soil, whether under water or on dry land. Runoff from agricultural fields is another source. Chemical farming is not an exact science and in practice only about half of the fertiliser applied is actually taken up by the crop plants. Most of the other half ends up in water bodies. The effect on the river ecosystem can be similar to fertilising a grassland field, giving an extra advantage to the competitor plants which tend to do well in these places anyway. Sadly these days many lowland river banks are covered in a continuous bed of nettles, thriving on the nutrients left behind by enriched flood waters.

A waterside plant which has become common in recent years is oilseed rape. As a cultivated crop it's a fairly new introduction but it's lost no time in getting established as a naturalised plant. Here in Somerset it's mostly confined to the banks of rivers and drainage channels. It's an out-and-out competitor and certainly appreciates the extra nutrients, both from the enriched river water and from the spoil that's dumped on the banks when the river is dredged. As a biennial, the bare ground formed by the dredging spoil may help it to compete with the perennial nettles by giving it some bare ground to seed into. But it hardly seems to need any help. With its large seed, fast growth and big, horizontal leaves it appears well able to hold its own against perennials as long as it gets the nutrients. In some low-lying parts of Somerset you can spot a river from a distance by a line of bright yellow rape-flowers, often mixed with the white of cow parsley, cutting across the deep green of the springtime grassland. (See photo 39.)

Himalayan balsam is another naturalised competitor plant which has colonised the riversides. It's one of a small band of exotic plants which does so well that it often crowds out native plants, replacing diverse communities with a monoculture of itself. Rhododendron and Japanese knotweed are two other prominent members of the band. But despite its faults Himalayan balsam can be a help to landscape readers at times. Its seed is spread by the water of the rivers it lives beside and sometimes the exact extent of a winter flood can be shown the following summer by the area where the large purple-pink flowers come up.

Water features are forever moving. Rivers constantly shift their course as erosion and deposition change the shape of their meanders. Ponds silt up over time, and though there are ponds which have been in place for hundreds of years that's not long on the timescale of nature. Sometimes a spring can dry up in one place an reappear in another, and I've had the opportunity to observe this process in my field.

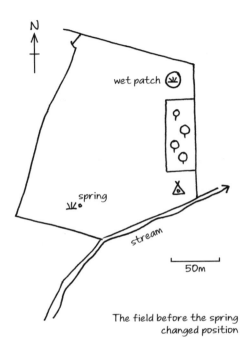

The field before the spring changed position

When I first came to the field there was a small spring, as shown on the first map (left). Though it ran for most of the year the flow was weak and the water only crept a little way down the slope before it seeped back into the soil. Further down the field, in the lowest corner, there was a larger wet patch. This wasn't a spring but simply a dip where runoff water collected in the winter. The only noticeable water plant in either place was rushes. There was a handful of them by the little spring and a thick growth in the wet patch at the bottom. The lack of other water plants was due to the grazing of the aftermath. Rushes, those knee-high tussocks of dark, spiky leaves, are much more resistant to grazing than other water plants. In winter they made ideal habitat for a pair of snipe which visited the wet patch at the bottom of the field every year.

After some years the spring suddenly dried up and reappeared at the bottom of the field, near the wet patch. There was no immediate cause for this change that you could put your finger on, but under the ground there lies evidence that the stream that runs along one side the field and any springs associated with it have steadily changed position over the long ages of prehistoric time. This evidence is a layer of tufa, some thirty centimetres below ground. Tufa is a limy substance which is sometimes deposited by a stream or spring flowing from a limestone catchment. In the stream you can see it forming as a thin surface layer on twigs and other debris in the water, a cream-coloured crust which makes the twig look petrified. Over time the organic nucleus is lost and the tufa gradually becomes a homogenous, granular deposit. Far from being confined to the present course of the stream, the underground layer of tufa is spread over several hectares, including all of the field and parts of its neighbours. All of this area must have been either stream bed or spring at one time or another.

In its new position further down the slope the spring flows a little more strongly and there's water on the surface throughout the year. It has created a new wet patch beside the old one but distinct from it. The two together

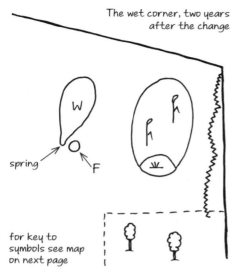

The wet corner, two years
after the change

spring

for key to
symbols see map
on next page

make up a sizable piece of land that's too wet to mow for hay. The change happened around the time I found it increasingly hard to get the aftermath properly grazed. Since then there have sometimes been a few cattle or ponies in the field but never a herd of voracious heifers keen to get in there and eat everything they could find. If the intensive grazing had continued the only water plants to appear around the new wet patch would have been rushes. But the lack of grazing allowed other plants to move in, and this has given me the opportunity to observe a bit of wetland succession as it has happened over the years.

Two dominant plants soon became established. One was the common reed. Over two metres high, with a feathery flowerhead which bends to the breeze, reeds are typical plants of shallow standing water. Much taller than rushes, they can easily out-compete them but they're palatable to grazing animals and soon die out if they're regularly grazed. With the decrease in grazing, they gradually took over from the rushes in the old wet patch. The other new plant was great hairy willowherb, almost as tall as reed and a cousin of the more familiar rosebay willowherb. It immediately colonised the area of standing water at the centre of the new wet patch. This was unexpected because it's normally a plant of moist rather than wet soil. Why the reeds shouldn't have taken over the new wet patch I don't know but somehow the willowherb got there first.

It's fourteen years now since the spring changed position. Since then I've made a sketch map of this corner of the field every year or two to record the changing vegetation. The two main characters in the story have always been reed and great willowherb and together they now dominate most of the area. The old wet patch has become drier. With no significant grazing to take away each year's production of plant material, the dip has gradually filled up with dead stems and leaves. This raising of ground level has dried it out a bit and allowed willowherb to come in and progressively take over from the reeds.

Meanwhile the new wet patch, fed by the little spring, stays wet all year round. Here you would expect the willowherb to give way to reeds, but it hasn't. Reeds have colonised the area surrounding the spring and become the dominant plant there. But in the very wettest part, where there's standing water for most of the year, there's still more willowherb than reed. Given reed's preference for wetter conditions than willowherb this is surprising. It seems that the willowherb is succeeding simply because it got there first and it's hanging onto what it holds. As ever in ecology, possession is nine-tenths of the law, and great willowherb seems to be a tenacious plant.

This is the general pattern of the vegetation but there are many other plants growing in the wet corner. Some of them are only clearly visible at certain times of year. Bindweed, for example, reveals itself in August when it has climbed far enough up the other plants to spread its leaves above them. Nettle and creeping thistle, which are comparatively short plants in this community, are most visible in autumn when the reeds begin to lose their leaves. Looking through the series of sketch maps I've made through the years also shows how much apparently random change there is from year to year. For example, the patch of rushes on the west side has been a pretty constant presence for the last ten years, but one year I noted that area down as 'mainly willowherb'. On the other hand the clump of golden-flowered fleabane, which was already there when I first got to know the field twenty-five years ago, has been a steady central point amid the ebb and flow of other plants. Blackthorn and bramble wait in the wings, steadily advancing from the edges as the ground progressively dries up. So far just one new willow tree has sprung up in the reedbed but this is surely the first representative of what must be the next stage in the succession.

The tall, undisturbed vegetation must provide more habitat for wild animals than the meadow it replaced, though apart from the tiny grassy ball of a harvest mouse's nest I haven't seen much sign of it. On the other hand there has been a loss. Since the vegetation grew tall and dense the snipe which used to come and stay each winter have stopped coming. Be that as it may, that piece of land obviously wants to be a marsh and who am I to stand in its way? When I said as much to my neighbour who farms the land all round, he turned to his farmworker and gave him a smile and a nod that said, "You see. I told you this bloke's a complete nutter."

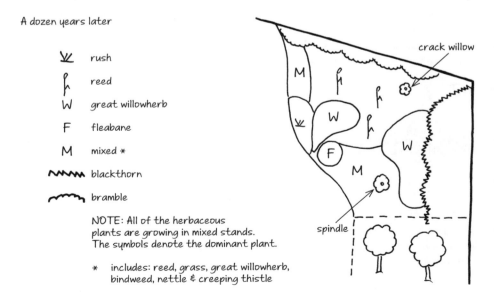

A dozen years later

У   rush

ſ   reed

W   great willowherb

F   fleabane

M   mixed *

⋀⋀⋀⋀ blackthorn

⌒⌒⌒ bramble

NOTE: All of the herbaceous plants are growing in mixed stands. The symbols denote the dominant plant.

* includes: reed, grass, great willowherb, bindweed, nettle & creeping thistle

From Source to Sink

Small springs like the one in my field are potentially useful. Even if water only flows in the winter there may actually be a usable yield all year round which is consumed by the surrounding plants. Clear away the plants and collect the water in a pipe and you may have enough to meet a modest domestic need.

In grassland, small springs may reveal themselves by a strip of grass which is brighter green than its surroundings, stretching downhill from the point where the water emerges. It's greener in summer because it's kept continuously moist and it's greener in the winter because the water from under ground is warmer than the cold winter air. Patches of green like this must surely have been the inspiration for the first watermeadows. (See pages 223-224.) In a frost the green stripe is even clearer as it stands out against the white rime. When spring comes and all the grass is green, the wet stripe may stand out brown if it gets poached by the feet of cattle, turned out after spending the winter inside. Later in the year the poached area may grass over again but, if the field hasn't been rolled, the pockmarked surface will give it away. On soils where buttercups aren't common the moist patch around a spring may be thick with them and stand out a brilliant yellow at flowering time. A stronger spring or one on a heavier soil is more often marked by rushes. In woods the creeping buttercup may indicate a slight spring while golden saxifrage usually indicates soil that's too wet to walk on. Larger springs are often marked by the presence of alder trees.

While small springs only affect their immediate surroundings, large ones can influence the landscape on a broad scale because they attract settlement. Before the days of mains water every village, indeed every farm, had to have a reliable supply of its own. Springs are also attractive as village sites because they're often situated on the boundary between hill and vale, and thus give the village access to both kinds of land. This happens where the hills are made of a permeable rock, such as limestone, and the valley of an impermeable one, such as clay. Rain which falls on the hills seeps down through the permeable rock till it comes to the impermeable layer. It can't go down any further so it stays there, saturating the lower part of the permeable layer and forming an underground store of water known as an aquifer. Where the junction between the two rock types meets the surface, which is usually near the bottom of the hillside, the water comes out in a series of springs. This is called a spring line. An example of a spring line on a very small scale can be seen in the sketch on page 232, where sand overlies clay. Notice the alder trees in the hedge that marks the boundary between the two. They indicate where the water is coming out of the ground.

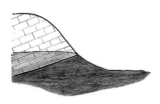

Structure of a spring line

On a larger scale, pretty well every range of chalk or limestone hills in England has a series of villages along the spring line at its foot. The territory of each village extended in a narrow oblong, reaching up onto the hills above and down into the clay vale below, giving each one its share of the different kinds of land: downs for pasture, clay vale for arable and riversides for meadow.

The boundaries of these territories have often been inherited by the modern civil parishes and you can trace them on the 1:25,000 Ordnance Survey map.

A disadvantage of the downland pasture was that there was no surface water up there and the grazing animals had to walk a long way to drink. A solution to the problem was the dew pond. These days the word is often used loosely to describe any pond out in the fields but a real dew pond is a specific kind of pond designed to collect rain water. Why they ever got called dew ponds I don't know, because the dew doesn't provide enough water to flow along the surface and fill a pond. A dew pond is a shallow, circular bowl at the centre of a catchment area which naturally drains towards it. Both pond and catchment area were lined with puddled clay to make them waterproof. On the Mendips the pond itself was also lined with flagstones, perhaps to prevent the clay lining being punctured by the hooves of the cattle as they jostled to get at the water. In the rainy maritime climate of the Mendips the ponds held water for most of the year, though they did need cleaning out regularly. They were made at the time of the enclosure acts, when the land was parcelled out into stone-walled fields and the cattle could no longer walk down to the valleys for water. These days they've been superseded by drinking troughs fed by the mains.

While a dewpond needs careful construction and regular maintenance, nothing could be easier than making a pond in low-lying clay country where the water table is near the surface. All you need to do is dig a hole and it will fill with water. But you do need to be sure that the water table is indeed near the surface. The best way to find out is by digging a narrow test pit and seeing what happens over twelve months. The twelve months are necessary because a pond which fills in winter may dry up in summer. In some places a test pit may not be totally reliable as one part of the site may be more leaky than another, for example where clay is interbedded with irregular layers of limestone. A pond which dries up in summer is not necessarily bad for wildlife. It will just have a different range of inhabitants from a perennial pond, including some creatures which can only survive in seasonal ponds because they're vulnerable to predation by fish.

A newly dug pond will become a thriving ecosystem of its own accord, as most plants and animals which inhabit ponds are well adapted to colonising new ones. They need to be good colonists because of the relatively short life-span of ponds. Water beetles can fly, newts can walk quite long distances, and sticklebacks have sticky eggs which hitch a ride on the feet of water birds. Some plant seeds also stick to the feet or feathers of birds, while other plants, like bulrushes and reeds, produce masses of light, windblown seeds.

People have made ponds for various reasons. Watering animals was the most usual one but soaking the wheels of wagons was another. The wooden parts of a wheel can shrink in dry summer weather, making the wheel weak and unstable, so every now and then they needed a good soaking. That's why the haywain in Constable's famous painting is parked in a pond. The first steps towards the mechanisation of farming actually increased the need for ponds. The steam traction engines used for ploughing needed a copious supply of water and in some places extra ponds were dug.

Steam was also responsible for some unintended ponds. Railway embankments were sometimes made simply by digging the material out of the ground beside the tracks. The resulting hollows filled with water and today they have become long dark ponds, shaded by willows and alders. At the same time, ponds were being formed by subsidence in coal mining districts and especially in salt mining areas. The meres of Cheshire, which now seem so much a part of the natural landscape, were formed by subsidence of the salt workings there. Even the Norfolk Broads are the product of an extractive industry. They were long assumed to be natural lakes till the botanist Joyce Lambert, while working on the ecology of the Broads, discovered that they had vertical sides and flat bottoms and were quite clearly old peat workings. Once she'd pointed this out, other clues to their artificial origin came into focus, such as the remains of the baulks that were left between one set of workings and another. They survive either as narrow peninsulas or lines of little islands in a straight line. But they weren't obvious before Lambert made her discovery.

It can be hard to tell an artificial pond from a purely natural one but natural ponds are probably quite rare. Ponds and lakes can't be formed by water erosion because water doesn't make hollows. It cuts down ever deeper along its course. In fact a stream flowing out of a pond will eventually drain it, though this will take a long time if the rock is hard. But ice can make hollows and the great majority of the natural ponds and lakes in Britain are the result of the Ice Age. The lochs of the Highlands and lakes of the Lake District lie in the bottom of U-shaped valleys carved out by glaciers. The hollows in which they lie were formed towards the end of the Ice Age. As temperatures rose glaciers melted at lower altitudes and only remained in the higher parts of the valleys. They continued to erode the valley floor downwards while the lower part of the valley, where the glacier was replaced by a small river, was eroded more slowly. Thus the upper part of the valley became deeper than the lower and when the ice melted altogether the upper part filled with water. Hence the characteristic long, narrow shape of lochs and glacial lakes.

Just at the point where the glacier becomes a river a dam of debris can form which adds to the depth of the lake. The debris consists of all the mineral matter, whether boulders, gravel or rock dust, which are carried in the ice and left behind when it melts. This kind of material is known as moraine and a dam at the end of a glacier is called a terminal moraine. Of course moraine is a relatively erodible material and the river which replaces the glacier can cut down into it and lower the level the lake. You can see this process halfway completed at Tregaron Bog in Mid Wales, which started out as a lake formed behind a terminal moraine. The River Teifi, which runs through it, has eroded a gap in the moraine deep enough to partly drain the lake and turn it into a bog.

Unlike blanket bogs, which form because the climate is extremely wet, Tregaron Bog is a valley bog, which has formed in a valley where drainage is restricted. The first stage in the formation of a valley bog is shallow water, shallow enough for emergent plants like reeds and bulrushes, which have their roots in the soil below the water and tall stems which emerge into the air above. Most plants couldn't live in such a waterlogged and therefore oxygen-deficient soil. The emergent plants manage it because they have vertical channels in their tissues, almost like internal pipes, which take air down to the roots. With this problem solved, they're able to put on prodigious growth every year, most of which dies back in winter and falls into the water. Since most of the micro-organisms which decompose organic matter need oxygen for respiration and still water contains little oxygen, decomposition is always incomplete in such wet conditions. Every year more dead organic material is added to the water than is decomposed. Gradually a layer of partly-decomposed material builds up in the water. This is peat.

Eventually the peat builds up to the level of the water surface and the character of the reedswamp begins to change. It's no longer a body of open water but a mass of very wet peat, interspersed with pools and channels. Now different plants take over. If the water is acid, they're the typical bog plants such as cotton grass, deergrass, bog myrtle and above all bog moss. But if the water comes from a catchment of alkaline rocks, such as limestone, the water will be alkaline and a different range of plants will grow on the peat. Prominent among them is the great fen sedge, or sword-sedge, so called because its leaves are so sharply serrated that they can cut your fingers to the bone if you clutch them and pull. They say it's better to fall over and get a soaking than to 'save yourself' by grabbing it! There are many other sedges, rushes and herbs that grow along with great fen sedge, but it dominates. Alkaline peatlands of this kind are usually called fens while acid ones are known as bogs, or mosses in Scotland and the north of England.

From this point a valley bog or fen can develop in two different ways. It can succeed to woodland or become a raised bog. Trees can take root if parts of the peat become dry enough, perhaps during a run of dry years. Once the first pioneers get established they pave the way for more trees because they reinforce the drying-out process. All plants pump water from the soil and lose it from their leaves in a constant stream. Trees, being much bigger than other plants, do this at a much greater rate and they can lower the water table significantly. Even so conditions remain wet and the main trees are the water-loving ones, alder, willow and white birch. This kind of woodland is known as carr.

A raised bog develops where rainfall is higher and the surface stays very wet. The absorbent nature of the valley bog or fen means that the rainwater can't flow away as it would on a mineral soil and rain-fed peat starts to form on the surface. This is similar to the formation of blanket bog but the very high levels of rainfall which are needed to get a blanket bog going on a mineral surface are not necessary where the surface is already made of peat. Unlike a valley bog or fen, which is fed by drainage water from the surrounding country, a raised bog is fed entirely by rain. Since rainwater is slightly acid, this means that raised bogs are always acid, regardless of whether the peat below is acid or alkaline. They contain the usual range of bog plants, with bog moss as the dominant one. Like blanket bogs,

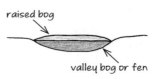

Profile of a raised bog

they can stop growing and dry out at the surface, usually in response to a long-term drop in rainfall. If this happens heathers take over from the bog moss and a heath or moor is formed, and if the grazing pressure is low trees can get established.

TYPES OF BOG		
Blanket	*Valley*	*Raised*
Forms in regions of extremely high rainfall; covers all flat and gently-sloping land with a layer of peat.	Forms in a low place where drainage is impeded, fed by drainage water.	Forms on top of valley bog, fed by rainfall. Always acid.
Always acid.	May be either: acid – bog or moss or alkaline – fen	

The progression all the way from open water to carr woodland is a process of succession similar to the one which goes from bare soil to woodland. In fact carr is not the end point, in theory at least. Alder, willow and birch are all pioneer species. They're able to cope with wet conditions but, by lowering the water table and by themselves adding to the depth of peat when they die, they eventually make conditions much drier and allow species like ash and oak to become established. In practice this probably didn't happen very often under natural conditions. The water level in places like the Fens of eastern England and the Somerset Levels has gone through cyclical changes over the ages, periodically knocking succession back to square one. Even without an overall change, individual places must have experienced cycles of wetness as rivers changed course, sandbanks were laid down and eroded away again.

These days there are very few lowland wetlands which have not been drained for agriculture. One of the few places you can get an idea of what a natural wetland might have been like is in the Norfolk Broads. On the margins of some of the less disturbed broads you can see concentric zones of vegetation, running from open water with floating plants, through marginal reedswamp, to willow scrub and finally to alder carr. What's happening here is that the broad is gradually filling up, from the margin towards the middle, as all ponds and lakes eventually do. The further you are from the centre of the broad the more advanced the succession.

Concentric rings of vegetation of this kind are a typical pattern on ponds and lakes. Although they do represent stages in succession, in the moment they simply indicate different depths of water or degrees of wetness. Photo 40 shows an example. In the middle of the pond where it's deepest there's open water, next comes yellow water lily, then bulrush in the shallow fringe, and in the wet soil round the margin there's yellow flag iris with some purple loosestrife and a young willow. The bulrush and yellow flag are hard to tell apart in the picture as neither is flowering but the purple loosestrife stands out.

Wetlands aren't always located at the lowest point in the landscape. The nature of the underlying rock, whether it's permeable or impermeable, can have its effect too. Once, out walking in Dorset, I saw a reedbed which lay on gently sloping ground a little way above the bottom of a small valley. A stream ran along the bottom of the valley but there was no marsh there. Later I had an opportunity to check the geological map. The reedbed lay neatly on an outcrop of Gault clay, below the chalk and above the greensand of the valley bottom. In effect it was a large, diffuse spring sited where the porous chalk lay over the impermeable clay.

Small wetlands like this have a valuable role in the landscape, not least in helping to even out the movement of water. This is especially noticeable if a stream runs through the wetland. Even a small marsh or bog will absorb a great deal of water when the stream is in spate and release it slowly afterwards. I lived for a few years just below the southern edge of Dartmoor, first in one hamlet and then in another. By chance both these hamlets lay on streams of similar size. One of the streams passed through a marsh on its way from the moor and the other didn't. Both of them could overflow their banks after torrential rain but the volume and violence of the floods were much less on the stream which passed through a wetland. Conversely, in dry times the water stored in a wetland is slowly released and this helps to maintain the flow of rivers and streams. You notice this effect especially in areas where streams are inclined to dry up altogether in summer. The New Forest, with its sandy and gravelly soils and comparatively dry climate, is just such a place, and New Forest streams which pass through valley bogs are much more reliable in the summer than the ones which don't.

Another factor which may have affected my two Dartmoor streams is woodland. The second stream passed thorough more woodland than the first and this may have contributed to its more even flow. The value of woods in slowing down and evening out the flow of water though the landscape is well known and appreciated but the mechanisms by which they do it are less well known. Although the woodland soil and the leaf litter do act like a sponge, they're not that much more effective than the soil of permanent pasture. The thing which really makes a difference is log dams. Dead branches and even whole trees fall into streams, the gaps between the sticks are filled by autumn leaves, and sand and silt build up behind the organic debris to make a water-holding dam. These dams slow down the water and allow it to seep into the soil on either side. As well as evening out the flow of water they add diversity to the structure of the stream, with shallow riffles and deep pools. This adds to the diversity of habitat for wildlife and fish populations are usually higher where there are more log dams. Eventually they rot away but new ones are always being formed.

Comparisons with relatively untouched woods in North America suggest that log dams were very much more common in the wildwood than today. Like so many other natural features in the countryside they're victims to the incessant human urge to tidy up.

Nevertheless woods are one place you can see what a natural stream may have looked like. Out in the fields and in most secondary woods the smaller streams have almost all been straightened and turned into ditches. But in some primary woods they're still more or less untouched. There you can see the pattern of interlocking spurs, so characteristic of young rivers and streams, on a miniature scale. (See pages 36-37.) You can also see how a stream can act as a natural drain. The place where it rises, if not a definite spring, is often a marshy area. But once the stream is concentrated in a single channel it starts to cut down below ground level. This has the effect of draining the ground in much the same way as a ditch which has been dug for the purpose. So the banks of a stream are not always, as one might expect, the wettest part of the landscape but often the driest. Where the banks are marshy it's either because there's virtually no fall on the stream or because its flow is obstructed, perhaps by a log dam.

CASE STUDY

THE SOMERSET 'MOORS'

Some landscapes are dominated by water. Many of these are highland land-scapes which are wet because it rains so much, covered with blanket bogs and drained by rivers which roar in spate after every heavy rainfall. Others are in low-lying, flat places where it may not rain very much but drainage is slow. Most of the latter kind have been drained for agriculture and are now far removed from their natural state. But the water is still there. Though it may be confined in artificial waterways and straightened rivers, it's ever ready to burst these bonds and flood the land like it used to. In these places you could almost say the landscape is not so much formed by the interaction between people and the land as between people and water. Rather than describe them in a general way, I'll end this chapter with a case study of one which I know well. Officially it's called the Somerset Levels, though traditionally it was known as Sedgemoor and local people always refer to it simply as 'the moors'.

The area has much in common with other wetland landscapes. Its structure is similar to that of the East Anglian Fens, though the land is in grass in contrast to the arable of the Fens. But I include this case study not just as an example of a wetland but also as an example of a complete landscape. Much of this book is about the elements which go to make up a landscape, even if some of them are quite big ones such as woods and heaths. Here I describe a larger area of land which has a definite boundary and try to show how the different elements within it fit together.

The moors are the heart of Somerset. They stretch back from the coast, not in a broad plain but in tongues of wetland separated by peninsulas and islands of

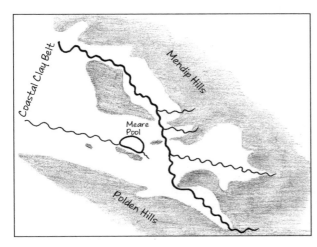

The study area before draining

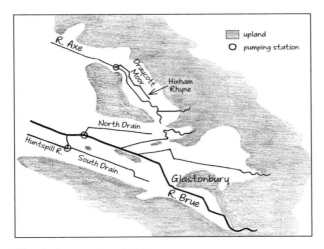

upland

○ pumping station

The study area after draining

upland.[†] Although there is a certain feeling of remoteness down on the moors, there's nowhere so far from the upland that you can't look up and see a hedged landscape dotted with cosy farms and hamlets or a glimpse of the familiar cone of Glastonbury Tor. The low Polden Hills cut right across the moors, separating the northern moors from the southern. I'm much more familiar with the northern part, comprising the valleys of the Rivers Brue and Axe, so I'll concentrate on them. (See photo 41.)

At first sight the moors look boring. They're flat, the roads are mostly straight and the fields are plain rectangles. Instead of hedges there are wet ditches which combine the functions of draining the land and keeping the animals in the fields. The moors don't have the immediate attraction of the upland countryside with its winding lanes, hedged fields and ancient villages. But flatness and straightness are superficial qualities. When you look beyond them you find a rich wildlife, a subtle variety in the character of the different moors, and a blend of natural and human influences which have led to a landscape which in some ways is upside down.

I've lived here for thirty years, travelled over the moors, walked on them, pollarded the willows that grow there, and for the past ten years looked out over them every day from my front window. But it was only a couple of years ago that I got to understand how this landscape really is upside down.

[†] Elsewhere in this book 'upland' means hilly country but in this context it means any land other than the moors themselves.

My 'aha!' moment happened one day when I was sitting on the edge of the Mendip Hills, looking down on the moors of the Axe valley.

Rodney Stoke, 22nd April 2005

On Draycott Moor I can see a drove with a thickly-treed hedge all along it. The ditches which lie at right angles to the drove are hedge-lined or dotted with willow trees near the drove. But the trees and shrubs soon peter out till, at the ends of the fields, where the ditches join the rhynes*, there are few. There's a distinct impression that the land gets lower further away from the drove. But the difference in height is so little that it's hard to be sure it's not just an optical illusion caused by the way the trees get smaller and fewer further away from the drove. It could also be an unconscious assumption of mine based on the knowledge that trees don't grow so well on waterlogged ground. On the map the drove is marked as Brook Bank, which seems to support the idea that it's higher than the surrounding fields. The watercourse at its side is marked as Draycott Brook.

*   Artificial drainage channels.

In fact my impression was right. The ditches do run away from Draycott Brook and into the channels which lie parallel to it. This is what I mean by the landscape being upside down: the rivers and streams, which you would normally expect to run along the bottom of the valleys, are actually slightly higher than the surrounding land. The ditches run away from the rivers and drain into artificial channels which run more or less parallel to the rivers. On the map on the previous page you can see this pattern on a larger scale where the River Brue is flanked on either side by the North and South Drains. The ditches flow away from the Brue and into the two Drains. An artificial channel which is bigger than a ditch and smaller than a drain is known as a rhyne, pronounced 'reen'. The two rhynes which run parallel to Draycott Brook flow into a larger rhyne, the Hixham Rhyne, which runs parallel to the River Axe. The brook itself flows straight into the river and in order to do that it has to be carried over the Hixham Rhyne by a culvert.

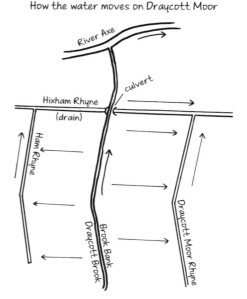

How the water moves on Draycott Moor

Before that day I'd seen the separate elements of this pattern on different parts of the moors: ditches which run away from rivers instead of into them; the embanked river Brue so high above its surroundings that its bottom is level with the surface of the fields, streams and even rivers carried over rhynes on culverts. But that day I saw it as a whole for the first time. When I went home I consulted the geological map and then I began to understand why it's like that. The map shows Draycott Brook running along a narrow band of alluvium while the moors on either side are peat. Looking at the moors as a whole, some areas are composed of peat and others of alluvium but all the rivers and streams are on alluvium. Wherever one of them crosses a moor made of peat the map shows it running on a strip of alluvium. These riverside bands of alluvium are the result of the rivers repeatedly overflowing and depositing their load of sediment. Most of the sediment settles out quickly, as soon as the water leaves the watercourse and slows down. So the sediment accumulates close to the river and over time the banks build up higher than the surrounding land. This natural embankment is topped off with an artificial one to stop the river flooding the low-lying land on either side and very soon you have a river running above the level of the surrounding land. The field ditches are below the level of the river and the only way they can flow is away from it. The rhynes and drains run parallel to the streams and rivers and take away the water from the ditches.

The problem with this layout is finding an outlet for the drains and major rhynes such as Hixham. There's very little fall on the rivers, so however far downstream you go it's hard to engineer a situation where the drains are higher than the river and can empty into them easily. Going straight to the sea might be an option but then you come up against another element of upside-downness in this crazy landscape. In most places you expect the land to become lower and lower as you get nearer the sea but on the Somerset Levels it gets higher. If you travel towards the coast along the road that runs across the wide, wet moors between the River Brue and the North Drain you can see this quite clearly because here the rise in the land happens quite abruptly. Although the difference in level is only a couple of metres it stands out plainly as you approach it across the flat, hedgeless moors. There's a sharp rise in the ground and on top of it are hedged fields. The combination of bank and hedge almost has the appearance of a defensive palisade but the change in level is in fact completely natural.

As you cross this boundary you move into a different landscape. The river, which was previously perched above the surrounding fields, now flows at the bottom of a trench and the land is much better drained than before. There are still big ditches between the fields but they're not always full enough of water to keep the cattle in the fields, so most of them are hedged. Curiously, the hedges often grow on both sides of the ditch rather than on one side only. They meet over the middle and when they're mechanically trimmed they make an enormous table-top several metres wide. At first the hedges are pure hawthorn and the fields rectangular. But as you move nearer the coast or towards the northern part of the coastal belt these give way to the mixed hedges and irregular-shaped fields of ancient countryside. Clearly people have lived and farmed here for a long time, in contrast to the low inland moors, where the fields are strict rectangles and houses are rare to this day. The hedgerow trees and scattered farmsteads distract the eye from the flatness of the landscape but the map confirms that it's dead level. In fact this is the area which is properly called the Levels. The moors got their name in historical times because they were wild and untamed while the coastal belt, though equally flat, was already rich farmland.

The coastal levels are higher than the inland moors because when the sea breaks its bounds most of the sediment it carries is deposited near the coast. In this way a layer of marine clay some three metres deep has built up on top of the peat and now it acts like a dam to the drainage of the inland moors. This is much the same process which has left the rivers on the inland moors perched above the surrounding fields, so both the upside-down elements in this watery landscape have fundamentally the same cause.

When the moors were comprehensively drained, around the beginning of the nineteenth century, there was no choice but to divert the North and South Drains into the River Brue. Cutting deep channels right through the coastal clay belt was just not an option. But after heavy rain, when the river was full with water which had flowed in from the surrounding upland, its level was higher than that of the drains. So the drains became useless just at the time they were most needed. Even today old people tell of their childhood on the moors when the whole family would move upstairs in winter because downstairs was flooded. It was only in the mid-twentieth century, when pumps were installed and the Huntspill river was dug, that the situation improved. But still the combined waters of the Brue and the North Drain have no outlet to the sea other than the lower course of the Brue, which was originally formed by a smaller river. (See map on page 271.) So the moors still flood regularly in winter, especially on the northern side of the river. Usually it's a controlled flooding. The water is let out of the river more or less onto the site of the former Meare Pool and pumped away as soon as possible. But for a few days we look out over sheets of water, a reminder of how it may have looked in times past – and maybe a preview of times to come.

The name Sedgemoor gives a clue to what the landscape was like before draining. The sedge in Sedgemoor was the great fen sedge and most of the moors were originally fen, watered by the alkaline runoff from the surrounding limestone hills and claylands. The former fens are now moors of alkaline peat. There are also moors of alluvium. These have formed in places where a large river has brought down heavy loads of silt from the uplands, such as alongside the Brue upstream from Glastonbury, and also on the site of Meare Pool, where the water was too deep for emergent plants to grow and thus form peat. The peat and alluvium moors are surprisingly similar in appearance, no doubt because both soils are alkaline. Not so the third kind of moor, which has formed where parts of the alkaline fens developed into acid raised bogs. This happened during a period of increased rainfall several thousands of years ago. More recently the climate became less rainy and the surface of the raised bogs dried out enough for heathers to become dominant and the bogs were transformed to heath.

The heath hasn't got a single name but you can trace it on the map by a whole series of Heath names, such as Ashcott Heath, Walton Heath and half a dozen others, each bearing the name of the village on the adjacent upland which owned it. The peat of the heaths was and still is a valuable resource because it's workable. The fen peat, which is formed mainly from the great fen sedge, is fibrous and hard to cut. But the raised bogs on which the heaths developed consisted mainly of bog moss, which yields a soft and homogenous peat. Historically the moss peat was cut for fuel and today it's dug for the horticultural trade. The modest hand cutting of the past has been replaced by heavy machinery and some parts of the peat moors have been reduced to a moonscape. At the same time, these acid moors are little use for farming and much of the land which isn't cut for peat has been allowed to succeed to woodland of alder, willow, birch and oak. Recently many of the worked-out peat diggings have been converted to nature reserves, either as broad reedbeds or as carefully-sculpted areas of alternating land and water, designed to give a variety of habitats.

It's a landscape of sudden contrasts. The woods have a deceptively primeval feel. Although they're young as woods go, still in their first generation of trees, their dank mossiness gives them a prehistoric, almost mystical atmosphere. But in a few moments' walk you can be in the industrial world of the active peat workings, and as soon again walking beside a wide expanse of reeds which would make a time-traveller from the Neolithic Age feel quite at home.

Both the coastal clay belt and the former heaths stand out as distinctive landscapes among the Somerset Levels as a whole. Although the rest of the moors are broadly similar, each one does have its own individual character. The former Meare Pool, for example, is notably treeless and bleak, no doubt because frequent flooding is bad for tree growth. Much of the Axe valley has a similar character to the coastal belt, with hedges as well as ditches on the field boundaries. This perplexed me till I looked at the geological map and saw that this part of the valley is on the same marine clay as the coastal belt, due to a major incursion of the sea in the past. But the differences between the other moors are less easy to explain.

Some are almost treeless except for the odd self-sown alder or hawthorn. Others have lines of pollard willows along the banks of the rhynes. Pollard willows used to be the emblematic image of the whole region but now they're less common than they were. Their economic value as a source of thatching spars has all but vanished and they get in the way of mechanical dredging. Where they survive they're often left unpollarded and some have grown into grotesque shapes. The road from Glastonbury out to Godney has a positively gothic feel to it on a winter's night, its gnarled old willows like a row of witches, silhouetted against the wind-scudded clouds. Some moors are well treed nonetheless. One or two even have short windbreaks of oak, willow, alder and pine, protected from browsing within a double ditch. I can't find any general reasons why one moor should be well treed and another not. Whether the moor lies on fen peat or alluvium seems to make no difference. Nor, it seems, does the date when the moor was first drained.

Maybe there are some reasons which I have yet to find. On the other hand this may be one of those cases which illustrate that landscape is not mechanically determined by natural and social forces. Human choice and even whim can leave their mark as well.

There's less difference between the fields themselves from one moor to another than there is between the trees and shrubs. The rectangular field shape is fairly universal, though in one place the rectangles are drawn out into what must be some of the longest and narrowest fields anywhere in Britain. These are mostly in the parish of Catcott, on a part of the raised bog which is not now dug for peat. But when the moor was first drained there was every expectation that it should be. The long, narrow fields stretch across the bog so as to give every small landholder a bit of the best moss peat in the middle with access from the road which runs along the edge.

The fields are almost all semi-improved grassland. There's little improved grassland because the ground is too wet to grow improved commercial strains of grass, let alone arable crops. There's even less semi-natural grassland because the moors are too flat to have escaped the spreading of a bit of fertiliser now and then. The orchids which used to tint the pastures pink in summer are now confined to the few grassland nature reserves. But some fields are still so full of common sorrel that the sward takes on the russet hue of its flowers when the fields are shut up for hay. Others are so full of rushes that they look dark green, sometimes almost black, from a distance. These are not just the wetter fields but also the ones which are more used for pasture than hay or silage, as rushes are unpalatable to cattle but are set back by mowing. Buttercups are another plant that cattle avoid and where the soil is moist rather than wet field after field turns deep yellow in springtime.

Almost every field has drainage gutters running across it. These are shallow furrows about ten metres apart which help to get the water off the fields and into the ditches after rain. Locally they're known as grips, pronounced 'gripes'. They only have open water in them for a short time but they do make a consistently wetter habitat for plants throughout the year. On buttercup fields they stand out vividly at flowering time as dark green lines striking through the yellow background. The effect is a series of bright yellow rectangles bordered in green, a vivid reminder that buttercups like the soil moist but not wet.

Most of the fields belong to farms which are based on the upland. The two kinds of land, upland and moor, complement each other. In high summer, when plant growth is limited by the lack of soil moisture, the moors come into their own. Then farmers whose holdings are mainly on the moors have the advantage. In spring they're at a disadvantage because the moors are cold and the grass is slow to start growing. They are in effect one gigantic frost pocket. Cold air from the surrounding hills flows down onto them and they're often covered in a thick layer of white rime while all the upland is frost-free. The frost pocket effect is compounded by the wetness of the soil. It takes more energy to heat up water than a dry mineral substance, so wet soil warms up more slowly in the spring than dry.

Until recently dairy farming was the mainstay of the moors. In summer the farmers would take small portable milking machines round their scattered fields as the cows ate off one field after another. But these days you need the economies of scale to stay in dairy farming. This means a big block of land within easy walking distance of a large, permanent milking parlour. But the landholding pattern on the moors is rather like the medieval open fields, with each farmer's holding more often scattered far and wide than concentrated in one block. So now most of the fields are grazed by beef cattle and sheep or mown for hay and silage. Recently I've even seen one or two abandoned fields, chock full of tall herbs which are gradually crowding out the grasses. Apart from the urban fringes, this is the only place where I've seen flat land let go of like this.

The open, water-filled ditches are universal as field boundaries on the moors. These narrow strips of water are the refuge of all the plants and animals which have survived from the days of the wild Sedgemoor. Not all of them have survived, indeed the great fen sedge which gave the region its name is now extinct in Somerset. Those which have survived are ones which can find a niche in the constantly-repeated succession which is driven by the periodic cleaning out of the ditches. This is an aquatic succession and the starting point is not bare soil, as in a dry-land succession, but open water; the direction of change is not so much from annual to perennial to woody as from water plants to land plants.

Just after a ditch has been dredged, when there's plenty of open water, the most prominent plant is the tiny floating duckweed. It can coat the surface so completely with its thin green film that an unwary dog has been known to take it for a lawn and try to run across the ditch! An abundance of duckweed is an indicator of nutrients in the water, as its minute roots have no access to the soil below. Before long rooted plants start to move in. One of the most characteristic is frogbit, with its bright, three-petalled flower and leaves like a miniature water lily. It indicates clean, unpolluted water. Arrowhead, which also has three petals but upright, arrow-shaped leaves, is a bit less common. While these truly aquatic plants occupy the centre of the ditch, marsh plants such as yellow flag iris, purple loosestrife and the scented meadowsweet grow along the banks. (See photo 42.) Bit by bit the ditch fills with vegetation, and the soil, which is really more of a viscous liquid than a solid, gradually wells up from below. Now the narrow-leaved, grass-like plants predominate right across the ditch. They include reedgrass, true reeds, bulrushes and sedges. Eventually a ditch can become dry land if it's left uncleaned for long enough. This doesn't often happen but where it does it usually leaves a tell-tale dip in the grassland with rather more rushes on it than there are on the original field surfaces at either side.

The ditches are also home to a wide range of animal life from insects up to the herons which stand like solemn sentinels on the banks, waiting stock still for the least sign of movement in the water. The bird life is the crowning glory of the moors' wildlife. You can hardly go down there without seeing swans.

Linger just a little longer and you'll surprise a pair of mallards on a ditch and see a moorhen or two scuttle away across the water, flicking their tails and bobbing their heads as they go. Barn owls are surviving well on the moors. Elsewhere road deaths are a major cause of their decline but there are few roads on the moors and those not much used at night. The sweet, seductive call of the cuckoo is heard a little more often on the moors than on the surrounding upland. On the moors the cuckoo lays its eggs in the sedge warbler's nest and this species is holding its own, in contrast to the less fortunate birds of ordinary farmland which host the cuckoo's eggs on the upland. The new reed-bed nature reserves must have given a boost to the sedge warbler's numbers.

The moors are famous for wild ducks and waders, both resident and winter migrants. The ducks appreciate the open water of the winter floods while the long-legged waders need moist soil for feeding. They find their food by probing the ground with their long bills and the soil must be soft for this. The most characteristic bird of the moors, the peewit or lapwing, is a wader. Its two names describe its plaintive call and the slow flap of its broad, round wings. The aerobatic display of the male birds in the breeding season is one of those things that brings a special joy to my heart. At the same time waders have an economic value in the ecosystem as many of the grubs they pick from the soil are pests which feed on the farmers' grass and reduce production.

Drainage is the enemy of both ducks and waders. People who remember the far-off days before the pumping stations were built speak of huge flocks of teal, widgeon and snipe coming in in the winter. Since then the moors have become progressively drier as more and more efficient pumps have been installed. There has also been pressure from a powerful minority of farmers who wanted more pumping so they could grow improved grassland. This wasn't so good for the other farmers who couldn't afford to plough and reseed their fields, because the native grasses are shallow-rooted kinds and adapted to a high water table. Nor was it good for the birds, whose numbers continued to decline. Now, when attitudes have swung more in favour of wildlife than food production, some sort of compromise has been reached. But the winter flocks of ducks and waders, though delightful, haven't gone back to the size which would take your breath away.

That distinction is reserved for the starling. Westhay Moor lies on a small raised bog to the north of the main heaths. It has extensive reedbeds on disused peat diggings, now largely a nature reserve. Every winter literally millions of starlings, from all over Britain and the Continent, come to roost there on the reeds. The mild western climate makes it a good place to spend the winter and the sheer scale of the roost must give them security. In the daytime they range out over the countryside far beyond the moors to feed. In the late afternoon you can see parties of them making their way back, flying in such tight formation that the flock seems to be one animal with one mind. Bit by bit the small feeding parties join together in ever bigger masses of birds. I sometimes fancy you can tell how close you are to Westhay Moor by the size of the flocks. Finally they come together in a murmuration of millions which wheels and swoops down to perch for the night on the reeds.

Hedges
and Other Field Boundaries

Hedges are like coppice woods in miniature. The shrubs take the role of the coppice stools and the hedgerow trees the standards. A regularly cut hedge is like an actively coppiced wood while a hedge which is left to grow has much in common with a neglected wood. Britain has less woodland than almost any other European country, but more hedges. Is this a coincidence or are the hedges to some extent a compensation for our lack of woodland? Although their main purpose was always to keep animals in the fields, hedges used to be an important source of firewood in districts that lacked woodland, and they also compensate for the lack of woodland both visually and ecologically. These days people plant hedges mainly for their visual and wildlife benefits. But in all previous ages, as with so much in the landscape, these benefits have been nothing more than the side effects of people going about the business of earning a living.

In some parts of the country the abundance of well-treed hedges can give the illusion that the landscape is well wooded, that somewhere not far in the distance there's a wood amongst the fields. But as you move through the landscape you may not find any woodland at all, just a succession of thick hedges and frequent hedgerow trees which combine to fill the view and hide the open fields that lie between them. Perhaps this is where the word 'woodland' comes from in the old distinction between woodland and champion countryside.

In fact hedges are more like continuous woodland edges than complete woods. Like woodland edges they're rich in shrubs which can flower and set fruit in the abundant light. They usually have a bank, or at least a strip of herbaceous vegetation at their foot which doesn't get farmed along with the rest of the field, in effect a narrow belt of semi-natural grassland. Many hedges also have a ditch, a linear wetland. So hedges are to some extent a microcosm of the semi-natural landscape. In many parts of the country they're about the only semi-natural vegetation left. Despite the grubbing out of thousands of miles of hedges during the past half-century, they're the largest single wildlife habitat left in lowland Britain.

Whether they're a net benefit to farming or not is a moot point. Their main function is to contain animals and that can equally well be done by a wire fence. The question is which does the job better and cheaper, a fence or a hedge. Both need maintenance: regular trimming for a hedge versus complete replacement every decade or two for a fence. Most hedges also need a bit of fencing to make them a hundred percent effective but this is often minimal. It's little indeed compared to the full paraphernalia of a stand-alone fence, which includes two strands of barbed wire, heavy wire netting, a row of stakes, straining posts at intervals and so on. The hedge probably wins on cost grounds, especially as it's usually already there. In terms of usefulness it also wins because it has the added benefit of providing the animals with shelter.

As evidence of the overall advantage of hedges, look at how few of them have been replaced by fences in the grassland areas of Britain. Although some hedges have gone in the west it's mainly been a matter of amalgamating small fields. But in the east of the country, where arable farming reigns supreme, the argument has clearly gone the other way. As farms have become more specialised and the animals have gone, most of the hedges have gone with them. Bigger fields mean tractors spend more time working and less turning at the end of a row, and many hedges have gone for this reason. But hedges also take up space which could be growing crops and they need maintenance, which is a cost. Although arable farmers may see hedges as worth preserving for visual amenity and wildlife they generally see them as a dead loss in productive terms. But hedges also play a part in the ecology of arable farming. They have both positive and negative effects and the question is which are the greater, the positive or negative.

You can see the negative side quite clearly in some arable fields. The hedge casts shade on the crop, especially if there are hedgerow trees, and competes with it for water and nutrients. Slugs live in the hedgebank and venture a little way into the field to eat the crops. The strip of crop affected by shade, competition and slugs can stand out clearly on the edge of the field. But the positive effects are spread more widely and less easily seen. The shelter effect increases crop yields less markedly but over a much longer distance downwind. Beetles live in the hedgebank over winter and range widely over the field in summer, eating pests as they go. Wild flowers in the hedge provide food for flying insects which lay their eggs on the crop, where their larvae feed on caterpillars and aphids. The hedge protects the land from erosion by wind and by water for a fair distance downwind and downhill, but the effect may only be revealed

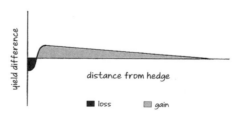

after the hedge has been removed. So the negative effects are concentrated and highly visible while the positive ones are spread out and not visible, though on balance the positive are greater.

The positive effects of hedges are also easily masked by the quick fixes of chemical agriculture. The benefits of pest control won't show up in a landscape which is regularly sprayed with pesticides. Nor will reduced soil erosion be appreciated where any loss of topsoil can be countered with an extra dose of fertiliser. But in a sustainable landscape, where chemical remedies aren't available and the health of both soil and crops is founded on biological diversity, hedges are an economic asset to arable farming.

Kinds of Hedges

Just like woodland, hedges can be ancient or recent. In the case of woods the division between the two is fairly arbitrary (see page 183) but with hedges there's a step change between the old and the new. The change came with the parliamentary enclosures of the eighteenth and nineteenth centuries, which transformed the open fields of the planned countryside into the hedged fields of today. Suddenly there was a need for millions of shrubs to make thousands of miles of hedges and the nursery industry was born. In truly modern fashion it churned out a monoculture of hawthorn, which is quick and easy to grow and has the great benefit of being thorny. Before that time the natural increase of the landscape provided enough plants for the occasional new hedge that was needed. People used whatever shrubs were available, probably digging up young plants from the woods rather than growing them from seed. So ancient hedges are usually a mix of shrub species while recent ones are mostly hawthorn. (See photo 44.)

There are some ancient hedges in the planned countryside, often on parish boundaries, and you may see the odd recent one in the ancient countryside. But the level of diversity of shrub species in the hedges is one of the prime clues to which kind of country you're in.

The most diverse hedges of all are usually ones which were carved directly out of the wood. In the ancient countryside they may go right back to the original clearing of the wildwood, though they may not stand out very clearly among the mixed hedges which are the norm here. But in the clear light of the planned countryside these woodland relict hedges contrast strongly with the straight hawthorn hedges of enclosure. They often came about when a wood, already enclosed with a bank and hedge, was grubbed out to make more fields, and what was the wood-hedge became the field-hedge. It usually has a crooked course and contains a similar mix of tree and shrub species to that in local woods. It may also have woodland wildflowers growing in it, such as dogs mercury, yellow archangel and bluebells. But these plants aren't sure indicators of a woodland relict hedge. They can also gradually spread into recent hedges which abut onto woodland or a former wood-hedge. It's a slow process and the clue may be that the woodland plants only reach part of the way along the hedge. Dogs mercury is often the first colonist and may reach further along than the other woodland herbs.

In ancient countryside woodland wildflowers are much more common in hedgerows generally, especially in the wetter west of the country. The most spectacular displays of wildflowers I've seen anywhere in Britain are on the high hedgebanks of the English West Country. In late spring and early summer they come alive with a mix of both woodland and grassland flowers. There are flowery hedgebanks in other parts of the country and drab ones in Devon and Cornwall but the overall contrast with the rest of England is dramatic. It's also puzzling. Why should there be such a difference? Is it because of the mild, moist climate? Is it that the high banks which are so characteristic of the region provide special conditions of microclimate and soil which favour wild flowers?

Or is it a matter of time? If the hedges of the West Country survive from a much earlier age than those of other parts it would mean they've had more time to develop that diversity which comes with age. Maybe it's all of these reasons combined or maybe it's some other which I'm unaware of. (See photo 43.)

The further west you go the bigger the banks tend to be and the bigger they are the less importance is attached to the hedge on top. In Cornwall the banks are massive and usually stone-faced, the shrubs on top may be nothing more than a bit of gorse or even heather, and the bank itself goes by the name of hedge. There are similar banks in many parts of Wales, especially the south and west. In fact Pembrokeshire and Cornwall have a very similar feel to them, with high, rocky sea-cliffs surrounding a country of small, banked fields, dotted with white-washed, black-slated farmhouses. The high banks of Wales and the West Country are a distinct regional variation but hedgebanks are found in other parts of the country though they're usually smaller. On the whole they're a sign of age. The older a hedge is the more likely it is to stand on a bank. Whether the hedge has a ditch or not is more a matter of the soil type. Ditches are usual on clay, which normally needs draining, but not on lighter soils.

Another clue to the age of a hedge is the course it traces. Ancient hedges may be straight or crooked but recent ones are almost always straight. The contrast between the small, irregular fields of ancient countryside and the large, rectangular ones of enclosure act countryside is usually quite clear. The c- and s-shaped curves of former open-field strips and the dog-legs of former furlongs indicate enclosure by agreement, probably during early modern times. (See pages 42-44.) But a curving hedge doesn't always have this origin. It may follow the course of a stream, sometimes a stream that no longer flows because the land has been artificially drained. It may also follow the break of a slope, which frequently follows a smooth curve. Hedges are often sited on the break of a slope because this puts steep and flat land into separate fields. A field which is flat enough to plough in one part but not in another is an awkward field to manage. Over the years lynchetts can form on these hedges and this accentuates the natural break of the slope.

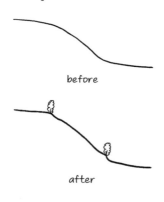

before

after

Hedges often follow the boundary between one soil type and another. This clearly makes sense as different soils need different treatment and they may be suited to different crops. My own field is an example. In the last chapter I described the layer of tufa which underlies it and parts of some neighbouring fields. (See pages 261-262.) The soil over the tufa is a medium loam, in contrast to the clay loam of the surrounding land. Although the boundaries of the field are fairly straight, they coincide with the change in soil type pretty closely. It took some effort to make it so. The stream which runs along one boundary of the field has clearly been diverted from its natural course in order to run along the boundary between the two soil types. Its present course runs across the slope while the natural run of the land would take it through the middle of the field.

A neighbouring field is half on the medium loam and half on the clay but I suspect it was two separate fields in the not-too-distant past. It was almost certainly two different furlongs back in the days of the open fields. When it's ploughed you can see the difference because the plough brings up a little of the tufa so the medium loam has a white cast to it which stands out against the dark brown of the clay loam. When a crop of grass or maize is growing there you can't tell the difference between the two soils by the condition of the crop, as the blanket applications of fertiliser and farmyard slurry obliterate the subtle distinctions in crop growth which were significant to our forebears. My field has not been ploughed now for many years but the tufa gave it its traditional name, the White Field, which I've taken for my surname.

Sometimes an unusual pattern of field boundaries on the map can alert you to an interesting piece of country that's worth a visit. One such place that caught my attention was St Braivels Common, on the edge of the Forest of Dean. The reason for its distinctive pattern soon became apparent when I visited it.

St Braivels Common, February 2000

This place has always looked fascinating on the map. It's a landscape of tiny fields with scattered farms and cottages, which contrasts with the larger fields and compact village of St Braivels itself. It has no unenclosed land at all, despite being called a common. What could be its history?

In a word, it's stones. The Common lies on a great dome of hard sandstone and has a sandy soil. Originally there must have been a lot of loose stones on the soil surface because all the fields are walled and the small size of the fields shows that there was a lot of stone to be got rid of. In fact there was so much that in some places the walls weren't enough to use it all up and there are neat piles of stones the size of a small house, made simply to get the excess stones out of the way. Compared to the loamy soil near the village it must have been an unattractive place to cultivate and was left as common land for long enough to acquire its name. Then at some point population pressure must have risen so high that the task of clearing the stones from the land became a necessity. Most of the walls now have hedges growing up beside them and the landscape has a mellow, mossy feel to it.

June 2003

Coming here in summer is very different. You can hardly see the stone walls now the hedges beside them are in leaf. What a beautiful little world it is! There's a remarkable number of flower-rich meadows, dominated at the moment by ox-eye daisies. Along the narrow green lanes there are foxgloves and lovely crooked-limbed oaks, both old and young. We came here to photograph the stone piles but with the trees in leaf and the herbaceous plants grown tall we couldn't get a good shot.

"This is now the preserve of the rich," said Cathy. Too true! Each of the cottages is probably on the site of a shelter thrown up by a family of land-hungry squatters but today these cottages are within commuting distance of three major cities.

See photo 45.

Later I heard that the residents have got together to buy an 'alpine' tractor, a tractor small enough to get down the little lanes and through the narrow field gateways. The meadows we saw are rich semi-natural grassland. One of them boasts eighty-two species of plants. As the fields are too small for present-day farming they're getting neglected and beginning to succeed to woodland. The little tractor means they can be mown again and maintained as grassland.

St Braivels Common is an unusually interesting and biodiverse corner of countryside. But even the comparatively plain hedges of the planned countryside have their story to tell. Those pure hawthorn hedges don't stay pure forever. Like any other ecosystem they become more diverse with time. The first woody colonists are usually elder, ash, blackthorn, rose and bramble. Hazel, dogwood and maple are slower to arrive and their presence in a hedge suggests that it's ancient, though they will come in more quickly if the hedge abuts on an ancient wood.

This extract from my notebook features two hedges which appear to have been colonised after planting, one recent and one ancient.

High Stoy, 23rd March 2005

There's a drove with hedges on both sides. One is shown on a map of 1616, the other isn't.

Old Hedge, pre 1616	New Hedge, post 1616
Shrubs:	Shrubs:
Mostly hazel with some sycamore, blackthorn and pussy willow.	Mostly hawthorn and blackthorn with some elder, sycamore and hazel.
Herbs:	Herbs:
Lots of bluebells, also red campion, lords and ladies, lesser celandine and bracken.	Mostly red campion with some lesser celandine, a little dogs mercury and very little bluebell and ramsons.

The drove is situated right on the edge of the planned countryside, on the northern rim of the Dorset chalk downs, which stretch southwards from here all the way to the sea. One end of the drove abuts onto the upper edge of the ancient woodland which occupies the scarp slope. From that high point you could literally throw a stone into the ancient countryside of the Blackmore Vale which lies below. The recent hedge appears to have been originally planted as a mix of hawthorn and blackthorn while the ancient hedge shows every sign of having been planted as pure hazel. Both have subsequently been colonised by three species of shrub, which gives the younger hedge a higher species count.

This is unusual. You'd expect the older hedge to have been planted as a mixture and it's certainly had more time to acquire new species than the younger one has. Even the list of shrub species from both hedges is fairly similar. What distinguishes them is the proportion of the different species in each hedge, one being mainly hazel the other mainly thorn. The same applies to the herb layer. The species list is much the same either side of the drove but the proportions are different. Bluebells, which dominate the ancient hedge, are true woodland wildflowers while red campion, which is dominant in the recent one, is more of an edge species and more typical of hedges than of woods. This was just a quick list I jotted down in passing. If I'd had the time it would have been interesting to compare the ends of the hedges nearest the wood with those furthest from it to see what effect distance from the wood has had on the distribution of shrubs and herbs.

The mixture of hawthorn and blackthorn was occasionally used in enclosure-act hedges instead of pure hawthorn. Blackthorn is a useful hedging plant as its suckering habit helps to keep the hedge thick at the bottom. Another variation on pure hawthorn was to include some standard ash trees. It's usually pretty easy to distinguish between a hedge planted with ash standards and one which has been colonised by ash after planting. The planted ash trees are now at the end of their mortal span. They're either over-mature or they've already been felled and have regrown in a multi-stemmed form like a coppice stool. They'll also be fairly evenly spaced along the length of the hedge. Self-seeded ash trees are likely to be younger and more randomly spread. Under the modern regime of constant trimming they don't often get a chance to grow into standard trees and usually merge in unobtrusively with the shrubs.

Some recent hedges are entirely self-sown, the hedgerow equivalent of recent semi-natural woodland. They grow up along field walls and fences, which make convenient perches for birds which shit out the seeds of hawthorn, bramble and other berry-bearing bushes. Ash and sycamore seeds, light enough to be carried by the wind but big enough to be able to get going in a grass sward, are also common. The level of diversity in the hedge is usually low at first and increases over time as new species colonise.

Whether or not a hedge self-seeds along a particular wall or fence depends on the frequency of grazing. A few years of continuous arable crops or silage-making can give the plants the break they need. There's always a narrow strip at the edge of the field where the plough and the mower don't reach. On a roadside the hedge is most likely to form on the side of the fence nearest the road, which never gets grazed. A self-sown hedge is a form of succession and as with all kinds of succession in Britain, the key influence is the amount and pattern of grazing.

You can often see self-sown hedges in the process of getting established. On the chalk downs of Wiltshire you may see hawthorn bushes scattered at random along the line of a roadside fence, neatly trimmed to a uniform height and shape as though they were a continuous hedge. On the Mendip Hills, which is stone wall country, a few self-sown hedges have now developed to the point that you hardly notice the original wall, reduced to a low pile of stones at the foot of a tall, thick hedge. At St Braivels Common the process has gone far enough that if you were to visit the place for the first time on a summer's day you might not realise there were walls there at all, though they're still quite visible in winter and still make an effective barrier to animals.

Hooper's Rule

Taking all hedges together, ancient and recent, planted and self-sown, there's a tendency for older ones to be more diverse. This is partly because of the historical change from mixed hedges to pure hawthorn ones and partly because the longer the hedge is there the more opportunity there is for new species to colonise. In fact it has been suggested that you can actually date a hedge by counting the number of tree and shrub species in it. This idea was the brainchild of Max Hooper and his colleagues, who were investigating hedges at Monks Wood Experimental Station in Huntingdonshire during the 1970s. By studying old maps and other historical documents they were able to date some of the local hedges. They compared these dates with the number of shrub species in the hedges and found there was a direct relationship between the two: one woody species per hundred years of age.

It wasn't an exact match, but close enough to deserve further investigation. The next step was to see if it held true in other parts of the country. They recorded hedges in different regions and most of the results they gathered supported the hypothesis. But there was one area in Shropshire where the hedges were known to be late but were nonetheless very diverse. Evidently there had been no hawthorn nursery in the locality at the time the land was enclosed and the hedges were planted with mixed shrubs in the old style. Dating these hedges by the one-species-per-century rule would make them hundreds of years older than they really are. When Max Hooper and his colleagues published their findings they emphasised that the idea should be "treated with caution and not used as an immutable universal law".[†] Nevertheless, it caught the imagination of professionals and amateurs alike and soon came to be known as Hooper's Rule. Many people gave it, and some still give it, much more authority than its originators ever claimed.

The method is simple. Choose three random thirty-yard sections of the hedge, or fewer if it's too short. They shouldn't include the ends of the hedge and

[†] *Hedges*, E Pollard, MD Hooper & NW Moore, Collins, 1974.

should be well scattered along its length. Try to resist the temptation to choose specially diverse sections which will give you a high count! Include all woody species except brambles and climbers such as ivy and old man's beard. Then take the average of your samples.

It can become quite addictive. Regardless of whether it can really give you an accurate date, it's a good way to get to know a hedge – and to brush up your tree and shrub identification skills. I always feel a sense of delight when I come upon a particularly diverse hedge. It's a joy to find such diversity however old it may be. I also feel a sense of wonder. A hedge like this is as much a link with the past as an old parish church or the gravestones in its churchyard. It's a living thread which connects my world with the world of the people who planted it. It gets me wondering about them. Who were they? What did they look like? Why did they plant it? What tools did they use?

Over the years, as more and more species counts have accumulated, it's become clear that Hooper's rule really can't be used with any confidence. Hedge planters haven't been as consistent in their habits as was first thought. More pockets of country have emerged where people went on planting mixed hedges into the age of parliamentary enclosures. Such places are mainly outlying areas of enclosure in predominantly ancient countryside, places where there perhaps wasn't enough demand to make a nursery business worthwhile and there was plenty of woodland to provide wild seedlings. Likewise, early hedges were sometimes planted with one species of shrub rather than a mixture, as the drove at High Stoy illustrates. Very recent hedges are sometimes planted with a rich mixture of native shrubs. Many of these are now old enough to have been laid and it would be possible to mistake one for an ancient hedge.

There are also several natural influences on diversity which aren't related to the age of the hedge. The acidity of the soil is one of them. Alkaline soils naturally support a much wider range of plants than acid ones, and shrubs like spindle, wayfaring tree and dogwood will never be found in a hedge on an acid soil, however old it is. The climate is another. The natural diversity of trees and shrubs is greater in warmer regions than cooler ones. This means the potential diversity is greater in the south of Britain than in the north, and at low altitudes than in the hills. The nearness of seed parents also has an effect. If the hedge abuts onto an older hedge or an ancient woodland it will usually have a higher count. The end of the hedge will be colonised first but in the end all of it will be affected. This is the reason for not taking the ends of a hedge as samples when you're making a count. Finally, some plants are so competitive that they can take over a diverse hedge and reduce it to a pure stand, which can give an old hedge a very low count.

I've known elder do this. It's one of those plants which inhibits the growth of its neighbours by putting out chemicals which are poisonous to them. It's also a competitive plant which responds very vigorously to a nutrient-rich soil. The combination of chemicals and competition can suppress and kill other trees and shrubs. The most extreme example I've seen was a windbreak of alternate elder and damson trees, planted around a vegetable plot with the aim of combining shelter with an edible yield from both species. Although damson is the toughest of fruit trees and often used in windbreaks, after only ten years it had all been killed, leaving a pure stand of elder. I would suspect any pure elder hedge of once having been more diverse.

More common is the pure hedge of suckering elm. There are two kinds of elm in Britain. One is the wych elm, which is a single species, and the other is a group of several species which all have a suckering habit. The wych elm is more common in woodland than hedgerows, reproduces by seed and never suckers. The suckering elms have lost the ability to reproduce by seed and are more characteristic of hedgerows than of woods. This is perhaps because they usually grow in more fertile soils, the kind of soils which have mostly been cleared for farming rather than left as woodland. Hedgerow elm trees used to be a characteristic feature of the clay vales of England. They were a constant background to my childhood in north Somerset. Like tall billowing thunderclouds, they gave the landscape a shady depth on a hot summer's day. When I was away in Africa in the early seventies I heard that they were all dying of Dutch elm disease. My first response was that I didn't want to go back to a landscape without them. Of course I did come back, and though the mature trees are now just a memory the suckering elms are far from extinct.

Elm leaf. Note the asymmetric base, which is typical of all British elms.

It's their suckering habit which has saved them. The fungus which causes the disease is spread by bark beetles which feed on medium- to large-sized elms. It inevitably infects a young suckering elm when its trunk diameter reaches about twenty centimetres. But plants smaller than this are safe and these young trees can produce more suckers without ever growing to mature size, so they survive indefinitely as virtual shrubs. The hedges are as full of elm plants as ever they were, it's only the big ones that have gone. It's this same suckering habit which has made elms such strong competitors with the other shrubs in a hedge. Reproduction by seed is not easy in a dense hedge, where the competition for light and water from the established shrubs is intense. A suckering species has a great advantage here. A new sucker has the support of all the established plants in the clone and this gives it the edge over a seedling. Other shrubs can co-exist with elm but they can't compete with it when it comes to reproduction. So, as all plants must die in the end, elm eventually replaces the others. There must be more to this than suckering, though. Blackthorn and dogwood are two other hedge plants which sucker but neither of them takes over in the way elm does. I suspect that elm also indulges in a little chemical warfare which just tips the scales in its favour.

Keeping a hedge well-trimmed seems to favour elm, as this extract from my notebook illustrates.

Butleigh, 18th July 2004

On the track to Broad Park there's a grown-out elm hedge. All the elms are dead, standing tall and straight above the living hedge. They're being replaced from below by blackthorn, hawthorn, elder, bramble and a little elm. Will the hedge ever go back to being pure elm? Elder and blackthorn may give it a run for its money. But I suspect that in the end the elm will win out. It usually seems to do so on these heavy soils.

The hedge on the other side of the track has been kept trimmed and is pure elm.

In the grown-out hedge there are no mature trees and shrubs other than the dead elms, which suggests that till recently it was pure elm. As it grew unchecked the elms would have become bigger but fewer through the same process of exclusion that goes on in a woodland at the thicket stage. (See page 207.) The apical dominance of these trees would also have discouraged the growth of suckers. (See page 171.) So when the trees reached the vulnerable size and died there weren't many suckers ready to take over and the other species were able to come in and fill the vacuum. Meanwhile the hedge on the other side of the track, which belongs to another farm, was kept regularly trimmed. This meant there were no dominant apical shoots to inhibit the growth of suckers. The elms suckered freely and no other shrub species could get a toehold in this crowded environment.

The bare poles of dead elms in untrimmed hedges have become a familiar part of the landscape, often giving an uncared-for and scruffy look to parts of the clay country. On a visit to East Anglia a few years ago I was surprised to see a row of suckering elms growing quite happily though they were well past the age at which suckering elms normally die. This wasn't the anomaly it seemed because the different species of suckering elm vary in their susceptibility to the disease. The so-called English elm, the commonest species over central England, is totally susceptible and always dies, while the East Anglian elm has some resistance. The non-suckering wych elm, by contrast, is quite resistant and seems to survive more often than it succumbs.

The difference is in the genes. Sexual reproduction, which in plants means reproduction by seed, allows a species to change its genetic makeup through time. Different plants have resistance to different strains of disease and sexual reproduction can throw up individuals which have resistance to new strains. There have been repeated epidemics of elm disease from Neolithic times to the 1930s. The present one is unusually severe but not unique. Through these long ages the wych elm has had the opportunity to develop new resistance while the suckering elms, restricted as they are to vegetative reproduction, haven't. They're stuck with the genes their ancestors had thousands of years ago. As for the difference between the English and East Anglian elms, it has recently been discovered that the English elms are all just one clone. Every single tree in the country is genetically the same individual, descended by vegetative reproduction from one seedling tree that grew at some time in the dim past. That individual happens to have no resistance at all to the current strain of the disease. Presumably the East Anglian elms have a bit more genetic variation.

Like grey squirrels, rhododendron and oak mildew, all of which have had significant effects on the landscape, the present epidemic of elm disease is a product of globalisation. It came to Britain on infected timber imported from North America. How it will end is hard to predict. The fungus which causes the disease is capable of genetic change and may become less virulent in time. It also has its own parasite, a virus, and this in its turn may become more virulent and act as a control on the disease. One way or another this epidemic is likely to pass just as all the previous ones have.

Hedge Shapes

Like any other semi-natural ecosystem, hedges develop under a blend of both natural and human influences. We manage hedges with the aim of keeping them compact and dense while they continually try to grow upwards and outwards. As they grow they don't just get bigger, they also move succession onwards. If they grow outwards they spread shrubs out into the fields on either side, moving the fields on from the grassland to the shrub stage. If they grow tall, the trees have the advantage over the shrubs and thus the hedge itself moves from the scrub to the woodland stage. Human action, as usual, aims to keep succession where it is. The result of these conflicting human and natural forces is a whole range of hedge sizes and shapes, from the wildly overgrown to the severely trimmed. Just as the species makeup of a hedge can tell us about its origin, its shape can tell us about its recent history.

A hedge needs to be high enough and dense enough to keep animals in the field, but no bigger than necessary because the more it grows the more it competes with the grass or other crops in the fields. As hedges grow there's a constant tendency for the upper branches to grow vigorously and shade the lower part, which makes it thin and gappy. So controlling the size of the hedge helps to keep it dense at the same time. Browsing by farm animals can also thin out the lower part of a hedge, so there's often a wire fence on one or both sides to protect it. Sheep browse harder than cattle and being smaller animals they do it at just the height where the hedge is weakened by shading. Sheep also make more use of any resulting holes than cattle do, as they can fit through smaller gaps and take more interest in escape.

A grown-out hedge. Note the browse line.

Maintaining a hedge costs time and money and in these days of expensive labour and financial squeeze some hedges are left to their own devices. Just how a hedge develops when it's no longer tended varies according to what trees and shrubs it contains. A common pattern is for the trees to become dominant. These include both standard trees and trees which were previously kept trimmed to shrub size.

Small trees like hawthorn and rowan can take on the role of canopy trees in parts of the hedge where there are no bigger species like oak and ash. The combination of shade and root competition from the trees weakens the shorter shrubs. They're usually finished off by browsing and by the animals barging their way from one field to another through the resulting holes. The hedge gradually turns into a line of trees with a distinct browse line on them.

From a distance a landscape full of grown-out hedges of this kind can look more attractive than one ruled by tight little hedges of trimmed shrubs. But these big hedges are dying. Like an overgrazed wood, they won't survive longer than the lifespan of the trees. If most of them are small, short-lived species like hawthorn this may not be very long. Even while they survive these lines of trees have already lost most of the useful functions of a hedge. They're no longer a barrier to farm animals. The nesting habitat for hedgerow birds is gone and the hedgebank habitat is severely degraded by trampling. As for shelter, a line of trees like this doesn't slow the wind but speeds it up. When the wind blows against the dense crowns of the trees some of it is deflected up and over them and some of it is deflected downwards, where it joins with the ground level wind to blow through the line of bare trunks with redoubled force.

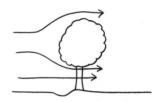

A neglected hedge with blackthorn in it can develop in quite a different way. Like elm, blackthorn is a suckering plant but unlike elm it's armed with formidable thorns. These enable it to spread out from the hedge into the pasture on either side, even in the face of browsing. Only mowing, topping or arable cultivation will stop it. It tends not to out-compete the established shrubs in the hedge in the way that elm does, so eventually the hedge becomes a sandwich, with blackthorn as the bread and the original hedge as the filler. The mass of blackthorn suckers keeps the hedge dense at the bottom. It becomes an excellent nesting habitat for song birds and a hedge like this which is both tall and dense provides just the shelter than bats and butterflies need in order to travel around the landscape from one piece of habitat to another. But the hedgebank habitat is degraded by shade. Reptiles such as adders make great use of sunny hedgebanks for basking but on a shady bank they would never accumulate enough energy to go hunting. Sun-loving wildflowers get shaded out, though woodland species such as primroses may survive and flower under the seasonal shade of the blackthorn. If the blackthorn is continuous along the hedge it can make an effective barrier to farm animals but it's a net loss to the farmer because it takes up so much land.

These are two distinctive ways a neglected hedge can develop. In practice there are many variations and intermediate forms. A variation on the blackthorn-fringed hedge is where a gap develops between the fringe of blackthorn and the original hedge. This happens most often where the main shrub in the hedge is hazel. When other trees and shrubs are neglected they grow mainly upwards but once hazel has reached a modest height it grows outwards, over the top of the shorter blackthorn. This is because it's naturally multi-stemmed and as the stems

age they grow away from each other. As time goes by they grow at a lower and lower angle, shading the blackthorn nearest to the hedge and weakening it, while the outer edge of blackthorn continues to spread into the field. Sooner or later cattle will find their way behind the outer belt of blackthorn and trample down the weakened plants behind. As they spend more and more summer days in the cool shade of the hazel they barge out a green tunnel between the original hedge and the belt of thorn.

A hazel hedge which is fringed with bramble rather than blackthorn can also develop this structure. It can happen without the help of animals as shade alone is enough to kill the brambles nearest the hedge. At Thatchers end (page 72) you can see an extreme case where the belt of brambles has been pushed even further from the hedge by the shade of planted trees. This has happened because the trees nearest the hedge have grown more vigorously than those further away.

The blackthorn sandwich hedge isn't nearly as common as the dying hedge which is becoming nothing but a line of trees. These days more hedges are lost from the landscape by neglect than by deliberate grubbing out. It's a constant process which goes on unobtrusively without the noise and fuss of bulldozers or the ugly gash of bare soil which proclaims that a hedge has been removed. Even if the farmer doesn't particularly want to get rid of the hedge, losing it saves the cost of maintaining it. On the other hand a blackthorn sandwich represents a lot of land lost to production, so they're usually taken in hand well before they get to that state. There are basically three ways of managing a hedge: coppicing, laying and trimming.

Coppicing means cutting the entire hedge down to the ground and removing all the cut material. You don't see it done very often but it can be the best remedy for a seriously overgrown hedge. It's difficult to either lay or trim a hedge composed of large-diameter stems, and coppicing also gives the best opportunity to plant new shrubs where there are gaps, because for one season at least there's no competing shade. Of course the hedge has to be fenced while the shrubs regrow but a laid hedge generally needs fencing at first too. Some weeding may be necessary in the first year after coppicing, as a vigorous growth of bramble or even herbaceous plants such as nettles can interfere with the regrowth of the shrubs.

A variation on coppicing which you sometimes see is what might be called high coppicing. This is cutting the overgrown shrubs at a normal hedge height, around a metre or so above the ground. I think people do this because they shy away from full coppicing, which looks too much like destroying the hedge. But on the whole it's not a very good idea. After years of competition and self-shading the remaining stems are rather far apart and largely twigless.

The regrowth comes from the top of the cut stems rather than the base and this leaves the bottom of the hedge gappy. It's hard to remedy this by planting because any new plants will suffer the shade of taller neighbours. This gives a hedge which is thin at the base and thick above, which is exactly the wrong way round. It's no barrier to animals, especially sheep; rather than give shelter it will intensify the wind at ground level; and it makes a poor habitat for wildlife. It's usually much better to take the plunge and coppice right down to the ground. The only exception to this is a hedge with plenty of suckering species in it. Elm, blackthorn and dogwood are the commonest of these. Suckers, with the support of the clone they belong to, can succeed in filling in the bottom of the hedge where seedlings would fail. High coppicing actually encourages the growth of suckers because it breaks apical dominance in the standing shrubs.

A laid hedge looks very different to a coppiced one. The result of the hedging operation is not bare ground but a compact, living hedge. But in fact the two methods have much in common. In both cases all the stems are cut at ground level so that new growth will spring from the base. The difference is that in laying not all of them are cut right through and removed. The over-large, crooked or awkward ones are cut out but a selection of medium-sized, straight stems are cut three-quarters of the way through and bent over to form the new hedge. This doesn't kill the stems. All the vital tissues are located just underneath the bark and enough of them remain for the stems to survive and put on new growth.

There's no job I enjoy more than laying a hedge. It's always done in winter, when the trees and shrubs are dormant. The winter has a quality of quietness, quite unlike any other time of year. It has a shy, easily-missed beauty, with a thousand subtle shades of grey. The brown-backed fieldfare and the russet redwing provide the accent, rather than the bright wildflowers of spring and summer. But it's not a time to sit under a tree and contemplate nature. It's too cold for that. There are other winter jobs out on the land, like ditching, pollarding, fencing and hedge trimming, but none of them gives the same satisfaction as laying a hedge. When you lay a hedge you're working with living plants to create a living structure which, though it conforms as closely as possible to a preconceived plan, is always a blend of your own intentions and the nature of the plants you're working with. Even in the simplest hedge there's some variation in the size, shape and species of the shrubs, and the more varied the hedge the more interesting it becomes.

There are, of course, regional styles of hedge-laying. I was brought up with the Somerset style, which produces a big, thick hedge laid between two rows of stakes. This is what I thought of as normal hedgelaying till, on a childhood visit to Gloucestershire, I had my first sight of the Midland style. I was very impressed. Midland style is much neater and more regular, with a single line of stakes down the centre and the living stems woven between them, almost like a wattle hurdle.

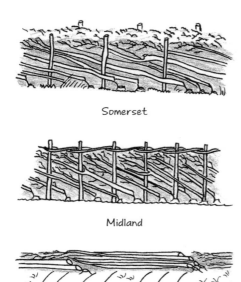

Somerset

Midland

West Country

But this neat regularity is much easier to achieve with the recent, single-species hedges of the planned countryside than with the anarchic ancient hedges of north Somerset. I still use the Somerset style, not just for tradition's sake but because it seems to me the best way to work with the hedges round here. (See photo 46.) Travelling the other way from my home, into the West Country and Dorset, you come upon a different style again. Here the freshly-laid hedge doesn't need to be a barrier to sheep and cattle as it grows on top of a bank which does that job quite adequately. So the hedges are laid very low, with a flat or semi-circular profile.

The Lancashire and Westmorland styles are similar to Somerset, with two rows of stakes, while the Cumberland style is similar to West Country. But over most of England the single-stake Midland style, or variations on it, is the norm. In Wales the tradition of hedgelaying has survived better than in England. Perhaps this is partly due to a cultural difference but it's also true that hedges and fences need to be better kept in Wales because of the preponderance of sheep. Most of the Welsh styles are broadly similar to the English Midland style. But in South Wales, from Monmouthshire to Pembrokeshire, the hedgebanks are higher and the traditional hedge is much the same as in the English West Country.

Sometimes these days you can see a style of hedgelaying that can only be described as nominal. Almost all of the stems are discarded except for a single line, laid flat on the ground so that the tip of one just reaches the butt of the next. Functionally this is coppicing rather than laying. It's much quicker and cheaper than real laying but presumably still gets the grant payment and perhaps that's why it's done.

In the first summer after laying a laid hedge starts to regrow. The shoots which spring from the base of the cut stems will eventually renew the hedge, just as in coppicing. But shoots also spring from the laid stems themselves and grow up vertically, giving the hedge a wispy upper storey over its dense lower layer. From this point on the hedge is normally kept in shape by regular trimming. Bit by bit the laid

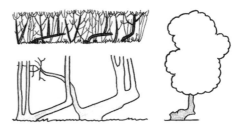

Clipped hedge, grown-out hedge and isolated tree all with signs of former laying

stems are superseded by new growth from below but they can persist for a long time, either alive or dead. In a regularly trimmed hedge they stay much the same size they were when they were laid and are hard to see among the dense twiggy growth except in winter time. But if the hedge is neglected and becomes gappy some laid stems will grow bigger and bigger, still lying at the tell-tale angle which shows that they were once cut and laid. Occasionally you may even see an isolated tree which bears the signs of once having been laid, clearly the sole survivor of a hedge which has since been lost. (See photo 47.)

Traditionally a hedge would be trimmed for a number of years and then, when it started to get thin and gappy at the bottom, left to grow tall so it could be laid again. These days that rarely happens. Laying is labour-intensive, while trimming is done by machine. As labour is so much more expensive than fossil fuels hardly anyone can afford to lay their hedges regularly and constant trimming has become the norm. It's done with tractor-mounted flail cutters, which not only cut the twigs but chop them up into little chips which can be left to gently decompose on the ground. This saves the labour of removing the trimmings. It may also gradually increase the level of nutrients in the hedgebank soil and thus reduce the diversity of herbaceous plants growing there, though I have no direct evidence that this has actually happened in any specific case.

Most hedges are trimmed to a rectangular profile. Some years ago attempts were made to interest farmers in the A-shaped profile. The main advantage of this is that the base of the hedge is wider than the top and thus doesn't get so heavily shaded, which means it will stay dense for longer under a regime of constant trimming. But the idea never really caught on. I think there's something solid and substantial about the rectangular shape that appeals to the sensibilities of farmers. "Let's get everything squared up," was a phrase often used on the farm where I first went to work. 'Squared up' meant tidy and it was the way things were supposed to be.

In fact a well-tended rectangular hedge can stay in good condition for many years. In central Somerset where I live there are plenty of thick, dense hedges which show no signs of deteriorating after several decades of trimming. They keep cattle in without the aid of a fence or at most with a single strand of wire. They stay dense partly because they're mixed hedges and include several shrub species which naturally grow from the base. The suckering shrubs, blackthorn, dogwood and elm, regenerate from the roots; hazel forms new shoots at ground level; and maple tends to branch lower down than hawthorn when it's regularly clipped. They also stay dense because of their size. Many of them reach up to head height and are broad in proportion, which gives the shrubs enough space to develop and maintain a healthy shape. Big hedges like these are also good for nesting birds. Some species like to nest well above ground level, so taller hedges will often have a greater diversity of birds, and all nesting birds are more secure from predators in a thick hedge than a thin one.

Most hedges are trimmed every year, at least here in the west of England, where they're a functional part of the farm. In some parts of the east, where hedges are more of an amenity than a tool of the farmer's trade, things can be different.

Huntingfield, Suffolk, 17th May 1997

In this part of Suffolk there are some huge fields with all their hedges removed, but they alternate with areas where there are still plenty of hedges. The hedged areas have a really good feel to them. As hedges have no economic use here they survive through the goodness of heart of individual farmers. The custom seems to be in most cases to trim the sides but not the tops, so they're very tall. The hedges are ancient and mixed, full of singing birds at this time of year. You get a feel of what Suffolk must once have been, though it does feel eerie without any farm animals.

Back to Somerset, 19th May

After the snowy scenery of Suffolk with the hawthorn in full bloom, Somerset is drab. Here the hedges are cut every year and it makes such a difference. As we drove out of the station I suddenly noticed it and it came as a shock.

Sometimes you have to go away in order to really understand your home landscape. I'd never realised before how little we allow the hawthorn to flower round here. Hawthorn only flowers on twigs in their second year of growth, so trimming every year means it never flowers. Come to that I didn't appreciate how much it was flowering in Suffolk till I got home and saw the contrast. The note I made in Suffolk makes no mention of the blossom.

It would be wrong to think of this entirely as an east-west split or one between arable and grazing areas. The Vale of York is also on the eastern side of the country and also purely arable but it looks very different from Suffolk. It's an area of planned countryside, with a regular grid of fields bounded by pure hawthorn hedges, in contrast to the irregular field shapes and mixed hedges of Suffolk. But the recent treatment is different too. The fields are big and many hedges must have been grubbed out to make them that size, but all of them are still hedged. I didn't see anything like the open 'prairie' parts of Suffolk when I was there. I didn't see much which resembles the hedged parts of Suffolk either, as the hedges are kept severely trimmed. Only on the fringes of the villages and towns are a few of them allowed to grow tall and blossom.

Small, neat hedges like those of the Vale of York often suffer from over-trimming. The smaller the hedge the cheaper it is to maintain because you don't have to make so many passes with the trimmer. Once along the top and one pass for each side is very economical but it makes a hedge that's too small to be viable in the long term. It also means that the hedge is cut to the same height every year so all the new growth is concentrated near the cut ends. After years of this treatment the hedge becomes little more than a row of sticks with tufts of twigs at the top. Then the shrubs start to die, leaving gaps, and eventually it becomes nothing more than an irregular row of isolated shrubs. Sometimes it can be hard to tell whether you're looking at an old hedge in the process of dying or a new one in the process of self-seeding along a fence line. I don't actually know a hedge that has been killed altogether by over-trimming, but it is the logical end of this process.

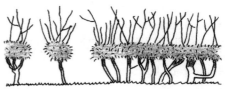

Uncontrolled sheep browsing can also turn the bottom of a hedge into a row of sticks. If the hedge is being assaulted from above by excessive trimming and from below by sheep it can deteriorate quickly, as in this example. The wispy twigs above are the current year's growth and will be removed next time the hedge is trimmed. A hedge can actually survive quite a lot of browsing if the upper part is given space to grow, though it may not be much use as a hedge. I noted this example in North Yorkshire.

Hambelton Hills, 9th May, 2004

Cross section of the hedge

The lambs like to get in under the hedge and the hedge bottom is one of the most tightly grazed parts of the field. All the plants in the hedge are hawthorn, except for one standard holly with its top untrimmed.

Although it looks substantial this hedge is pretty useless. It doesn't keep the sheep in the field. That job's done by the wire fence. Far from providing any shelter at ground level, where the lambs would appreciate it, it actually intensifies the wind. (See the drawing on page 291.) As for wildlife, the hedge-bottom habitat which is crucial for most hedgerow plants and animals is completely missing. The field is used for sheep grazing year after year and to survive such a monoculture the hedge needs fencing on both sides. Otherwise sheep could be alternated with cattle, silage or arable, which would give the hedge regular breaks in which it could regrow. But now it's too late for these remedies. Bereft of active buds and shaded by the bushy upper part, the lower stems won't grow more twigs if the field is rested from sheep. The only real solution would be laying or coppicing.

Trees and Walls

The only plant which had been allowed to grow as a standard tree in that hedge was the single holly. Hawthorn will make a small tree if allowed to and trimmed holly makes an excellent hedging plant but someone had chosen to let the one holly grow and none of the hawthorn. There's an old belief that it's unlucky to cut down a holly tree and this includes holly stems which raise their heads above the top line of a hedge. It seems to be a widespread belief. It was firmly held in Somerset when I was young, and in East Anglia, I'm told, it has saved some hollies from the general destruction of hedges and hedgerow trees. The uncut holly in the hedge I saw on the Hambelton Hills suggests that the belief extends to the north of England too. You do see holly hedges which are kept neatly trimmed. No doubt the belief is waning. In any case it's only when a stem is quite definitely proud of the surrounding shrubs that it becomes unlucky to kill the King of the Trees, as he once was known.

Most hedgerow trees have been allowed to grow for less spiritual reasons. In the past they were grown either for timber or as pollards. A bit of firewood, some tool handles and the occasional piece of timber to repair a house were useful products in the rural economy. Whenever a hedge was laid the hedger would leave a few young saplings to grow on, just as a woodsman would leave a few standards when he cut the coppice. As mechanical trimming has taken over, fewer and fewer trees have been left. Machines don't discriminate in the way a hand worker can and they're designed to work fast, without interruption. It takes time to locate a suitable sapling, stop the trimmer just before you get to it, start again just after it, then go back and trim the shrubs round it by hand. It's so much simpler just to keep driving. And why would anyone bother? The small yield of timber or poles which were useful in the past are irrelevant in the modern mass-production economy.

Photographs of hedged landscapes from the 1950s or before have a distinctly antique look these days. It's not just that they're in black and white. There's something about them that makes you feel you're looking at a subtly different country, softer and at the same time more solid. The difference is in the abundance of hedgerow trees. If you're able to take the photo back to the place where it was taken you'll probably see a stark contrast between the old view and the new. Some people say it was the Dutch elm epidemic of the early 1970s that transformed the landscape. Indeed, in areas where elm was the dominant hedgerow tree there was a sudden change at that point, a change which has never been reversed. But the same process went on just as surely though more slowly in areas where other trees are dominant. It's not the death of trees that's changed the landscape but the fact that they're no longer being replaced. A sudden change like Dutch elm disease is obvious but a gradual change slips by unnoticed.

The trees in the old photos aren't just more abundant they're also more varied in age. Today almost all hedgerow trees are mature or old. A hedge where saplings have been left to grow into trees is a rare and pleasant surprise in the landscape. Sometimes, on closer inspection, the trees turn out to be planted beside the hedge rather than promoted from it. This may seem unnecessary when the hedge must surely be full of potential trees which only need to be allowed to grow. But not every hedge contains species such as oak, ash and maple which can grow into full sized trees. Even if it does, after years of trimming there may not be any stems left in sound enough condition to grow into healthy trees. Wherever people do promote or plant new trees it's a blessing to the landscape, not just for visual reasons but also for biodiversity. Trees are important wildlife in themselves and, by increasing the structural diversity of a hedge, they increase the number of niches available to other creatures. For example, song birds will only nest in a hedge where there's a song perch for the male, however suitable it may be in other respects.

Growing trees in hedges has always been a bit of a compromise between the growth of the tree and the growth of the hedge. A tree weakens the hedge beneath it both by shading and by root competition. You'll often see a gap in the hedge underneath a broad-boughed tree. So the interests of the hedge had to be balanced against the value of the tree. In the past shade was kept somewhat in check as pollards were regularly cut and timber trees were pruned. But these days both pollarding and pruning are very rare. Hedgerow trees also reduce the yield of crops in the fields, mainly by shading. A tree in a north-south hedge doesn't have as much effect as one in an east-west hedge because its midday shade falls on the hedge itself rather than on the fields. Looking around I see no evidence that farmers in former times took this into account when choosing which trees to promote, but it's one of the many tools which can be used now to design a landscape which is both biodiverse and productive.

Despite the negative effect on the hedge most farmers keep their mature hedgerow trees. It's not just a matter of beauty and wildlife but also of local character and distinctiveness. A neighbour of mine said he'd never cut down all the trees on his farm like his cousin had done because "It doesn't look like Somerset any more." His is certainly the majority view. Most of us feel it's worth paying a modest price to maintain that recognisable quality which makes a place not just any place but our home. Even when the hedge itself is grubbed out the trees are often spared, though they do compete with the crops and make tractor work more difficult. In some fields you can trace the course of a former hedge by a line of old trees dotted along it. This is more characteristic of the ancient countryside than the planned as there are more old trees worth saving and more of a tradition of trees in the landscape.

Trees are never more than a minor component of a hedge but shelterbelts and windbreaks are composed entirely of trees. A shelterbelt is made up of several lines of trees and a windbreak only one. Shelterbelts aren't very common in the British landscape. They take up a fair bit of land, which farmers and landowners may be reluctant to 'lose' in the sense of planting trees on it. Nor is there much return in terms of timber. The outer rows of trees aren't drawn up, so they become too branchy for timber unless they're pruned. But pruning wouldn't be worthwhile economically and would make the belt pretty useless for shelter. (See the drawing on page 291.) As with hedges, the negative effect on the crops is concentrated near the shelterbelt and is often quite visible, while the positive effect is spread out over the field and doesn't show up. The gain in yield outweighs the loss but it doesn't look like that. (See page 280.)

Single-line windbreaks are almost as rare as shelterbelts. They need a little more care and attention than a multi-row shelterbelt because the loss of even a single tree would create a gap and a gap forms a wind tunnel which intensifies the wind rather than slowing it down. On the other hand they take up less land. For both these reasons they're more suited to high-value crops, where management is more intensive and land more valuable. In this country they're almost exclusively used for orchards, which are particularly vulnerable to wind, both at the blossom stage and when the fruit hangs heavy on the bough.

If you see a windbreak in the landscape you can be almost sure that there's an orchard at its foot, or that there was one until recently but it's been grubbed out.

The trees used most often in windbreaks are poplars and alders. Both grow fast and straight and there are varieties available which keep their lower branches well, to avoid a thin, gappy bottom. Conifers are less often used because a permeable barrier has a more calming effect on the wind than a solid one, which causes gusts and eddies. Where conifers are used they're usually wide-spaced so as to allow some of the wind to pass between them.

In many hilly areas hedges give way to dry-stone walls. Stone wall country is often a sign that the ground was originally so stony that people couldn't start farming till they'd got rid of the stones. St Braivels Common is an example of this. Not every place had such a surplus of stone and sometimes you'll see small field quarries which were used once the surface stone had run out. Another reason why walls are typical of hill country is the climate. It can be difficult to establish a good strong hedge in the cold, windy and wet conditions of the hills.

Different ranges of hills have their distinctive styles of wall and these are partly determined by the nature of the local stone. The sleek, smooth walls of the Cotswolds are the product of the thin, even-shaped stones that the Cotswold limestone yields. Mendip walls are much more chunky. You couldn't make a neat Cotswold wall with the irregular lumps of Mendip limestone. No doubt there are also cultural reasons for regional variations. One may be the date the walls were built. The parliamentary enclosures brought the values of the new industrial age to the countryside. Just as the hedges of the time were pure hawthorn and the field layout strictly rectangular, the walls tend to be uniform and regular. Stone-walled fields which were enclosed by agreement in an earlier age not only follow the characteristic curves and dog-legs of the former furlongs, they also tend to be built in a more informal style, with more variation in height, width and shape.

Obviously walls don't need regular cutting like hedges but they do need maintenance. Stones fall off from time to time and sometimes whole sections can come crashing down – as when an unsuspecting rambler thinks that a dry-stone wall will support their weight just like a wall built with cement. Repairing them is a slow and skilled job. It's not too much of a burden if you keep up with it but once they've been let go it can be a huge job to get them back in good condition. I know a small farm of about a hundred acres on the Mendips where the new owners were quoted a price of thirty thousand pounds to put all the walls in order. There's no way that present-day farming can support that kind of cost. A strand of barbed wire alongside the wall is a cheaper solution and may also save the wall from further damage by discouraging cattle from rubbing themselves on it. In some parts of the country the walls are fenced even where they haven't broken down. This is mostly where the local walling style is tall and narrow, which makes a wall more vulnerable to the odd push from an itchy cow.

Just like hedges, walls can disappear from the landscape if they're neglected, but it's a slower process. Bit by bit they tumble down till there's nothing there but a long, low pile of stones. Over the years any inert object lying on the soil surface will eventually be buried by the earthworms. They constantly bring soil up from below and deposit it as worm casts on the surface and this gradually raises the soil level relative to any solid object like a stone. Where the soil is deep whole buildings can be buried a metre or more below ground level. This is why archaeologists spend so much time digging. But on the thin hill soils where most field walls are found there's not enough depth to bury even a low heap of stones and even after hundreds of years a slight grassy bank may remain. On hills made of base-poor rocks such as granite, where the soil is too acid for earthworms, the ruins of walls may stay on the surface more or less forever.

On the whole walls haven't been lost so much through deliberate destruction. Even with modern machinery it's much more expensive to shift the thousands of tonnes of stone which go to make up a wall than to grub up and burn a hedge. In any case stone walls aren't often found on the high-value arable land where most of the hedge-grubbing has happened.

Walls are challenging environments for wildlife as they have little or no soil, experience wide fluctuations in temperature and are very dry in summer. But some plants and animals can adapt to these conditions and even take advantage of them. Lichens can live on the surface of the stones. A lichen is a combined organism, half fungus and half alga. The alga provides food by photosynthesis while the fungus protects it from the extreme conditions of microclimate which prevail on rocks and the bark of trees. Without any soil, they're dependent on the air for both moisture and mineral nutrients. They're also very sensitive to air pollution and the presence of lichens is an indicator of unpolluted air. This varies according to species. On the whole the ones which form a flat crust on the surface they inhabit are more tolerant than the ones which branch like miniature bushes. Lichen experts can read the prevailing level of air pollution quite accurately by noting which species are present and which are not. With such sparse supplies of the necessities of life, and only part of the plant able to photosynthesise, it's no wonder that their growth rate is infinitesimally slow. They couldn't compete with other plants and are mostly found on surfaces where nothing else can grow. Mosses are only a bit more competitive than lichens and also benefit from the absence of vigorous plants but they need more moisture. They're more common on walls in the constantly damp climate of the west or where a wall is shaded by trees.

Some of the plants which survive the extreme dryness of summer on walls are exotics from the south. Ivy-leaved toadflax, with its little purple-and-yellow flowers which seem to wink at you in the summer sun as you pass by, is from Italy. Pennywort or navelwort is also a Mediterranean plant but its native range includes the south and west of Britain. It lives happily on the walls and stone-faced hedgebanks of the West Country. Both its common names come from the shape of its leaves, which are circular with a soft dimple in the middle. They grow mostly in the moister wintertime and are thick and fleshy so as to store water into the drier days of summer. Stonecrop is another succulent-leaved plant which lives on walls. Its little star-like flowers are white or yellow according to the species but all stonecrops have the same thick, teardrop-shaped leaves, each one a tiny water tank.

For basking reptiles the dry heat of a wall is an asset. The hard, pale surface of the stone reflects the energy of sunlight onto any body resting on it. By contrast, the darker, more complex surfaces of vegetation absorb the sun's energy rather than reflect it. Lizards seem more at home on walls than snakes. They climb better and can hide more easily in the holes between the stones. Snakes are more often seen just at the foot of a wall, where they benefit from the shelter as well as the reflected heat.

Roads
and Paths

There's nothing more enticing than a track over the moors or a path through the woods that curves away from you into the unknown beyond the bend. Even a tarmac road can hint at untold stories which once unfolded between its winding green banks and suggest new vistas awaiting over the brow of the next rise. It may seem prosaic by comparison to pore over the same paths and roads on a map but maps can tell another story, one which I find equally fascinating. Just as the pattern of fields and hedges can be easier to grasp from the birds-eye view of a map, so can the network of roads. Maps can complement what you see on the ground, though they can never act as a substitute for being there.

In some places, a combination of mapwork and direct observation can show how the roads have changed over the ages, how an old road has been superseded by a newer one. The earlier road often takes a more hilly route than the modern one. Steep slopes weren't so much of a problem in the days when walking and riding were the main means of transport, and the most economical route was usually the straightest rather than the flattest. In fact valley bottoms, with their clay or alluvial soils, were often impassable in the days of unsurfaced roads. But as wheeled transport became more common the most economical route was more often one which avoided the steepest slopes even if that meant a detour. The older, straighter roads may show both their age and importance by the way the field pattern is formed around them. Roads which cut across the field pattern must be younger than the fields.

An example of the contrast between old and new is the route from Buckfastleigh to South Brent in Devon. From the hamlet of Dean the old road goes directly up the hill and heads more or less straight for South Brent, with just a jiggle in its line where it drops down to ford a brook. Its age is witnessed to by the way the ancient field boundaries conform to it. Although most of this road is still in use as a tarmacked lane, the steep section just south of Dean is now only a track giving access to the fields. It's been superseded by a more recent stretch of road which gently curves round the side of the hill, using the contours to lessen the gradient and avoiding the hilltop, which the old road marched right across. This in turn has been replaced by the modern dual carriageway, which ignores the old road altogether, taking the lowest possible route, which is further flattened out with cuttings and embankments.

Old roads don't always go straight. Less important ones, linking individual farms and hamlets, are often crooked, going around the boundaries of the ancient fields. They conform to the field pattern rather than influence it. An example is the old lane for which Ragmans Lane Farm is named. It has been superseded by a more modern road which snakes its way from the River Wye up to the hilltop village of Ruardean by the least punishing route. The old lane has fallen out of use but it's still there, a sunken green lane between high banks. A disused lane which is not so sunken and which falls out of use may gradually disappear. First one of its hedges is grubbed out and, though still a recognisable track, it now becomes part of the grazing area when the field is in grass. Then, when the field goes into corn, it may get ploughed up along with the rest of the field. If it was originally straight it may survive as a ghost on the map, an alignment of hedges which seems to be going somewhere. But if it followed the irregular boundaries of the fields it will be lost without trace. Ragmans Lane survives because it would never be worth the effort to fill it in for the sake of gaining a little extra land.

New roads have more often been made by upgrading the old than by taking a completely new line. Even so you can see the changes on short sections, especially in steep country where the old straight route up a hill has been abandoned in favour of a gentler, curving line. On the ground this is often recognisable by a deliberate cutting on the uphill side of the curve, crisp and clear compared to the gradually-formed banks of older roads. The old straight section may survive as a lane, track or footpath or perhaps just a straight hedgeline with the new road curving around it. At Nettlebridge in Somerset the Roman Fosse Way cuts straight across a steep valley. The present main road follows the Roman route on the flattish land to the north but in the valley it loops around it, first on one side then on the other.

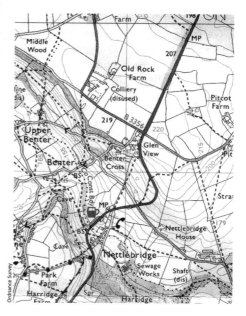

Many of the improved roads were turnpikes. Built mainly during the eighteenth century, they were financed by tolls. Though the toll gates have long gone you can still see the distinctive toll houses beside many country roads.

The three-sided facade, with one side facing the road and the others at forty-five degrees to it, enabled the toll-keeper to see what was coming in either direction. They're usually sited at a road junction, where they could catch more than one stream of traffic.

The toll house at Westhay, Somerset

Turnpikes were specifically designed for wheeled traffic, which wanted a hard surface and was prepared to pay for it. Travellers who didn't need it and didn't want to pay avoided them if they could. So the improvement of the main roads may have helped to define the drove roads, a network of green lanes dedicated to the long-distance transport of meat on the hoof. Welsh and Scottish drovers walked many hundreds of miles behind their cattle on the way to the markets in England. Within Wales, farms which offered hospitality to the drovers and their herds would plant a clump of Scots pines by the roadside as a sign. Some of these clumps are still there, often on droves which have now become part of the motor road network. Most of the present trees look too young to have been there in the days of droving and must have been replanted. Even today you can occasionally see a new group of young pines planted beside a mature clump. I've also heard it said that the best blackberries in Wales are to be found along the drove roads. The bramble is an unusually variable plant, with some three hundred subspecies recognised in Britain alone. There's a great difference between the best fruiting ones and the worst. The drovers would pass by many different types on their travels and could afford to be choosy. So as they went on their way, shitting out the seeds of what they'd eaten, they spread the very best blackberries along the sides of the drove roads. At least that's the story I heard.

The age of turnpikes overlapped with the age of parliamentary enclosure. In the planned countryside the network of village lanes was swept away along with the open fields and replaced by new straight roads among the new rectangular fields. Enclosure-act roads are sometimes so straight that people take them for Roman roads. They usually have very wide verges. Those broad strips of green on either side of the tarmac might seem strangely wasteful for an age which was so keen on productivity. Roads and their verges were part of the common land of the parish, so they could be used as common grazing. Verges were the poor person's pasture and many a cottager would keep a goat or even raise a couple of bullocks on 'the long acre'. But the spirit of the times was against common land and this was just the kind of self-reliance the enclosers were keen to stamp out. They wanted the poor as dependent labourers.

In fact the reason why enclosure roads are so wide is that they were not planned to be hard surfaced. The clear distinction between road and verge only came later. In wet weather a narrow unsurfaced road would quickly become an impassable quagmire, while the traffic on a wide road would be spread over a wider area and the road would remain usable in all but the wettest winters. Wide roads were a mark of efficiency rather than a waste of land.

When you're travelling by car, the first clue that you're in planned countryside can sometimes be the roads rather than the layout of the fields, especially if your view is confined by tall hedges.

Blackdown Hills, 8th November 2006

Driving along a straight road with wide verges and hedges of almost pure beech, I'm clearly in enclosure-act country. Then suddenly the road curves and becomes slightly sunken. The hedges are now mixed and they come in to the sides of the road, leaving no verge. In a moment I'm in a village.

On the hills of the eastern part of the West Country – the Blackdowns, Exmoor and the Quantocks – pure beech hedges are a sign of planned countryside. Beech is not native to these parts but for some reason it was considered the ideal hedging plant by the enclosers. As the new hedges were always planted on top of high West Country banks, perhaps they saw less need for a thorny species which would act as a barrier to animals. And since a well-trimmed beech hedge keeps its leaves in winter they may have felt it would give better shelter than other species.

The straightness of the road and its wide verges were also diagnostic. By contrast, the stretch of curving, slightly sunken road with no verge and mixed hedges was quite clearly ancient. It survives in the middle of the planned countryside because it's the village street and in England the villages themselves were left alone by the process of enclosure. Emerging from the other side of the village I was in planned countryside again till the road suddenly took a dip, heading down off the Blackdown plateau. Immediately I was back in the ancient countryside of winding lanes and mixed hedges.

One thing about that village street which is very characteristic of old roads is that it's sunken. Roads wear away through time. Look carefully at almost any village street and you'll see that the level of the houses and gardens is higher than that of the street itself, if only slightly. In fact sunken roadways are usually the most visible sign of a deserted medieval village in what is now a grassy field, though other earthworks may be present too. But it's not only in villages that roads are sunken. Broadly speaking, the older the road the more likely it is to have been worn away below the level of the surrounding fields. So sunken roads are characteristic of ancient countryside, where the lanes follow the courses they've taken since time immemorial. In some places a road can be so sunken as to form an obvious gully and this is known as a holloway.

Age is not the only factor. The steepness of the slope and the kind of underlying rock also affect the formation of holloways. The slope is important because holloways are caused by soil erosion. The impact of feet and wheels compacts the soil in the roadway and wears away the vegetation. When rain falls it can't penetrate the compacted soil so it runs along the surface. Where water runs on the surface, especially a surface unprotected by vegetation, it takes the soil with it. The steeper the slope the faster the water flows and the more soil it carries away with it. Look at any holloway you come across and you'll see how closely its depth matches the steepness of the slope. As the slope increases so does the depth of the holloway and as it levels out again the holloway peters out.

Sandy soils are more erodible than clays, so the sandier the soil the less slope is needed to form a holloway. Holloways are typical of sandstone hills, which have both the slope and the soil type which favour them. In fact there are some sandstones which are almost as erodible as the soil itself and the holloway cuts down through the bedrock just as it did through the soil. On the Yeovil sands of south Somerset you can see deep, vertical-sided gullies on land which only slopes very gently. The famous diarist Gilbert White described the same thing on the greensand at Selbourne in Hampshire. On less erodible rocks holloways usually erode down to a solid base and then stop. On chalk this may be a layer of flints, consolidated by the traffic of the ages. On limestone it may be the top surface of the bedrock. In the old Ragmans Lane there's an exposed slab of limestone bedrock and running through it is a neat, narrow groove cut by many generations of iron-tyred cart wheels, or maybe the sledge runners of an earlier age. This holloway probably formed quite rapidly after the lane was first laid out and then stayed at its present depth for centuries. Many holloways are now tarmacked, which of course stops any more downward erosion, but doesn't stop erosion of the banks. Sometimes they slump in heavy rainfall, especially where the wheels of passing vehicles have bitten into the base of the bank, wearing away the vegetation. The slumped soil is washed away from the surface of the tarmac, so the holloway can still get wider if not deeper.

Roadside Vegetation

Holloways tend to be very shady. It seems to me that the hedges are left unmanaged more often on a holloway than on other roadsides. Perhaps this is because the steepness of the terrain makes them less accessible. Even on quite wide roads the branches from either side often meet over the top and little lanes can become green tunnels that seem to lead you back through time. The characteristic plants of these dark places are ivy and the hart's tongue fern, with its shiny, undivided fronds which do look like huge tongues. Mossy tree roots are often exposed on the steep sides, buttressing the trees against the steep sides of the holloway. Beech roots in particular can grow into fantastic shapes, giving a Tolkeinesque feel to these places which always have an air of being just a little removed from the rest of the landscape.

True holloways shouldn't be confused with West Country lanes which, having a tall hedgebank on either side, feel as though they're sunken even if they're level with the surrounding fields. The great diversity of wildflowers which is typical of West Country banks can make these lanes places of wonder in the late spring and early summer. The plants are a mixture of woodland and grassland species together with hedgerow specialists like cow parsley and red campion. The distribution of the plants is affected by light and shade. You can find different mixes of plants on the north- and south-facing banks of the same lane, and between shady lanes with overgrown hedges and sunny ones where the hedges are kept well trimmed. This extract from my notebook gives an example.

16th May 1994

> *An east-west lane near the sea in south Devon, with very low hedges. The vegetation is similar on both sides – cow parsley, nettles, alexanders, red campion, bluebell, herb Robert etc – but alexanders is dominant on the south-facing bank and nettles on the north-facing. In a similar lane nearby with high hedges, nettles are dominant on both sides.*

The distribution of nettles in these hedges may not be related to the abundance of plant nutrients in the soil, as it so often is. As well as responding to high levels of nutrients they're also quite shade-tolerant plants and here the difference in shade is enough to give them the competitive edge over alexanders. Alexanders is a naturalised plant from the Mediterranean, originally introduced as a pot herb. It belongs to the cow parsley family but instead of the familiar flat plates of white flowers it has pom-poms of greenish-yellow flowers. It mostly grows near the sea. As you get within a mile or two of the south Devon coast you start to see it in the hedgerows, typically in a ratio of two thirds alexanders to one third cow parsley. Perhaps it needs the milder winters which you get beside the sea.

The distribution of plants isn't always so neatly explained by light and shade, as this sketch reveals.

30th May 1998

A lane near Horsebrook, south Devon.

mainly ferns

mainly ransoms

Ramsons is one of the most shade-tolerant of plants and so are most ferns. The distribution here probably has nothing to do with light levels. It may be due to chance but more likely to some ecological interaction I'm not aware of.

Fascinating as they are, these Devon lanes aren't typical. Most roadside verges are much less colourful and the few wildflowers which grow there tend to be those competitive plants which readily respond to high levels of nutrients. Roadsides receive extra nitrogen from the exhaust gasses of passing vehicles and, more importantly, plant nutrients accumulate there over the years because they're not removed by grazing or mowing. If a verge is mown to improve road visibility the mowings are left to lie there and they enrich the soil as they decompose. The typical mix of plants is tall, coarse grasses with nettles, hogweed, cow parsley, goosegrass and bindweed. The occasional splash of colour is added by meadow cranesbill, common mallow, rosebay or great hairy willowherb. Sometimes you'll see a verge overgrown with bracken and that's usually a sign that you're on a sandy soil.

There are a few verges here and there where a more delicate and diverse grassland has survived. In fact some local authorities have recognised them as mini nature reserves and put up discreet signs to that effect. Somehow these verges have avoided the increase in nutrients. The reasons why this has happened probably vary from place to place. A soil which is naturally low in bases is a good starting point but the fact that these verges haven't been enriched over time is probably more significant. They're often on little-used lanes where grazing may have continued well into the age of motor traffic, in which case nutrients would have been constantly removed in the milk and meat of the grazing animals. The lack of traffic will also have reduced the deposition of nitrogen from vehicles. In some places the landform may help. The diversity of the West Country banks may in part be due to their very steepness which allows plant nutrients to be leached out of them more easily by the rain.

As well as the nitrogen from traffic, roadside soils can be affected by the salt that's applied when snow falls. On major roads it can build up to the point where coastal plants colonise the verge, even far inland. You may have noticed a swathe of small white flowers along the central reservations of dual carriageways and motorways in early spring. This is Danish scurvygrass, a native plant despite its name, which isn't a grass but a small member of the cabbage family. There are several species of scurvygrass, which are all edible and have the strong, sour taste which goes with a high level of vitamin C, the cure for the sailors' chronic illness. They normally grow by the sea and so appreciate a bit of salt in their soil but it's not clear why this particular member of the tribe has expanded across the motorway network so successfully.

Another roadside curiosity is the occasional apple tree which, like scurvygrass, suddenly springs into focus at blossom time. These aren't crab apples or even planted domestic apples. They've grown from the cores tossed out of the windows of passing cars by people eating on the move. Domestic apples don't come true from seed so it's very unlikely that the fruit of these trees will be good to eat. But they add a little colour to the scene and a distinctive character to the roadside landscape. You can see them by railway lines too.

Roadsides can also act as a refuge for plants which have lost their niche in the wider countryside. The common poppy, for example, is an annual which needs disturbed soil in order to germinate. It used to grow in cornfields because the annual disturbance of ploughing gives just the conditions it needs. Wheat fields red with poppies were a favourite theme of the impressionist painters towards the end of the nineteenth century. Not long afterwards the cornfields of northern France were torn apart with high explosive in the First World War and this brought up the biggest flowering of poppies ever seen. Its seed stays viable in the soil for a long time and the massive disturbance of the war brought up many years' poppies at one time. As a symbol of peace and hope amid the hell of war the poppy is still with us, but it's largely gone from the cornfields because it's easily killed with herbicides. Every now and then some roadside works on the site of a former cornfield brings up a batch of buried seed and the blood-red blooms splash their colour along the verge as they used to over barley and battlefields alike.

Verges aren't completely independent of the fields around them and sometimes you can see how the neighbouring land has affected the vegetation of a verge. I have an example in my notebook from a walk I took on the Cotswolds.

River Windrush, 14th June 2003

The river is a ribbon of wildlife running through the arable monoculture of the Cotswolds. I saw a wealth of birds, mostly ducks and other water birds, though only one butterfly and few wild flowers. The narrow flood plain and steep southern bluffs are grassland, grazed by a few beef cattle and horses or shut up for hay. Everywhere else as far as the eye can see is continuous arable. In a farmyard I saw the remains of a milking parlour, just the pit where the cowman used to stand with all the above-ground parts removed – a relic of the days of mixed farming. The adjacent milk room is now an artist's studio. I didn't see a single working farmyard, only modernised farmhouses and cottages.

The profile of this little valley reflects the lie of the rocks underneath. The sharp bluff on the south side is a little scarp slope while the gentle rise to the north is the dip slope of the underlying strata.* I walked upstream by the river and came back along the lane that connects Burford and Little Barrington. The disordered vegetation of its verges is a relief from the arable on one side and dull grassland on the other. The northern verge is full of meadow cranesbill, knapweeds and other wild flowers. The southern verge, which lies just downhill from the arable land, has some meadow cranesbill but is dominated by nettles, cow parsley and goosegrass. Presumably this contrast is due to the runoff of nutrients from the arable land.

* See page 34.

Apart from the river itself the lane was the most interesting and attractive part of that intensively farmed landscape, its verges cut off by stone walls from the boring grass on one hand and the monoculture of wheat on the other. This is often the case when a road passes through a modern agricultural landscape. Where a road or track crosses unenclosed land there's less contrast and the verge may seem to merge seamlessly with the surrounding grassland, heath or moor. But if you look carefully you can often see a strip of vegetation which is influenced by the presence of the road. In many cases the effect seems to be related to the use of limestone chippings, which reduce the acidity of the roadside soil. One dry summer's day in a nutrient-poor, unfertilised hay meadow I noted how the grass was much taller and thicker in a band on either side of a stone-surfaced track which cut across it. Then a car drove along the track and raised a small dust cloud which spread and settled either side of the track, covering just the same width as the belt of taller grass. A tarmacked road doesn't give off dust but lime can be slowly leached from the surface chippings by rainwater and deposited in the soil beside the road. Where a road crosses a heath the effect can sometimes show in the composition of the plants. For example, if there's a mix of heather, bracken and gorse over the heath as whole there may be a definite concentration of gorse beside the road. Herbaceous plants may be affected too. The little yellow flowers scattered over the heath may be tormentil, a common plant on acid soils, but those near the track may be creeping cinquefoil, which also has small yellow flowers but grows in alkaline soil.

On unsurfaced tracks the effect is physical rather than chemical. Any plants which grow there have to contend both with compacted soil and the direct damage of trampling. The centre of a well-used path, where both effects are severe, is usually bare. A little further out, or on a less well-used path, there's a suite of plants which are highly resistant to compaction and trampling. These are much the same ones as are found in gateways. (See page 230.) Both greater plantain and pineapple weed are common. Further out again there's a zone dominated by plants which are moderately tolerant of compaction and trampling. They include white clover, dandelion and silverweed, whose yellow flowers could be mistaken for buttercups at first sight, though the silvery sheen on its feathery leaves is quite unmistakable. These plants survive trampling because they're low growing and have tough, elastic tissues which resist physical impact. Nonetheless they often get trampled to death so another adaptation they have is the ability to regenerate rapidly by seed. Where trampling is severe they can't reproduce in situ and are replenished by seed from outside, often brought in on feet, hooves and wheels.

It's not always a matter of survival, though. Some of these plants are actually favoured by the conditions they find on a path. Soil compaction means poor drainage and both greater plantain and knotgrass need wet soil for germination. White clover, although it grows well enough in well-drained soils, spreads more easily where the surface is wet because its runners take root more readily. Silverweed is another moisture-lover. Although it's most common in the compacted soil of paths it also grows in soils which are not compacted but just wet. All the path plants are short species and the lack of tall ones, which would shade them out, is another advantage of the path habitat for them.

Woodland paths are usually bare, as very few of the plants which tolerate compaction and trampling also tolerate shade. One exception is the lesser celandine, with its little rounded leaves and flowers like yellows stars. It's a tough, ground-hugging plant which readily regenerates from its pea-like tubers which, as many gardeners know, are almost indestructible. It's a common component of ancient woodland, either mixed in with the other flowers or favouring the wetter patches of soil. Where a path goes through an ancient wood the rich mix of wildflowers on either side of it may dwindle down to just this one on the path itself. Where cattle have access to a wood and cause a lot of compaction celandine may cover quite wide areas in a pure stand. (See page 161.)

Lanes, paths and hedgerows are one element of the countryside which can be carried over into the urban landscape. Woods and even meadows sometimes get built around and survive amid a sea of bricks and mortar but they aren't so much part of the town as bits of encapsulated countryside. Lanes and paths, on the other hand, can actually become part of the structure of the town.

You can see this in the neighbourhood where I live now, an area of mostly council housing that was built on the edge of Glastonbury in the 1960s and early 70s, known as Windmill Hill. If you compare the historical map from the beginning of the last century with the current Ordnance Survey map you can see that the old network of lanes and footpaths is still there. You might have thought that a network which was made to serve the needs of the rural landscape would have been swept away and replaced. But in fact it's quite intact and has been added to rather than superseded. The neighbourhood is a flat hilltop and there was one lane which zigzagged across it, made up of five straight sections joined more or less at right angles. It was typical of those crooked lanes which originally wound their way between the furlongs of the old open fields and were preserved when the fields were enclosed by agreement. (See page 94.)

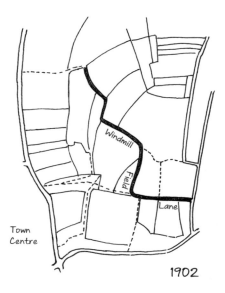

1902

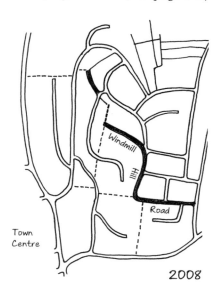

2008

It's marked on the old map as Windmill Field Lane. The first three sections of it are now called Windmill Hill Road and are streets of houses. It looks odd that three streets which lie almost at right angles to each other have the same name, especially as two of them now have straight extensions which bear different names. But when you know the story it all begins to make sense.

The fourth section has been demoted to a footpath, running between private gardens. It still has the old mixed hedges that were there when it was a country lane, except for one part which has been replaced with a typical garden hedge of lonicera. In spring the path is white with cow parsley and if you half close your eyes you can imagine yourself still in the country. There are other footpaths like this on the hill, some of them with a surviving rural hedge on one side, others completely urbanised. Most of them are on the slopes, connecting the flat hilltop with the old town at the bottom. The old map shows that they were almost all there a hundred years ago, just where they are now. It also shows several 'nurseries' on the hilltop, probably meaning market gardens. Gardening is labour-intensive and the paths must have been used by people from the town walking to work and back. Today they're mainly used by schoolchildren who live on the hilltop and go to the secondary school at the bottom.

In the past footpaths probably came and went according to need. I suspect that most of the paths on Windmill Hill only came into use when the nursery business developed and workers needed to come up from the town. When the land was built on the original purpose of the paths was swept away, but not the paths themselves because by then they'd become legal rights of way. The Rights of Way Act of 1949 fossilised the path network as it was in that year, just at the dawn of the car age. Or rather it fossilised those parts of the network which landowners admitted to. Many of them, then as now, would much rather not have had the public on their land at all. The result is that the footpath network we have now was originally created for quite different purposes from the one it's used for now, which, in the countryside at least, is mainly recreation.

The last section of what was once Windmill Field Lane is detached from Windmill Hill Road by the footpath and goes by another name. But it's origin is hinted at by one of its hedgerows, now a garden hedge. The hedgerow shrubs are gone but two old oaks and a pollard ash still stand among the exotic conifers. Until recently there was also an old native maple on the line of another hedge nearby but it started dying back and was removed as a safety hazard. Most of that hedge had survived more or less intact till a couple of years previously. It gave this corner of the housing estate a pleasant, slightly rural feeling and a connection with the countryside which surrounds it. It gave us elderflowers in summer and blackberries in autumn. In springtime blackbirds would nest in it and the cock bird would sing that melancholy evening song that never fails to tug at my heart. Then one day a gang of workers arrived and ripped it out. Only after the event I found out that some of the neighbours had asked the council to remove it because they thought it was unsightly. Now those neighbours have a view of a row of concrete garages, enlivened with a little graffiti.

The urbanising of the vegetation acts like a ratchet. The relics of the rural landscape die one by one and are never replaced. New trees may be planted but they're urban trees, often of exotic species and lacking the historical meaning or the wildlife value of the old rural ones. Meanwhile the trees which started life when the houses were built grow bigger and impart more of their character to the landscape. On Windmill Hill a patch of ash, sycamore and conker trees, planted just after our house was built, is now a tall woodland, where children have carved dens into the steep bank and woodpeckers can be seen feeding in the treetops. The urban landscape is by no means boring but it is distinctively urban and develops a vegetation which has little connection with what went before it.

So the long-term legacy of the rural landscape is much more in the layout of streets and paths than in the vegetation. Windmill Hill is not unusual in inheriting so much from the rural past. Although many urban landscapes are laid out as though on a blank sheet of paper, many more are not. Very often the irregular pattern of streets and houses which seems so illogical preserves the shapes of fields and furlongs in the boundaries of housing developments, and country lanes in busy urban thoroughfares.

Afterword

During the time I've been writing this book things have changed. The world price of grain, which had been falling in real terms for thirty years, has doubled in the past year.

For most of my adult life there's been a surplus of food worldwide. Though people have gone hungry it's been because food was, and still is, unevenly distributed, not because there wasn't enough. But now we're moving into a time of real shortage. Demand goes on rising but the supply of arable land is limited and most of the suitable land on the planet is already being cultivated. Significant increases in yield per hectare are not likely either. The 'green revolution' of the past half-century has already taken crop plants quite close to their genetic ceiling. Genetic engineering may fiddle about at the edges but it can't alter this basic fact.

On a world scale this is an enormous challenge and one which is central to my work as a permaculturist. Its specific effects on the British landscape are hard to predict because it will put us in a completely novel situation. For the first time since the 1870s growing food will become really profitable, as opposed to being propped up by subsidies. But since then there have been enormous changes in technology, population and lifestyle. It won't be a return to the past, and who would dare forecast the future? Nevertheless one thing which is almost certain is that the emphasis in farming will change from producing food as cheaply as possible, which has been the general aim since the end of the Second World War, to producing as much food as possible. For half a century labour has been expensive compared to machinery and chemicals, so the number of people employed on the land has fallen and fallen. We've come to associate this trend with 'productivity'. Yet productivity per person isn't the same thing as productivity per hectare. The truth is that smaller farms with more people working on them produce more food per hectare than large, mechanised ones. In a world that's short of food the total output from the available land is likely to be the top priority and this will mean more people employed on the land. A landscape with more people living and working in it would certainly look different. Would it be more appealing and more sustainable than the one carved out by heavy machinery over the past half-century? I can't help feeling that it would.

Timber prices are rising too, mainly in response to rising demand from China. It remains to be seen whether this is a permanent trend or simply a blip which will disappear once the logging industry catches up with the surge in demand. There really is a lot of timber in the former Soviet Union. The area of land between Vilnius and Vladivostok is unimaginably vast and much of it is forested. If those forests are harvested without restraint they can surely match world demand for many years to come. But if the price rise is permanent the small plantations of lowland Britain, which now largely stand unharvested, may become an economic proposition and a functional part of the rural economy. Those dark, derelict rows of conifers may become hives of productive activity.

Orchards are beginning to experience a similar revival already, though in their case it's not so much a matter of hard economics as a change in attitude. For thirty years or more fruit growers have been grubbing out their trees as supermarket buyers have increasingly gone abroad for cheaper fruit. But now customers are becoming more aware of food miles and are starting to demand home-grown fruit. The buyers are responding, and for the first time in many years some British growers are planting new fruit trees.

Just as food supply is falling behind demand, so is the supply of oil. There's probably as much oil left in the ground as ever came out of it but the oil which was easiest and cheapest to extract was taken first. What remains is more difficult to get at and more costly. As demand goes on rising production will steadily fall. We are now at, or very near, the point known as peak oil, when oil production reaches its all-time maximum and starts to decline. A similar peak for natural gas is not far behind. Never again will fossil fuels be so readily available or so cheap. In the short term, peak oil on its own may not drastically reduce the amount of fuel available to farmers. Food is one thing we simply can't do without, so farming will probably go on getting all the fuel it wants. But the rising price will make labour relatively less expensive, which can only add to the increase in the number of people working the land.

At the same time it will have a drastic effect on the economics of the commuting lifestyle. Cheap oil has made a forty or fifty mile drive to work a negligible expense for someone on a good salary. This has driven up the price of rural houses to the point where only the rich can afford to buy them. A house with a few acres where horses can be kept is even more expensive. But soon the commuters may have to move back to the city, closer to their jobs. Meanwhile there's a sizable body of people who are eager to go smallholding but are held back by the sheer cost of buying small parcels of land. The general price of agricultural land will probably go on rising in line with food prices. But the ridiculous figures presently paid for pony plots, which keep genuine smallholders out of the market, will probably soon be a thing of the past. Once again the country may become a place for people who work on the land.

I realise I'm sticking my neck out by making any predictions about the future. If there's one thing such predictions have in common it's that they usually turn out to be wrong. Maybe there's a touch of wishful thinking in my vision of a repopulated countryside, with more hedges and orchards, revived woodlands and new smallholdings. But a positive vision gives us a direction to aim for, if not an exact description of what to expect.

Glastonbury, Spring 2008

Glossary

Words in italics are defined elsewhere in the glossary.

Aftermath: grass which regrows after a hay crop has been taken.
Alga: singular of algae.
Alkaline: opposite to acid.
Alluvium: soft mineral sediment deposited by rivers.
Annual: plant which completes its life cycle in one year. See *biennial* and *perennial*.
Arable: land used for crops such as cereals rather than for grass.
Aspect (of a slope): the direction it faces relative to north, south, east and west.
Bases: a group of chemical elements which include many plant nutrients and which tend to make a rock or soil which contains them more alkaline.
Basic: containing a high proportion of bases.
Biennial: plant which completes its life cycle in two years. See *annual* and *perennial*.
Biomass: the overall weight of plant and animal material, living and dead, in an ecosystem.
Biotic: caused by a living thing, plant or animal.
Brown earth: a fertile soil type, neither excessively leached nor habitually waterlogged.
Browse line: line on trees or shrubs below which the leaves and twigs have been eaten by grazing animals.
Burn (Scots): stream.
Climax: the supposed final, stable stage of natural succession.
Clone: group of plants which result from vegetative reproduction from one parent and are thus genetically identical.
Combe (pronounced 'coom'): small, narrow valley.
Deciduous: tree which loses its leaves in winter.
Dip slope: usually gentle slope which follows the angle of the underlying rock *strata*.
Emergent: erect plant which has its lower part in water and its upper part in the air.
Exotic plant: one which is not native to the country where it grows.
Fallow: arable land which is temporarily resting, usually for one year, with no crop grown.

Herbaceous: plant with no woody parts, including herbs, grasses etc.

Invertebrate: animal without a backbone, e.g. insects, worms, molluscs.

Leat: artificial water channel.

Loam: soil containing a mixture of sand, silt and clay particles.

Loam, medium: soil containing an equal mixture of sand, silt and clay particles.

Lynchett: terrace formed on a hedge line by soil erosion on the downhill side and deposition on the uphill. See also *strip lynchett*.

Maiden: tree which has been neither coppiced nor pollarded.

Mast year: year in which a tree produces abundant seed.

Monoculture: crop in which all the plants are of one species.

Omnivorous: animal which eats both plants and animals.

Osier: willow grown for basket-making, coppiced, with thin, vertical stems.

Perennial: plant which lives for more than two years. See *annual* and *biennial*.

Poaching: baring and compaction of the soil by trampling during wet weather.

Podsol: an infertile soil type, acid and excessively leached.

Primary wood: land which has been wooded continuously since the end of the last Ice Age. See *secondary wood*.

Pure stand: group of plants, either wild or cultivated, all of the same species.

Rhizome: root-like structure of herbaceous plants, usually running horizontally through the soil.

Scarp: steep slope whose profile cuts across the *strata* of the underlying rock.

Secondary wood: land which is now wooded but has been cleared and recolonised by trees at some point in the past. See *primary wood*.

Semi-natural ecosystem: one in which all the plants are self-sown but the structure has been determined by human activity such as grazing, mowing or coppicing.

Spar, thatching: long wooden staple used by thatchers to fix the thatch to the roof.

Species: a distinct group of plants or animals which can interbreed with each other successfully, e.g. apple, cow. See *variety*.

Strata: natural layers of rock.

Suckering: formation of new shoots from the root of a parent plant.

Tarn: upland pond formed by erosion during the ice ages.

Variety: a distinct subdivision of a species, e.g. Cox's orange pippin, bramley. See *species*.

Vegetative reproduction: plant reproduction by means other than seeds, e.g. *suckering*.

Windthrow: uprooting of trees by the wind.

Woody plants: trees and shrubs.

Further Reading

General

The Hidden Landscape: a journey into the geological past, by Richard Fortey, Jonathan Cape 1993. A readable account of the rocks of Britain.

Plantatt, by MOHill, CDPreston & DBRoy, Centre for Ecology and Hydrology, 2004. A table of all the native and naturalised plants of Britain, including their soil preferences, based on the work of Heinz Ellenberg in central Europe. It has its limitations. Firstly, plants don't necessarily behave in the same way in Britain as they do in central Europe. Secondly, it's not possible to express all the nuances of plant behaviour in a table. For example, some plants may tolerate a wide range of soil moisture and others may be restricted to a narrow range but both kinds have their soil moisture preferences expressed by a single number on a scale of one to twelve. Nonetheless, the 'Ellenberg numbers' are a valuable resource for reading indicator plants.

The History of the Countryside, by Oliver Rackham, Dent, 1986. The best book on the history of the landscape.

The Illustrated History of the Countryside, by Oliver Rackham, 2003, Weidenfeld & Nicolson. A much shortened version with the addition of colour pictures and some illustrated walks in various parts of the country.

The Ecology of Urban Habitats, by Oliver Gilbert, Chapman & Hall, 1991. A fascinating account of how the urban landscape works.

Arkenfield: portrait of an English village, by Ronald Blythe, Penguin, 1972. The human element of the landscape comes alive in this book about country people at a time of great change.

The Wild Places, by Robert Macfarlane, Granta Books, 2007. An account of a very personal relationship with the landscape.

Field Guides

These are books which help you identify wild plants and animals. The most useful for landscape reading are ones for wild flowers and trees. When choosing a guide the main points to bear in mind are:

The key

This is the first step to identification. It usually takes the form of a series of questions with multiple answers which progressively narrow the possibilities down to a single species. Some are easier to use than others. Trees are usually keyed by leaf shape but a useful feature is a second key for use in wintertime based on twig shapes. Wild flowers are usually keyed by the blooms but in general they're more difficult to key than trees and some wild flower books lack an overall key. The best way to assess the effectiveness of a key is to use it, so an opportunity to borrow a guide from a friend to try it out can be useful.

Illustrations

In general, hand-painted illustrations are better than photographs. The illustrator can paint a typical example, avoiding both the idiosyncrasy of individual plants and confusing backgrounds. Tree guides often have really awful illustrations, hardly more than a series of lollipops with 'oak', 'beech' and 'walnut' under them. The overall shape of a tree is variable and usually more affected by its environment and history than its species. The most useful illustrations are detailed ones showing leaves, twigs, buds and flowers.

Size and range covered

A smaller book is easier to carry around with you but if it's too small there may not be enough information on each plant for a secure identification. The wider the range of plants covered by the guide the less space there will be for each entry, or the book will be inconveniently large and heavy. Many guides cover parts of the continent as well as Britain, which results in a lot of unnecessary entries if you're only going to use it here. Some tree guides include a lot of species which are only found in parks or arboretums and have no relevance to landscape reading. A useful range of trees is: all the native and naturalised species; the most commonly planted forestry trees, both conifer and broadleaved; fruit trees; and a few very common ornamentals. This would cover over 99% of the trees in Britain but less than half the species count. The native shrubs should be included too. Many of them are useful indicators and they're an important part of the landscape in their own right.

Distribution maps

These show the geographic range of each plant. They can save a lot of time in identifying a plant because a quick glance at the maps can often rule out several species from the range of possibilities. They're much more useful for herbaceous plants than for trees because trees have been so widely planted.

My favourite guides are:

Wild Flowers by Colour, by Marjorie Blamey, A&C Black, 2005. The ideal book for beginners, designed with easy identification as the top priority.

Wild Flowers of Britain and Ireland, by Marjorie Blamey, Richard Fitter and Alastair Fitter, A&C Black, 2003. A more advanced book, suitable for people with some previous knowledge of wild plants. As well as wild flowers it covers trees, grasses and ferns, so you only have to carry one book, but the entries for these latter groups are brief compared to those in a specialist guide.

Trees and Bushes in Wood and Hedgerow, by Helge Vedel and Johan Lange, Methuen, 1960. Still far and away the best tree guide, though sadly out of print. Second hand copies are available on the internet.

How to Find and Identify Mammals, by Gillie Sargent and Pat Morris, The Mammal Society, 2003. Not really a field guide because it's A4 size and wouldn't slip into your pocket, but good for identifying the tracks and signs of wild animals.

Geological Maps

These are available from:
British Geological Survey
Keyworth, Nottingham NG12 5GG
0115 936 3241
www.bgs.ac.uk

Identifying Trees
at a Distance

Before the leaves come out in spring the swelling buds can give a distinctive colour to the canopy of a wood. Birch buds give a rich red and oak a purple haze. An ash wood has a grey, ghostly cast, not because of the buds, which are black, but from the grey-barked twigs and branches. As ash comes into leaf later than other trees, usually in the second half of May, this grey look persists when the surrounding trees are already in leaf.

When oak leaves first come out in mid May they have a bright orange colour. For a while an oak wood shows a variety of oranges, reflecting the individuality of the trees, some coming into leaf faster than others. Young beech leaves are a bright yellowy-green and the trees stand out boldly against other kinds with darker hues. Sycamore is mid green while willow and hawthorn are a darker green. When the wild cherry trees blossom, in April or May, they're unmistakable, standing out like great white sugarloaves amongst their less dramatic neighbours.

By the end of May ash has a feathery appearance, with pale green leaves, oak is like a teddy bear's fur, medium green, while sycamore is a dark, dark green. The native maple can stand out with a splash of orange to purple as the leaves are slow to shake off their bright infant colours. When the hawthorn flowers, its white blossom is unmistakable.

By mid June the trees are getting less easy to distinguish from a distance, merging into a general green. Ash is the most distinctive, with a very pale green, while sycamore is darker than other deciduous trees. By July even these distinctions can be hard to spot. But sometimes you can see a difference between the pale greens of ash and willow and the brownish greens of alder and beech.

Beech is usually the first tree to start turning with the approach of autumn. By the end of August some beeches stand out brown against the green of other trees. But generally in the autumn there's more difference between individual trees than between species.

In September most trees other than beech still have the uniform dark green of late summer but some individuals from a range of species may turn yellow. These are often younger trees. Trees which are carrying a lot of seed, including lime, sycamore and ash, may have a pale appearance, as the seed is lighter-coloured than the leaves. But this varies between individuals and from year to year rather than between species. It's a hindrance rather than a help in identification.

Some of the smaller trees and shrubs do become easier to distinguish in September. Hazel is generally a more yellowy green than other trees at this time, and where it forms an understorey it may show up on the edge of the wood as a paler band below the dark green of the taller trees. Hawthorn can often be identified by the red cast of its berries. Some shrubs start to reveal themselves by the autumn colour of their leaves, such as dogwood with its characteristic deep purple. The climber, old man's beard also stands out at this time, with a silvery-grey sheen rather than the pure white of winter which gives it its name.

The ash is one of the first trees to lose its leaves, and once again those grey twigs and branches contrast with the leafy oaks, now tinted with the glory of autumn. Oaks are variable, both in the shade of autumn colour and in the timing of the colour change and leaf fall. The native maple stands out clearly in autumn with striking colours ranging from bright yellow to rich gold, contrasting with the more sombre oranges and browns of the oaks. The rich yellow of birch and the russet red of wild cherry are also quite distinctive. In a good year beech turns a fiery orange at the height of autumn, a colour so intense it almost seems to have its own internal source of light. Once they lose their leaves the hawthorns are unmissable in most years, a mass of bright red berries which stay on them till the fieldfares and redwings come down from Scandinavia in the winter and strip them bare.

Index

Inspiration for Self Reliance

Other books by Patrick Whitefield

The Earth Care Manual

The definitive book of permaculture for Britain and other temperate climates. Contains in-depth analysis and detailed advice relating to soil, climate, water, energy, buildings, gardens, farming, woodlands and the design process.

£39.95. 480pp. 250 photographs & many diagrams/illustrations

Permaculture In A Nutshell

The perfect introduction to the ideas behind permaculture, which applies to so much more than just food growing. Patrick, one of Britan's best known practitioners and teachers, outlines the principles, and how to apply them – in the city, the garden, on the farm and in the community. Over 15,000 copies sold to date!

£5.95. 96pp. 8 photographs, 14 line drawings

How To Make A Forest Garden

The book of forest garden design and one of the best examples of applied permacuture ever published. This guide contains everything you need to know in order to create a 'maximum output for minimum labour' design. Suitable to fit any space. A best-seller!

£16.95. 192pp. 45 photographs, 65 illustrations

available from all good bookshops
or direct from Permanent Publications at:

www.permaculture.co.uk

Patrick Whitefield is a contributing editor to this leading magazine

Permaculture Magazine **offers tried and tested ways of creating flexible, low cost approaches to sustainable living.**

It enables you, your family and community to:

- Discover the inspiring and exciting world of permaculture
- Meet the world's most creative and innovative green pioneers
- Learn practical, cutting edge methods for self reliance
- Find interesting courses, contacts and opportunities
- Save money and have fun

Permaculture Magazine is published quarterly for enquiring minds and original thinkers worldwide. Each issue gives you practical, thought provoking articles written by leading experts as well as fantastic eco-friendly tips from readers!

**permaculture, organic gardening, eco-building, renewable technology, transition towns, agroforestry, sustainable agriculture, ecovillages, downshifting, community activism, human-scale economy...
and much more!**

Permaculture Magazine gives you access to a unique network of people and introduces you to pioneering projects in Britain and around the world. Join the 100,000+ readers in 77 countries and share their knowledge!

Every issue of *Permaculture Magazine* brings you the best ideas, advice and inspiration from people who are working towards a more sustainable world.

Permanent Publications
The Sustainability Centre, East Meon, Hampshire GU32 1HR, UK
Tel: 0845 458 4150 or 01730 823 311 Fax: +44 (0)1730 823 322
Email: orders@permaculture.co.uk Web: www.permaculture.co.uk

THE QUEEN'S AWARDS
FOR ENTERPRISE:
SUSTAINABLE DEVELOPMENT